Saudi Arabia: An Environmental Overview

T0136071

Saudi Arabia: An Environmental Overview

Peter Vincent

CRC Press
Taylor & Francis Group
Boca Raton London New York

CRC Press is an imprint of the
Taylor & Francis Group, an **informa** business
A TAYLOR & FRANCIS BOOK

CRC Press
Taylor & Francis Group
6000 Broken Sound Parkway NW, Suite 300
Boca Raton, FL 33487-2742

First issued in paperback 2019

© 2008 Taylor & Francis Group, LLC
CRC Press is an imprint of Taylor & Francis Group, an Informa business

Typeset in [Sabon] by Vikatan Publishing Solutions (P) Ltd., Chennai, India

No claim to original U.S. Government works

ISBN-13: 978-0-415-41387-9 (hbk)
ISBN-13: 978-0-367-38781-5 (pbk)

Library of Congress Cataloging-in-Publication Data

Saudi Arabia: an environmental overview/Peter Vincent.
    p. cm.
Includes bibliographical references and index.
ISBN 978-0-415-41387-9 (hardcover: alk. paper)
1. Saudi Arabia–Environmental Conditions. I. Vincent, Peter.
GE160.S32V56 2008
363.7009538–dc22

2007024276

**Visit the Taylor & Francis Web site at
http://www.taylorandfrancis.com**

**and the CRC Press Web site at
http://www.crcpress.com**

Contents

List of Figures

List of Tables

Preface

This book is about a country I first visited in 1983 and thereafter more or less on an annual basis for the next twenty years or so. My first introduction to the country, however, took place much earlier when I was a graduate student at Durham University in the 1960s. At that time there was a large contingent of students from the Middle East and I became friends with Taiba Al Asfour, a coastal geomorphologist from Kuwait, and Asad Abdo who, in 1968 became the first Saudi geographer to obtain a Ph.D. As a young geomorphologist, I found it exciting to listen to Abdo and Taiba and learn about a region so far away. I was determined to go myself one day and particularly to Saudi Arabia. Ironically, my own Ph.D. was in glacial geomorphology.

That day came some fifteen years later and I vividly recall the wall of heat hitting me as I stepped out of the plane at Jiddah's old airport near the docks. The thrill of that moment is with me still, as are memories of camping expeditions into the edge Empty Quarter and of gazing up at the most wondrous starlit skies imaginable. Then there was a hugely exciting trip to the Farazan Islands in the Red Sea, of my first sight of raised coral reefs, of pearl divers and my time with the most hospitable of villagers. How they laughed at my fishing exploits and the 'shark' that I thought I had caught turned out to be nothing more than a hook on a fellow fisherman's line. As I tugged he tugged back. For a moment it was a very large shark indeed! Above all there was sand, and more sand, shaped into the most glorious curves by a wind which seemed like a blow torch on my white European face. Of course, I then travelled by jeep or car, but more recently by helicopters, aided by GPSs and satellite phones. I ate at one of the many transport cafés now dotted along the Kingdom's magnificent new roads – I recommend the chicken, though the lamb or goat cooked slowly in hole in the ground is very tasty once in a while. I still marvel at those intrepid explorers, like Doughty and Thomas who navigated by compass and the stars and whose needs were little more than sweet water and a few dates. How did they do it?

Doubtless in such a broad book as this, experts will find many things to quibble with. Writing about Saudi Arabia is not the easiest of tasks and it is only in the last twenty years or so that computerized information has become available. Not everything that one would want to know is in the public domain and the geographical spread of environmental data is very uneven to say the least. With all these excuses in mind I have tried to paint a picture of a fascinating country with a long tradition of environmental awareness and sensitivity; a country whose development is pitted against some of the harshest environments on earth.

The Latinization of Arabic is a linguistic minefield and in writing this book I have attempted to be consistent (though doubtless I have failed here and there). A particular difficulty concerns the transliteration of place-names and throughout the text I have used as my authorities either the *Kingdom of Saudi Arabia – Geographic Map of the Arab Peninsula* (1984) at the scale of 1:2,000,000 published by the Ministry of Petroleum and Mineral Resources: Deputy Ministry for Mineral Resources, Jiddah or the *Atlas of the Kingdom of Saudi Arabia* published by the Ministry of Higher Education (1999). Occasionally, I have modified the spelling of a place-name cited in a reference so as to match the standard spelling provided by these map authorities. In using these map sources I have made no attempt to employ the diacritic marks. They are not widely understood by non-linguists and even for linguists the system is confusing and highly irregular.

There is no capitalization in Arabic and I have followed convention for capitalization of words in English. The Arabic definite article has been retained for personal names, such as Al-Jerash, and where I have used it for geographical names its use is based on the map authorities cited above. In general *al* is not hyphenated in place-names. Arabic has just three vowels, transliterated as: a, i, u. This can cause spelling confusions in English transliteration. For example, the traditional areas of land put aside by villagers and tribes for conservation purpose should properly be transliterated *hima* but is often spelled *hema*. Similarly, there are no Arabic place names starting with the letter 'o'. Thus, Hima Oneiza cannot be found on any maps even though mentioned in Draz (1985). As it turns out, 'o' here is properly sounded as a long 'u' and transliterated as 'u'. Oneiza thus becomes Unayzah, which is on standard maps. In short, one often has to do a little bit of detective work in order to sort out the muddle and I hope that readers will excuse me if mistakes have crept in. I have not corrected these transliteration irregularities in any of the references cited and I have retained the author's original spellings. Similarly, modern Arab authors writing in English spell their names in a variety of ways and when cited as a reference I have kept to their original spellings. Thus the same author may appear in several spelling guises.

Where necessary, I have updated location names cited in references to correspond to those in my map sources. In one or two cases it has simply not been possible to locate a place cited in a reference because of the non-standard transliteration and lack of detailed cartographic information. I have not, however, up-dated the actual citation in the bibliography and these are presented as in the original publication.

Radiocarbon dates have been quoted as given by their authors and have not been calibrated. Islamic calendar dates are based on the *hijra* (Arabic: migration) when, in 622 AD, the Prophet Muhammad left Makkah for Yathrib – later to be renamed Madinah. Those followers of Muhammed who migrated from Makkah (*muhajirs*), together with believers in Madinah (*ansar*), became the nucleus from which Islam was born. The calendar was formally introduced under the second legally constituted caliph, Umar (634-644 AD). The Islamic year is lunar and based on 354 days with eleven leap years of 355 days in each cycle of 30 years. Thus there are approximately

103 *hijri* years to the Gregorian century. Throughout I have used the Gregorian calendar but should the reader so wish the simple formulae given by Hattstein and Delius (2000) provide approximate conversions from and to the Gregorian calendar (G) and the *hijra* (H):

$$G = ((32/33) * H) + 622$$
$$H = 33/32 * (G - 622)$$

The Permanent Mission of the Kingdom of Saudi Arabia to the United Nations, New York, maintains a very useful information Web site with links to calendar conversion information at http://www.un.int/saudiarabia/index.htm [April 2007].

A central purpose of writing this book is to provide non-Arabic readers with a broad overview of this fascinating, and very important part of the Middle East. In choosing which journal references and books to cite I have kept almost entirely to material written in English. Much of this material is generally available in university libraries in the Western world and Saudi Arabia's scientists, many of whom trained in either the United States or the United Kingdom, are now very much part of the global community of scholars who publish in international journals.

Throughout the text I freely use the terms Saudi Arabia and Kingdom interchangeably. By Kingdom I mean the Kingdom of Saudi Arabia. Had I more expertise it would have been natural to deal with the whole of the Arabian Peninsula but this would have been a major undertaking. Nevertheless, where appropriate, I have used maps for the Peninsula as a whole, and made comments on circumstances in neighbouring countries.

Information on Saudi Arabia's environment is widely scattered and much is only available in unpublished Government and consultant reports. A helpful bibliography covering these sources and journal citations prior to the 1970s is provided by Batanouny (1978). There are many useful Web sites relevant to the topics covered in this book. Some come and go very quickly and in referring to sites I have chosen only those which, in my judgment, are reasonably permanent.

The text is organised as ten more or less self-contained chapters dealing with most aspects of the Kingdom's environment. I have deliberately not included a separate chapter on marine environments but they are mentioned where relevant. When I first thought about the structure of the book it was suggested to me that I include an introductory chapter for those readers who might need some background information before diving into the technical material. I also thought it interesting to say something about the discovery of the Kingdom by European adventurers – the first tourist, many of whom made interesting observations during their travels. These first two chapters are deliberately wide ranging and meant to set the scene.

Peter Vincent
November 2007

Acknowledgments

Producing this book has been no easy task for many reasons and would have been quite impossible without the advice and encouragement of many colleagues, particularly those in Saudi Arabia. First I should like to thank my former student, Dr Ahmad Sadah Al Ghamdi now at King Abdul Aziz University, Jiddah, for many memorable excursions mostly using his family car. How the suspension must have suffered. I am also indebted to another former student, Dr Ahmad Al Nahari, now at King Khalid University, Abha, for introducing me to the south and the Farazan Islands – what a magical place. At the National Commission for Wildlife, Conservation and Environment in Riyadh I received much encouragement from Professor Abuzinada and invaluable help from Eugene Joubert and Othman Llewelyn. Dr Philip Seddon, then at the NCWD, and now at the University of Otago kindly provided me with a copy of his MA thesis on protected areas which has proven invaluable. There are many colleagues at the Saudi Geological Survey, Jiddah, who have given freely of their time and knowledge. I should like particularly to thank Peter Johnson, John Roobol, Fayek Kattan and H.E. Dr Zohair Bin Abdul Hafeez. Simon Chew, Lancaster University, drew all the maps and is warmly thanked.

I would like to acknowledge the assistance of Dr Janjaap Blom and Lukas Goosen at Taylor & Francis The Netherlands/Balkema publishers during the production of the book.

Chapter 1

An environment discovered

Contents

1.1 INTRODUCTION

By way of introduction to a book dealing with the environment of Saudi Arabia I thought it of interest for the reader to learn about some of the more important, and mostly European, travellers and explorers who were drawn by the mysteries of the Orient, and specifically to the region now known as Saudi Arabia. Several of these desert adventurers made important environmental observations and their tales of travel in the desert encouraged others in their footsteps. Indeed, their exploits seem to have caught the imagination of modern-day adventurers who wonder, no doubt, how anyone could possibly manage without a 4 × 4 truck (Barger, 2000). But manage they did, and even though the golden age of exploration was soon over, and oil men in their planes had surveyed the whole country for black gold, much still remains to be discovered about the Kingdom's environment and adventures are still to be had. Of course, the view of the environment presented here is decidedly Western and Orientalist and the commodification of the environment has become a pervasive theme of the Kingdom's development trajectory. How this harsh desert environment is perceived by the declining numbers of *badu* is, as yet, an unresearched question – the other side of a fascinating coin. The term "bedouin," is actually an Anglicisation of *badawi*, or *badu* in the plural, and is used to distinguish the nomadic Arab of the desert from those in villages and oases.

As the Kingdom progressively becomes the focus of the modern tourist gaze one can only marvel at the exploits of these early explorers, armed with rudimentary maps, often with no knowledge of local customs, and as Christians in a Muslim world. Mostly they were tolerated and some, such as Lady Anne Blunt almost revered, and now part of folk memory. Indeed, the Blunts' crossing of the great northern sand desert of the Kingdom, An Nafud, and their stay in Jubbah is now part of the tourism literature for the Ha'il region.

1.2 EARLY EUROPEAN EXPLORERS

For more than a thousand years, since the time of the Prophet Muhammad, Arabia was hardly known to Europeans. The difficulties of desert travel, and Arab caution towards non-Muslims, deterred exploration of Arabia by Europeans until the eighteenth century, and for nearly two hundred years the writings of these explorers remained the only sources of environmental information available to the outside world.

Prior to this period one or two adventurers had made their way to Makkah but seemed not to have been in interested in collecting scientific information. One of the first Europeans to visit the holy city was the swashbuckling Italian Ludovico de Varthema who joined a troop of Ottoman soldiers escorting a pilgrimage from Damascus in 1503. Instead of returning to Damascus, Varthema smuggled on board a boat and set sail for Aden. There his luck failed him and within a day of his arrival he was put in chains. His escape was arranged by the Sultana and he left Aden having spent about ten months in Arabia.

The first Englishman to visit Makkah was Joseph Pitts who was born in Exeter in about 1663. His life seems to have been full of adventure having been captured by an Algerine pirate during a sea voyage back from Newfoundland and then sold into slavery and bought by Ibrahim, a wealthy Turk. After a good deal of torture – mainly by being beaten on the feet with a stick, Pitts was persuaded to become Muslim. Pitts was sold for a third time to a kinder owner who took him to Jiddah, Makkah and Madinah on pilgrimage. Pitts was finally given his freedom on return to Algiers, and enlisted first in the Turkish army, and then the Turkish navy. He finally managed to get back to England in 1693 (Bidwell, 1976). Pitts wrote one of the first detailed accounts of the annual pilgrimage to Makkah, the *hajj*, and also published a plan of the city's great mosque (Radford, 1920).

Another half century passed by before the first team of scientists visited the region when, in October 1762, a small expedition of six, relatively young men funded by King Frederick V of Denmark, arrived in Jiddah on board a pilgrim ship from Egypt. This ill-fated expedition was led by Carsten Niebuhr, a German surveyor, and also included the botanist Pehr Forsskål, a Swede by birth and a pupil of the great taxonomist, Linnaeus. Niebuhr made a map of the city and also recorded the temperature at hourly intervals for some days. As the expedition members were not permitted to go to Makkah they decided to sail south along the coast to Yemen. Niebuhr and his friends were treated to wonderful hospitality saddened by the fact that several of the expedition members died of a mysterious disease – possibly malaria. Niebuhr returned to Copenhagen in 1788 with a wealth of cultural and scientific information. Although his expedition travelled little beyond the environs of Jiddah he should be regarded a truly great pioneering explorer and the first western scientist to visit the Kingdom.

1.3 THE NINETEENTH CENTURY

At the beginning of the nineteenth century Arabia was still poorly known to the outside world and for the most part it was still a land of myth and legend. Still less was known about the physical environment. European interests seem to have been stimulated by Napoleon's ill-fated invasion of Egypt in 1798. Along with his army of 30,000 soldiers there were nearly 1,000 civilians – mainly administrators, but also botanists, zoologists, artists, poets, surveyors and economists whose purpose was to document and describe all they came across. On their return home they had collected enough information to produce the twenty-three volume *Descriptions de L'Egypte*, which became the authoritative text on Egyptology for generations. All sorts of pharonic and Islamic treasures were plundered and sent back to Paris where they intrigued and fascinated all who saw them. As Victor Hugo noted in his preface to his volume of poems entitled *Les Orientales* (1829), "In the age of Louis XIV everyone was a Hellenist. Now they are all Orientalists."

European exploration of the Middle East in the nineteenth century seems to have been almost, but not entirely, a British affair as, indeed, it was in East Africa and the search for the Nile's source (Simmons, 1987). Doubtless Horatio Nelson's victory over Napoleon in the Battle of the Nile on August 1798, was a turning point in the history of the region. One exception to the British dominance was that of the professional Swiss explorer Johann Ludwig Burckhardt (1784–1817) who travelled in the region and converted to Islam. As Ibrahim ibn Abd-Allah he went on a pilgrimage to Makkah in 1814 convincing his challengers by his outstanding knowledge of the Qur'an and its wonderful poetic language. He mapped the city and also visited Madinah for several months. Burckhardt achieved much in his short life, including the discovery of Petra in Jordan. He died of dysentery in Cairo at the early age of 33. At about the same time the Spanish traveller, known as Ali Bey (1766–1818), also posing as a Muslim, set out from Cádiz in 1803, and travelled through North Africa, Syria, and Arabia. On reaching Makkah he fixed the town's position astronomically. His exploits are recorded in *Voyage d' Ali Bey en Asie et en Afrique* (1814).

While most early travellers in Arabia seemed genuinely to like the people and the landscapes Captain George Forster Sadlier of the 47th (the Lancashire) Regiment of Foot loathed both. In 1819 the British authorities in India were concerned that marauding pirates were a threat to British trade in the Gulf. Sadlier was sent to negotiate help from Ibrahim Pasha who had captured the Wahhabis' eastern capital of Dariya a year earlier after a siege of three months. In an attempt to meet up with the Pasha, Sadlier travelled from Qatif to Hufuf and then across the Nadj and onwards to the outskirts of Madinah and finally to Yanbu and Jiddah (Figure 1.1).

Sadlier had become the first European to cross Arabia and it would be another century before this intrepid journey was repeated. Sadlier scrupulously recorded the details of his journey in his diary and although of interest to contemporary geographers the crossing seemed to have interested him very little (Sadleir, 1977 – see note in bibliography regarding this misspelling).

In spite of Sadlier's journey across the Nadj there remained some confusion as to where this region was precisely. As I shall note, several later explorers described their journey to the Nadj when in fact they were often to the north of it. One of the first Europeans to visit the northern Nadj was the Finn, Georg August Wallin who, in 1848 disguised as a learned Muslim Sheikh called Hajji Abdul-Moula, travelled from Jiddah to Ha'il and then on to the Euphrates. A *hajji* is someone who has undertaken the *hajj* (Arabic: pilgrimage) – one of the five Pillars of Islam. Unlike Sadlier, Wallin was interested in the geography and sociology of the region. A description of his journey was read in London to the Royal Geographical Society in 1852, and a rather

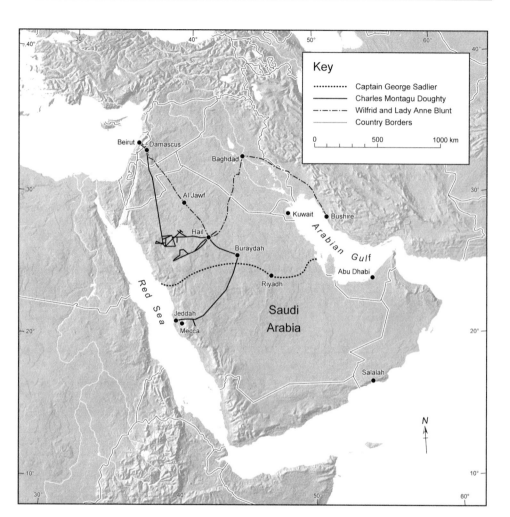

Figure 1.1 The Arabian journeys of Captain George Sadlier, Charles Montagu Doughty and Sir Wilfred and Lady Anne Blunt.

meagre account published two years later in the Society's journal. A year earlier Wallin was awarded the Royal Geographical Society's Founder's Medal, "*for his interesting and important travels in Arabia.*"

In 1853, the British adventurer and polymath, Sir Richard Burton (1821–90) arrived in Arabia and, disguising himself as a Pathan – an Afghanistani Muslim, went to Cairo, Suez, Madinah, and then on to Makkah. Though not the first non-Muslim to visit and describe Makkah, Burton's was the most sophisticated and the most accurate. His book *Pilgrimage to El-Medinah and Mecca* (1855–56) is not only a great adventure narrative but also a classic commentary on Muslim life and manners, especially on the *hajj*. Burton had also wanted to explore the Empty Quarter and had obtained the support of the Royal Geographical Society. However, when he asked the permission of the East India Company it was withheld much to Burton's disappointment (Thomas, 1931). Burton made some attempt to look for gold along the Red Sea coast north of Jiddah but was apparently not very successful.

One of the most enigmatic British travellers to the region at this time was William Gifford Palgrave who was born in London in 1826. After conversion to Catholicism Palgrave joined the Society of Jesus and in 1855 was posted to a Jesuit mission near Beirut. Palgrave fell in love with the Middle East and notified the Jesuit authorities that he wished to go as a missionary to central Arabia where few Europeans had been. His plans were approved and details conveyed to Napoleon III who funded the expedition in return for a full intelligence report on the details of trading possibilities, and of Wahhabi political organisation. Palgrave crossed the red sands of An Nafud, which he seemed to have hated, and then journeyed south to Ha'il and onwards to Buraydah and Riyadh, the capital of Nadj.

Palgrave had little time for the *badu* (collective *badu* : singular *badawi* – the word bedouin is not used in Saudi Arabia) whom he thought treacherous and promiscuous brigands. He did, however, have much praise for the town-dwellers who were both helpful and kind to him. In 1865 his *Narrative of a Year's Journey through Central and Eastern Arabia* was published with a dedication to Neibuhr who had first opened up the possibilities of travel in Arabia to Europeans. There is doubt in some quarters as to whether Palgrave's account of his journey is true. Philby, who later travelled in much the same area, thought much of it pure fiction. However, Philby had his own agenda and more recently some of Palgrave's descriptions, for example some large stone circles north of Riyadh, which Philby had dismissed as figments of Palgrave's vivid imagination, have been found by archaeologists from King Saud University.

The first explorer to make a serious photographic study of the Hijaz and the holy cities of Makkah and Madinah was the Egyptian General, Mohammed Sadek Bey (1832–1902). During his first visit to Arabia in 1862 he was involved in military land surveys between the town of Al Wadj on the Red Sea coast, and Madinah. Sadek's outstanding photographs were some of the very first to be seen by Europeans, and he was awarded a gold medal at the Venice Exhibition of 1881.

A year later, in 1863 a Levantine of Italian origin by the name of Carlos Guarmani went in disguise to the Jabal Shammar area, east of Ha'il, to procure horses, and perhaps spy, for Napoleon III. Guarmani started his journey in Jerusalem and visited Tayma, Khaybar, Unayzah, Buraydah and Ha'il. He evidently crossed parts of An Nafud but did not record any physical features. He wrote a lively and interesting account of his journeys which was published in 1866.

European travellers to Arabia in the nineteenth century were a colourful bunch by all accounts and the diminutive Lady Anne Blunt – she was just five feet tall, and her lecherous handsome husband, the poet Sir Wilfred Scawen Blunt, certainly lived up to that reputation. Lady Anne was a good shot and spoke excellent classical Arabic as did her husband though it is doubtful if this was much good when trying to converse with the locals. To her goes the honour of being the first European woman to visit the Arabian Peninsula (Winstone, 2003).

The Blunts were indefatigable travellers and had explored much of northern Arabia before travelling south to make a crossing of An Nafud in early January, 1879 (Figures 1.1 and 1.2). Their prime purpose was to see the famous horses at Ha'il belonging to Ibn Rashid and to bring back breeding stock and establish a blood line in England. The Blunts were very aware of the physical environment and while

Figure 1.2 The little palace in Jubbah where the Blunts rested during their crossing of An Nafud (Photo: author).

crossing the dunes made original observations about the long lines of horse-hoof shaped hollows which *radi*, their guide on the crossing, called *fulj*.(pl: *fuljes*) The Blunts measured some of the larger fulj at almost 80 metres deep and 500 metres wide and pondered their origin. Were they formed by wind or by water? Sir Wilfred leaned towards the water explanation – erroneously.

The Blunts arrived in Ha'il on January 24th having journeyed several hundred kilometres across terrain that would challenge even the hardiest modern traveller and certainly defeat a 4-wheel drive. Lady Anne published details of the journey in her book *A Pilgrimage to Nedj* in 1881. Sir Wilfred contributed an interesting appendix on physical geography including some original observations on the dunes (Simmons, 1987). It has to be said, however, that he was no lover of desert animals and took every opportunity to shoot at more or less anything with four leg or two wings. On the other hand, he was a romantic and seems to have fallen under the spell of Arabia much more so than his rather snobbish wife (Tidrick, 1989).

The Blunts survived their visit to Ha'il but could not explore further south into dangerous Wahhabi territory. On the February 1st 1879 they left with a group of Persian pilgrims returning from Makkah to Baghdad on the ancient pilgrimage road, the *Darb Zubaidah* built during the Abbasid caliphate in the ninth century.

At much the same time Charles Montagu Doughty (1843–1926) ventured to Arabia and spent two arduous years (November 1876–August 1878) travelling almost totally alone, often penniless and reliant on the hospitality and good will of his *badu* hosts (Figure 1.1). Doughty was unlike any other British explorer. His quiet dignity and unflinching Christian faith, frequently tested in what were often dangerous and demanding situations, have given us a most sensitive and detailed knowledge of the Arabian peoples and their land (Jellicoe, 1989). As T. E. Lawrence noted, "When his trial of two years was over he carried away in his note-book the soul of the desert."

Doughty's epic travelogue, *Arabia Deserta*, deliberately written in a difficult, archaic, English prose, was published in 1888 and is prefaced by the telling phrase, "*The Book is not Milk for Babes*" – a poignant reminder of the raw, often harsh, times he endured. In 1878 Doughty left Jiddah for Aden and returned to England. We have Doughty to thank for the first descriptions in English of the Nabatean rock tombs, akin to those at Petra in southern Jordan, at Madain Salih (Figure 1.3) which both Burkhardt and Burton failed to reach.

Doughty had a keen eye for rocks, though not that of an expert geologist. He constructed a general geological map along the routes he took during the period 1875 to 1878 and separated the granites from the basaltic lavas between Ha'il and Makkah.

Figure 1.3 Nabataean tombs at Madain Salih first described in English by Charles Montagu Doughty (Photo: author).

He wrote, in his usual difficult style, the following in *Arabia Deserta* (p.419, v.1) about the conditions he found on the lava plateau, Harrat Uwayrid:

> Viewing the great thickness of lava floods, we can imagine the very old beginning of the Harra – those streams upon streams of basalt, which appear in the calls of some wady-breaches of the desolate Aueyrid. Seeing the hillian are no greater we may suppose that many of them (as the Averine Monte Nuovo) are the slags and the powder cast up in one strong eruption. The earlier over-streaming lavas are older that the configuration which is now of the land:- We are in an amazement, in a rainless country, to see lava-basalt pans of the Harra, cleft and opened to a depth of a hundred fathoms to some valley-grounds as Thirba. Every mass is worn in grooves in the infirmer parts by aught that moves upon it; but what is this great outwearing of "stones of iron," indomitable and almost indestructible matter. We see in the cliff inscriptions at Madain, that the thickness of your nail is not wasted from the face of soft sandstone under this climate, in nearly 2,000 years.

Doughty's observations about the effectiveness of wind erosion are mostly correct at Madain Salih (Arabic: the place of the saint Salih) although he failed to notice that the bottom of many tomb entrances have been undercut by as much as a metre, the reason being that most sand particles travel in the bottom few metres of the wind column. Doughty also examined in some detail the *qanat* irrigation system in the town of Al Ula (Figure 1.4), south-west of Madain Salih, and made expert drawings of their construction. In 1912 Doughty was awarded the Founder's Medal of the

Figure 1.4 Al Ula as it was in Doughty's time – now abandoned (Photo: author).

Royal Geographical Society for his remarkable exploration in northern Arabia and for his classic book in which the results were described.

By the end of the century much of Arabia north of a line roughly from Makkah to Hufuf had been explored. To the south, the mountainous Asir, the southern Najd and the Rub' al Khali were almost totally unknown to the outside world. Elsewhere, no European had penetrated more than a few hundred kilometres from the coast except to Jawf and Najran (Kiernan, 1937).

1.4 TWENTIETH CENTURY EXPLORERS

Exploration of Arabia by Westerners was still mostly a British affair in the twentieth century although there were a few notable exceptions. One was the Czech explorer Alois Musil who travelled through northern Hijaz mountains and the Najd during the period 1908–1915 having first spent some time wandering in northern Arabia with the Rualla *badu*. He mapped the topography, collected a large number of plants and in 1911 helped make observations that led to the first general sequence of the Phanerozoic geological succession of north-west Arabia. On the face of it, this is very stuff of exploration but in fact Musil had been sent by the Austrian government and filed reports to the Institute of Military Geography in Vienna.

In the first half of the century Arabia attracted a quite remarkable and colourful group of British travellers and adventurers. Captain William Shakespear (1878–1915) – what a lovely name – fits the bill perfectly: a soldier, diplomat, amateur photographer, botanist, geographer and not least, an adventurer. He was the second British explorer to cross Arabia from east to west and seemed to have done it in some style. As a Political Officer from the India Office in Kuwait he saw the advantages of persuading the British Government to help the Arabs rise up against the Ottoman Empire and to trust Ibn Saud, the only ruler who the Arabs would follow.

Shakespear was an intrepid explorer and in 1914 made a long journey by camel from Kuwait to Riyadh and on to Buraydah and Jawf. Apart from his huge plate camera and a collapsible bathtub he also carried a supply of wine and whisky discreetly stowed away so as not to offend his *badu* guides. Much of his journey was in un-mapped territory and little known to the outside world. This was recognised in 1922 by the Royal Geographical Society who awarded him their Gold Medal for his contribution to desert geology and geography. Sadly, he was killed in a tribal skirmish at the early age of thirty-six. Shakespear's personal album of photographs and letters from the 1914 trans-Arabian journey is now housed in the archives of Al-Turath, a cultural foundation in Riyadh dedicated to Saudi heritage (Harrigan, 2002).

Not all explorers were what they first seemed, even though they travelled under the aegis of institutions such as the Royal Geographical Society. Captain Gerard Leachman, for example, was a geographer/botanist who made several impressive journeys in Arabia between 1910 and 1913 partly funded by the Society. In fact, he was a British government spy whose main mission was to negotiate with Abdul Aziz Ibn Saud in Riyadh.

At about the same time as Shakespear was making his crossing of the Peninsula so too was of one of the most remarkable women ever to venture into Arabia – Gertrude Bell (1869–1926). Bell (Figure 1.5) became fascinated with Arabia and its peoples

Figure 1.5 Gertrude Bell (Photo: PADUA).

after visiting friends in Jerusalem in 1899–1900 (Wallach, 1996). Her passion was for archaeology, and such was the respect for her knowledge that she became Honorary Director of Antiquities in Bagdad. Later in her career she was appointed Oriental Secretary to the High Commissioner in Bagdad and played an influential rôle in the boundary definitions of the new state of Iraq after the collapse of the Ottoman Empire.

Gertrude Bell only made one journey into northern Saudi Arabia in the spring of 1914. Her journey was all the more remarkable because she travelled as a lone Englishwomen accompanied by a few guides. She never crossed the central Nafud desert on her way down to Ha'il but she saw enough of the desert to make some interesting observations which she documented in her diaries and photographic collection, now kept at Newcastle University, UK. Much of this material is online at: http://www.gerty.ncl.ac.uk/

On the 11th February 1914 she noted in her diary:

> "We have come so far south (the khabra was but a day's journey from Taimah [Tayma']) in order to avoid the wild sand mountains (tu'us they are called in Arabic) of the heart of the Nefud and our way lies now within its southern border. This great region of sand is not desert. It is full of herbage of every kind, at this time of year springing into green, a paradise for the tribes that camp in it and for our own camels."

Her diary entry for the following day has more detail about the sand hills:

> "The tu'us are heads of sand with no herbage on them. They stand up pale yellow out of the Nefud and have a sort of moat round 3 sides, the 4th being generally a deep depression. The sand banks run mostly in an east-west direction, more or less. The valleys between them have a way of dropping suddenly into a very steep hollow called a Ga'r which on these edges of the Nefud has sometimes jellad at the bottom."

Clearly Bell had a good eye for landscape and her vast collection of photographs is now an invaluable archive of environmental information. She was awarded the Royal Geographical Society's Founder's Medal in 1918 for her important explorations and travels in Asia Minor, Syria, Arabia and the Euphrates.

Accounts of Harry Saint John (*Sheikh Abdullah*) Bridger Philby (Figure 1.6) often note that his grave in Beirut bears the inscription, 'Greatest of Arabian explorers. In many ways this is true, and to this very day few other explorers have travelled so widely in the Peninsula (Bidwell, 1976; Monroe, 1973).

In many ways this is true, and to this very day few other explorers have travelled so widely in the Peninsula (Bidwell, 1976; Monroe, 1973). Philby no doubt learned much from Gertrude Bell when the two served together in Basra and Bagdad from 1915 to 1917. Like Bell, he was strong-willed and rather egocentric, but unlike Bell he seemed to have been extremely short-tempered and undiplomatic.

Figure 1.6 Harry Saint John Bridger Philby (Photo: PADUA).

Philby's adventures in Arabia started in 1917 when, as an official in the Indian Civil Service, he landed in Bahrain and travelled on to Riyadh to meet the Sultan of Najd (later to become King Ibn Saud of Saudi Arabia). At that time the British were anxious to enlist the help of the Sharif of Makkah in their efforts against the Turks and were aware of the rivalry between the Sharif and Ibn Saud. Philby and Ibn Saud soon became good friends and their friendship was strengthened when, in 1930, Philby became a Muslim.

Philby crossed the Empty Quarter, ar Rub' al Khali, once west-east in 1932, and once north-south and back in 1936 (Figure 1.7).

On his first journey he wrote detailed and accurate accounts of his travels and sent plant and animal finds back to the British Museum. During his expeditions into the Rub' al Khali he discovered freshwater lake deposits and eggshell fragments attributed to a species of giant ostrich not dissimilar to finds in northern Africa (Bannerman, 1937). This Arabia-Africa faunal link has developed into one of the main themes in

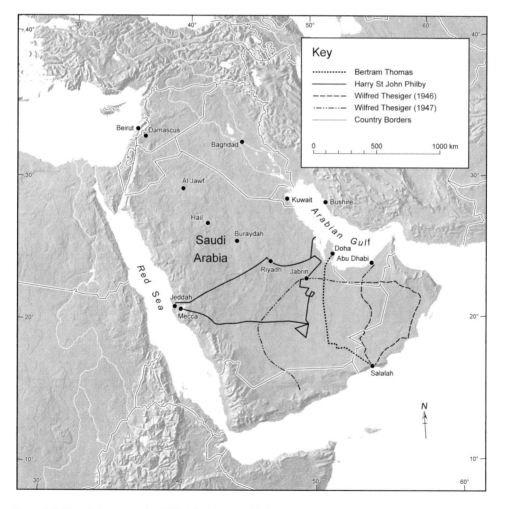

Figure 1.7 Travels into ar Rub' al Khali by Thomas, Philby and Thesiger.

Arabian biogeography. On his 1932 expedition to the Empty Quarter Philby became the first European to see the legendary Wabar impact crater. In his book, *The Empty Quarter* (1933a) Philby provides a wealth of geological and biological information and it remains a valuable source to this day. He also wrote for learned journals and published a number of interesting papers in the Royal Geographical Society's journal (1920a, 1920b, 1933b, 1949).

It was, however, Bertram Thomas (1892–1950), not Philby, who became the first European to cross the Rub' al Khali (Figures 1.7 and 1.8). This apparently quiet, unassuming, British civil servant, who at the time of his crossing was Wazir to the Sultan of Muscat, seems to have been the stuff of adventure yarns, par excellence. On October 6th 1930, and without the Sultan's knowledge, he secretly left Muscat harbour at midnight and boarded a homeward bound oil-tanker, the *British Grenadier*, which dropped him off near the village of Salala on the Dhofar coast of Oman. On December 10th, after a few months delay, his camel party set of for the *Sands,* as it was known by the local tribes, and arrived at Doha in Qatar on February 5th 1931, having made the 1,000 km crossing in fifty-seven days.

Thomas' crossing was not merely adventure. His aim was to determine the geological structure of this part of Arabia and to record its physical features and the wildlife. He was also interested in collecting details of the southern *badu* culture and languages. Thomas had a keen eye and made original observations on the massive dome dunes between Shu'iat and Shanna which he estimated to be as high as 230 metres. During his crossing of the *Sands* he collected some 400 specimens of mammals, reptiles and insects for the Natural History Museum (London). Of particular interest was Thomas' observation that the south Arabian fauna had African rather than Oriental affinities – a point I shall return to later in this book. In 1931 Thomas was

Figure 1.8 Bertram Thomas who made the first crossing of the Empty Quarter in 1931 (Photo: Library of the Faculty of Oriental Studies, University of Cambridge).

awarded the Royal Geographical Society's Founder's Medal for geographical work in Arabia and his successful crossing of Ar Rub' al Khali.

Oil brought other explorers on to the scene in the 1930s, this time American oil men from California chosen by the founder of the Kingdom, King Abd al Aziz (ibn Saud) in preference to British engineers not only because the King mistrusted the British, and their record of colonisation, but also because the Americans paid more. These oil men were not explorers in the ordinary sense but geologists who had a much more systematic approach to discovery and the lie of the land. One name that springs to mind is that of Karl Twitchell who produced an important geological and mineral survey of the Kingdom for the King.

Whilst American geologists were looking for gold and oil, others of a more romantic vision were challenged by the endless beauty of the desert, not least the great British explorer and self-professed nomad, Sir Wilfred Patrick Thesiger (1910–2003) who first read about the desert Arabs in Doughty's "*Arabia Deserta*" and T. E. Lawrence's "*Seven Pillars of Wisdom*."

Thesiger (Figure 1.9) encountered *badu* tribes for the first time as a member of the British S.A.S. behind enemy lines in Libya in the 1940s and was captivated by the timeless beauty of the desert and its people, and their astonishing ability to survive in such a harsh environment. When he discovered that the Locust Research Organisation was looking for people to travel the deserts of Southern Arabia in search of locust outbreak centres he jumped at the opportunity, and from 1945 to 1950 spent much of his time in the company of nomadic camel herders in Southern Arabia.

Thesiger's admiration for the *badu* is beautifully described in his classic travelogue *Arabian Sands* (1959) which details his two journeys across the Rub' al Khali, both feats of great endurance. His first crossing started out in October 1946, and penetrated

Figure 1.9 William Thesiger (Photo: PADUA).

the eastern *Sands* from Salala, in present-day Oman. His second, and more remarkable journey, was made in 1948 when he crossed the western Empty Quarter from Wadi Hadhramaut to the As Sulayyil oasis at the southern end of the Tuwayq Mountains. At that time, As Sulayyil was the home of one of the last *Ikhwan* (Arabic: brethren) communities in Arabia. This fanatical Wahhabi sect had no liking for Christians and Ibn Saud, angry that Thesiger had not sought permission to travel, ordered his arrest whereupon he was promptly locked up in the governor's castle. Philby, who was living in Riyadh at the time interceded on Thesiger's behalf and his party was able to continue eastward on the twelve hundred kilometre trek eastward to Abu Dhabi.

Thesiger made few scientific observations but his descriptions of the Arabian landscape and hazardous environments are beautifully crafted. In 1948 Thesiger was awarded the Founder's Medal of the Royal Geographical Society, *"for contributions to the Geography of Southern Arabia and for his crossing of the Rub' al Khali desert."* Thesiger's obituary in the Geographical Journal (Maitland, 2003) captures the essence of this remarkable explorer, traveller, writer and photographer.

Two other great Arabists, T. E. Lawrence – also known as Lawrence of Arabia (1888–1935) and Dame Freya Stark (1893–1993), made useful contributions to our knowledge of the Arabian environment. Lawrence, in helping Ibn Saud against the Turks, and the fall of Aqaba, never actually crossed the dunes of An Nafud – in spite of the wonderful story in the eponymous Hollywood blockbuster (see the map of his Hijaz compaign travels at: http://www.lawrenceofarabia.info/). Nevertheless he, more than anyone, brought world attention to wonderful desert environments of Saudi Arabia.

In two interesting papers Brookes (2003a, 2003b) has deconstructed Lawrence's magnum opus *Seven Pillars of Wisdom* (1926) to reveal remarkable geographical sensibility and landscape cognition. Indeed, Lawrence applied this geographical knowledge to strategic planning and the tactical execution of operations against Turkish forces. Lawrence was a trained archaeologist and must surely have observed the abundant evidence of early cultures in the desert terrain but presumably he had other things on his mind whilst planning his raids on the Turks.

The intrepid Dame Freya Stark, who was also very knowledgeable about archaeology and such matters, journeyed into Wadi Hadhramaut in the Yemen but a serious angina attack, and a timely rescue by the RAF, prevented her from travelling much further north into the Kingdom's southern Empty Quarter (Geniesse, 2000).

Quite why so many European and particularly British explorers were attracted to the mysteries of the Orient, and the vastness of its deserts, is somewhat puzzling. No doubt some were adventurers plain and simple, and others fascinated by the home of Islam and stories of *badu* hospitality and endurance. Whatever the answer, the draw is still there and much of Saudi Arabian environment remains as spectacular and fascinating now as it was then.

Chapter 2

The Kingdom

Contents

2.1 INTRODUCTION

In writing a book about Saudi Arabia's environment I am aware that many readers might have little knowledge of this part of the world. My aim in this chapter is to provide a very brief background to the Kingdom's geography and history. Contrary to popular belief, Saudi Arabia is not just a land of endless sand dunes, magic carpets

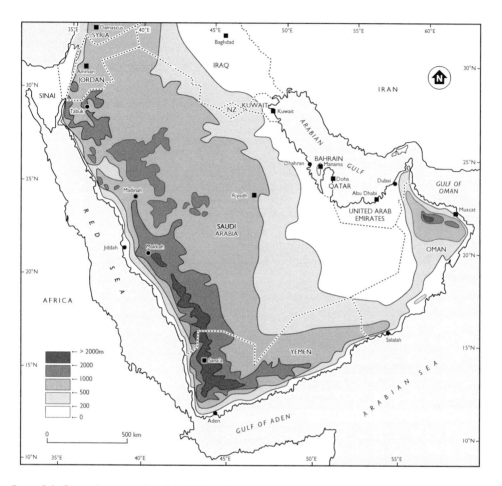

Figure 2.1 General topography of the Arabian Peninsula.

and *Tales From a Thousand and One Nights*. It is, however, a land of spectacular scenery and contrasts, from the verdant mountain valleys of the high Asir region, to the endless dunes of the Empty Quarter and the black, basaltic, lava fields of the Arabian Shield.

Saudi Arabia is a land of rapid economic and social change. The Empty Quarter is no longer quite so empty as oil companies seek the 'black gold' and gas buried in the rocks beneath the dunes. Many of the Kingdom's once small villages of mud-brick houses lying on either side of dusty caravan tracks have been transformed into thriving towns connected by modern highways and telecommunications.

Saudi Arabia occupies by far the larger part of the Arabian Peninsula which is sometimes just called Arabia, though this usage is not agreed upon unanimously. For example, the United States Geological Survey's Arabia maps include all countries north of Saudi Arabia up to the Turkish border. In ancient Greek and Latin sources, and often in subsequent sources, the term Arabia includes the Syrian and

Jordanian deserts and the Iraqi desert west of the lower Euphrates. The ancient Greek geographer, Ptolemy of Alexandria, described the following threefold division of Arabia in his famous '*Geography*': *Arabia Petræa*, miscalled 'stony Arabia', but really so-called from its chief city, Petra – essentially Jordan and Palestine; *Arabia Felix* ('happy Arabia') – the Asir province and Yemen and *Arabia Deserta* (meaning the interior) – probably the Nadj region of central Saudi Arabia and southern Iraq.

2.1.1 The physical setting

Saudi Arabia, which lies astride the Tropic of Cancer, has an area of 2,149,690 km^2 and occupies about three-quarters of the Arabian Peninsula. To get the country's size into perspective it is roughly nine times the size of the United Kingdom or about two-thirds the size of the European Union. The Peninsula is essentially an uplifted and tilted plateau that very gently slopes towards the Arabian Gulf, and contains some of the largest sandy deserts in the world (Figure 2.1). The climate is arid apart from the extreme south-west of the country where south-westerly Monsoon (Arabic: *mawsim*, season) winds are forced over the 3,000 m high Red Sea escarpment. Few places in Saudi Arabia receive more than 200 mm of rain a year and in the interior several years may pass without any rainfall at all. In summer the Peninsula bakes in temperatures which can reach into the 50s °C and on the high interior plateaus winter temperatures can drop below freezing. It is no wonder that this harsh environment is one of the most sparsely populated regions of the world.

2.2 MAIN PHYSIOGRAPHIC REGIONS

The combined influence of topography and geology has given rise to a number of distinct physiographic regions which are briefly describe here so as to provide the context for many of the chapters which follow. There is lack of agreement both for the names and precise definition of these regions and in Figure 2.2 I have used the definitions provided by the Saudi Geological Survey which are based on Brown *et al.* (1989).

2.2.1 The coastal lowlands

Both the Red Sea and Arabian Gulf coasts are fringed by coastal lowlands developed mostly on late Tertiary and Quaternary sediments. *Sibakh* (singular: *sabkhah* – a dry, salt encrusted, former wet zone) are common and evidence of sea level changes in terms of raised marine terraces is widespread.

2.2.1.1 The Red Sea coastal lowlands – Tihamah

A coastal plain runs more or less the whole length of the Red Sea but north of Jiddah it becomes narrow and discontinuous. From Jiddah south to Jizan the plain, or *Tihamah*, as it is often called, gradually widens to a maximum of about 40 km. Inland parts of the plain have been formed as a result of pedimentation of the foot slopes of the Precambrian scarp mountains and the coalescence of Quaternary

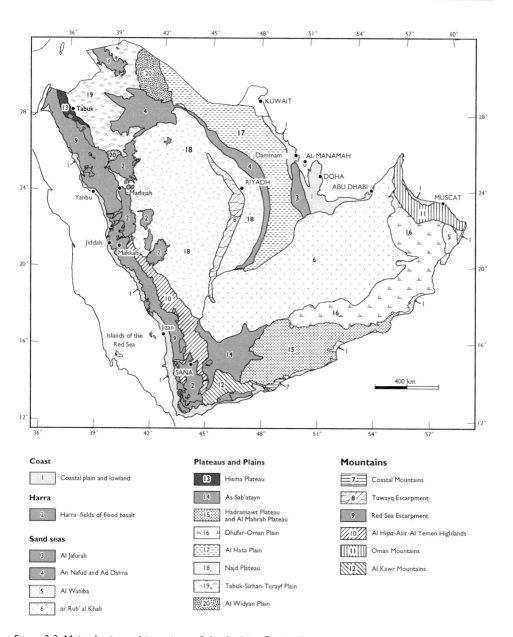

Figure 2.2 Main physiographic regions of the Arabian Peninsula.

gravel fans which have buried a young coralline platform. From Umm Lajj north to the Gulf of Aqaba portions of the plain comprise surfaces of raised Quaternary coralline reefs generally up to 30 m asl. (Figure 2.3) but where there have been neotectonic disturbances, as in the Haql area, the reefs have been raised up to 300 m asl. In places the coast is cliffed and high energy gravel fans debouch directly on to beaches.

Figure 2.3 Flight of raised beaches north of Umm Lajj (Photo: author).

South of Jiddah the *Tihamah* runs for several hundred kilometres until it narrows at Al Birk where it is terminated by the lava flows of Harrat Al Birk. This section of the coastal plain is known as the Tihamat ash Sham. South of the *harra* the plain widens to Jizan and is referred to as the Tihamat Asir. Much of the plain is hot, dry and infertile but in the Baysh-Jizan region the thick loessic alluvial silts in lower Wadi Baysh and Wadi Sabya, coupled with abundant groundwater, have given rise to significant agricultural development. In the Jiddah area high saline water tables are associated with *sibakh*. Air photo studies have shown that these *sibakh* have enlarged over the last thirty years or so and in Wadi Fatimah, south-east of Jiddah, farms have been abandoned as a result of the rising saline water tables.

2.2.1.2 The Gulf Coast plain

The coastal plain fringing the Arabian Gulf extends from the Kuwait border south to the United Arab Emirates. It is a zone of young Tertiary and Quaternary limestones and is mostly flat apart from a few low raised beaches and resistant reef patches forming hills 50 m or so asl. These are often draped by thick calcretes which form a protective crust. *Sibakh* are common and south of Dammam, barchan dunes of the Jafurah Desert sweep much of the salty surfaces clear of sand. Inland the coastal plain rises imperceptibly to the eastern edge of the Summan limestone plateau where it is often abruptly delimited, as at the Hufuf oasis, by fossil, possibly marine, cliffs, caves and lines of springs.

Figure 2.4 Young and old lavas on Harrat Khaybar north-east of Madinah (Photo: author).

2.2.2   *Harraat*

Basaltic lava flows (Arabic plural: *harraat*: singular *harra*: but before a named place: *harrat*) cover large portions of the Shield and give rise to high plateaus whose black surfaces, and low albedo, induce some of the highest air temperatures in the Kingdom. Old lava surfaces are weathered and can be traversed easily by jeep, but younger surfaces are a chaotic jumble of angular blocks and very difficult to cross (Figure 2.4). Although poorly vegetated, apart from the *wadi* floors which dissect the edges of the *harraat*, there is much evidence in the form of stone circles and linear constructions with unknown symbolic meaning, that in the past this forbidding terrain was once well-populated – perhaps in the Neolithic wet phase. In fact there are many springs on the *harraat* and conditions might not be as bad as perceived for those with local knowledge. The occasional camel can been seen grazing on these rocky wastes though the terrain is mostly too harsh for sheep and goats.

2.2.3   Sand seas – ergs

2.2.3.1   *Al Jafurah*

This small sand sea starts north-west of Dammam and widens southward to join Ar Rub' al Khali. Most of the dunes are mobile, crescentic, barchans blown south by the *shamal* winds often cover extensive *sibakh*. The dunes pose some hazards to industrial developments and airfields. Generally the sands are tan coloured but in places are a beautiful light cream colour. On the coastal edge of the main dune belt there are zones of low, horseshoe-shaped, parabolic dune anchored by vegetation though many sites have now been completely destroyed by infrastructure developments and especially those associated with the booming tourist industry. Overgrazing has re-activated barchans in several places and mobile dunes can be seen burying smaller parabolic dunes. Some of the inland *sibakh* probably formed when sea levels were higher. Deep below ground, buried in the young sedimentary rocks lie some of the largest oil and gas fields in the world.

2.2.3.2   An Nafud – Ad Dahna

The Nafud basin is part of a great plain, some 375,000 km^2 in area, that occupies north-central Saudi Arabia. An Nafud, or the Great Nafud, sand sea occupies some 57,000 km^2 of the basin between Jawf and Ha'il and has an average elevation of 900 m asl. Chapman (1978) describes the shape of An Nafud as being like a giant hand with long fingers pointing towards the east. The desert is without *widyan* or oases though the presence of wells, particularly in the west indicates a shallow ground water table fed by percolation. The village of Jubbah, about 80 km north of Ha'il, lies under the shelter of a small inlier of reddish Saq sandstone and is almost totally surrounded by dunes though there is a magnificent new Tarmac road south to Ha'il which cuts through the desert. Jubbah has been a settlement site for several thousand years and the sandstone petroglyphs found nearby are the most famous in the Kingdom (Figure 2.5).

Ad Dahna is one of the most distinctive physiographic units on the Peninsula. This long, arcuate belt of reddish sand dunes some 80 km wide stretches some 1,300 km from An Nafud southward to the Rub' al Khali. The sands are stained with iron oxide like those of An Nafud. The southerly moving mobile barchan dunes of Ad Dhana are a hazard where the main east-west highway from Riyadh to the Gulf cuts through the dune field.

2.2.3.3   Ar Rub' al Khali

The Rub' al Khali (the Empty Quarter) is locally known as *Ar Ramlah* (The Sands). It is the largest continuous expanse of sand desert in the world and has an area of some 640,000 km^2. The western region consists of fine, soft, sand and has an elevation of 600 m asl. The elevation drops steadily to the east where sand dunes are interspersed with extensive salt flats. Linear dunes are very common in the Rub' al Khali but zones of sand mountains up to 300 m high are also found. After rains the vegetation provides surprisingly good grazing. The eastern region of the Rub' al Khali has now been opened up by oil companies with a major oilfield being developed at Shaybah near the border with the United Arab Emirates.

2.2.4   Mountains

2.2.4.1   Red Sea Escarpment mountains

Ramping and faulting associated with the uplift of the Shield has developed a southwest facing mountain range some 40 to 140 km wide, known as *As Sarawat*.

Figure 2.5 Petroglyph at Jubbah (Photo: author).

Sets of faults running more or less parallel with the Tihamah demarcate the western foothills of the range which rise steadily towards an escarpment face which has eroded inland of the faulted zone since the mid-Tertiary uplift (Figure 2.6).

The highest point of the escarpment in Saudi Arabia is at Jabal as Sudah (also spelled Sawadah or Souda – the Black Mountain) to the north-west of Abha. From here the escarpment generally declines in altitude northwards and is about 1,000 m asl in the Madinah area. From Madinah northwards, through the greater part of the Hijaz, an erosional escarpment is hardly developed. It is not entirely clear why this is the case but apparently uplift along the escarpment is still active south of Taif.

Figure 2.6 Red Sea Escarpment west of Abha (Photo: author).

Where well-developed the escarpment is characterised by very steep *widyan* draining towards the Red Sea. Slopes are bare of any soil cover and small, dissected, gravel fans are common. Some *widyan* have eroded headward and have captured parts of the Tertiary drainage system which formerly flowed in a more north-westerly direction. Several *widyan* are spectacular. The canyon of Wadi Lajb (N17°35', E42°54') some 40 km north-east of Baysh is one of the most spectacular landforms in Saudi Arabia. In places the walls of the canyon are up to 400 m high and overhang the *wadi* floor which is only 3–20 m wide in places (Figure 2.7).

2.2.4.2 Tuwaiq – Al Aramah Escarpments

The central Nadj region is dominated by several prominent Jurassic cuestas (dip slope plus escarpment) that extend for some 1,600 km in an arc from An Nafud in the north to the Rub' al Khali in the south. The 800 km long, west-facing, limestone escarpment of the Jabal Tuwaiq is the largest of the escarpments and rises some 240 m above plains to the west. Its southern section, fringing Ar Rub' al Khali, has an average altitude of 800–1,000 m asl. The dip slope components of these cuestas are very gentle indeed and rarely slope more than two degrees to the east (Figure 2.8).

The Jabal Tuwaiq escarpment is breached by a number of major *widyan*, such as Wadi ad Dawasir and Wadi Birk, which drain from the Shield towards the Gulf. North-east of Riyadh is the 250 km long Al Aramah escarpment. This has an elevation

Figure 2.7 The spectacular canyon of Wadi Labj (Photo: author).

Figure 2.8 Tuwaiq Escarpment north of Wadi ad Dawasir looking north (Photo: author).

of about 540 m but is less prominent as it only stands about 120 m above the plain to the west.

2.2.4.3 Hijaz-Asir Highlands

The Hijaz-Asir Highlands (Arabic: *Al Hijaz* – the barrier) is a wedge-shaped crystalline plateau east of the Red Sea escarpment which extends from the Yemen northwards towards the Taif - Makkah – Madinah region where it is finally buried by the lavas of Harrat Rahat. Inland, the plateau comprises a mosaic of weathered, pedimented, basins and inselbergs in part buried by extensive lava fields such as the Harrat Khaybar – north-west of Madinah, and the Harrat Rahat between Madinah and Makkah. From Al Bahah south towards the Yemen boarder the plateau rises in elevation to more than 2,500 m asl. with many peaks in excess of 2,500 m. This rugged southern section of the plateau is commonly known as the Asir (Arabic: *Asir* – difficult) and the spectacular terrain continues into Yemen where the mountains rise to 3,760 m. Some small permanent rivers drain from the higher mountains such as Jabal Batharah but none flows for more than 50 km or so before disappearing into the *wadi* floor.

2.2.5   Plateaus and plains

2.2.5.1   Hisma Plateau

The Hisma Plateau is a region of Permo-Triassic sandstone mesas and buttes in the north-west of the Kingdom to the east of Tabuk. The plateau is about 100 km wide and, like the Hijaz to the south, is a ramped block sloping gently towards the north-east. The southern section of the plateau is overlain by the lavas of Harrat ar Rahah.

2.2.5.2   Tabuk Plain

The Tabuk Plain is a very shallow basin developed in Palaeozoic sandstones and is surrounded on three sides by mountains. At an average altitude of 800 m, the area around Tabuk city is a major cereal growing area with extensive use of groundwater for irrigation.

2.2.5.3   Nadj Uplands

The Najd (Arabic: *Najd* – highland) is a vast high plain stretching from the centre of Arabian Shield and eastward onto the sedimentary cover rocks as far as the Dahna dune belt. To the west of Riyadh the Nadj is interrupted by the cuestas of the Jabal Tuwaiq. To the north it is fringed by the Nafud basin and to the south by the sands of the Empty Quarter. The Nadj is dusty and hot in summer, and cold in winter. At its maximum the Nadj is about 580 km wide and gently slopes eastwards and south-eastward reflecting the configuration of the buried basement rocks. To the west the Nadj is a rocky expanse of coalescing gravel-strewn etchplains, pediments, inselbergs and *harraat* (Figure 2.9). To the east, on the sedimentary cover rocks, the Nadj is draped by vast, almost imperceptibly sloping, gravel fans emanating from the Shield and is broken by the presence of the Tuwaiq and younger escarpments.

2.2.5.4   As Summan

As Summan is a narrow, arcuate, mainly limestone plateau lying to the east of the Dahna dune field. The plateau is barren and extensively corroded and eroded. To the north, it is covered with Quaternary gravels of the Ad Dibdibah fan which spreads north-eastward into Kuwait. In the west the plateau has an elevation of c.400 m asl. and slopes gently eastward to the Arabian Gulf where its elevation is about 250 m. The whole plateau is karstified and some of the Kingdom's largest caves are located in the As Summan to the north-east of Riyadh. A number of oases such as that at Hufuf, are located at spring sites and the karst zone is an important zone of shallow groundwater recharge.

2.2.5.5   Al Widyan Plain

This high plain some 500 m asl. dips gently north-eastwards from An Nafud into south-central Iraq. These pedimented rangelands, covered with desert pavements of cherty gravels, are dissected by thousands of shallow *widyan* oriented down dip (Arabic: *widyan* – plural of *wadi*).

Figure 2.9 Crystalline Nadj uplands 100 km west of Unayzah (Photo: author).

2.3   A JOURNEY THROUGH TIME

2.3.1   Out of Africa – earliest human presence

The geographical dispersal of early hominids (*Homo erectus*) into the Arabian Peninsula from eastern Africa is thought to have taken place no later than 2.0 to 1.9 Ma. This is based on finds of *H. erectus* both in Java and central China dating from about 1.8 Ma and the time needed for *H. erectus* to migrate eastwards out of Africa into eastern Asia. *H. erectus* made simple bifacial tools such as hand axes and archaeologists class these implements as belonging to the Lower Palaeolithic Acheulian stone tool industry.

Whalen and Schatte (1997) suggest two possible migration routes for these early hominids: one route down the Nile and across the Sinai and on into Arabia; the other, and more direct route, across the Bab al Mandab strait into Yemen (Masry, 1977). This latter route conforms to a dispersal trajectory of early hominid sites from Olduvai in East Africa north-east to Djibouti and then on into western Asia (Figure 2.10).

In an interesting observation Whalen *et al.* (1989) note that the distance across the Bab al Mandab is only about 28 km and from the shoreline of either continent one may see the coastline of the other. They suggest that the crossing could have been made on rafts using the small island of Perim, near the Yemen coast, as a stopover. Once on the Peninsula two migrations routes are probable. One was northward along the Tihamah (Zarins *et al.*, 1980; Zarins *et al.*, 1981) and also the well-watered Asir

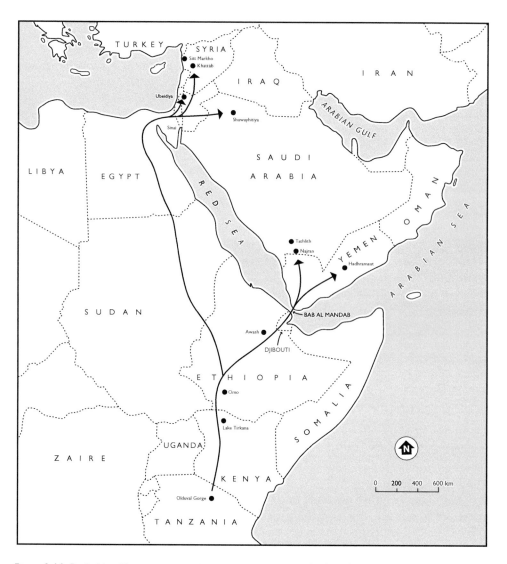

Figure 2.10 Probable *Homo erectus* migration routes into Arabia (Modified from Whalen and Schatte,1997).

highlands to Najran where there is a site of possible site of Acheulian age. A second route was eastward along the coast towards Oman.

Quartzite artefacts made by *H. erectus* entering Arabia via the northern route have been found in Wadi ash Shuwayhitiyah and possibly date from about 1.4 My. Morphologically the artefact types correlate well with Acheulian tools from sites in the Olduvai Gorge, East Africa.

It is now widely believed that all hominid lines which had descended from these original migrating *Homo erectus* populations became extinct. Templeton (2002) suggests a second major 'out of Africa' wave occurred between 840,000 and 420,000 BC when *Homo erectus* was evolving into early archaic *Homo* types.

A final 'out of Africa' migration took place across the Bab al Mandab into southern Yemen about 80,000 BC. Sea levels were low and exposed reefs and numerous islands would have made the journey relatively easy. At this time the northern Nile Valley route out of Africa would have been arid and much less likely. Mitochondrial DNA evidence indicates that perhaps as few as 250 early *Homo sapiens* made the crossing and their descendants were eventually to disperse throughout the world. Geneticists, who are able to link the information in the DNA to dispersal routes suggest that we are all African descendants of a so-called mitochrondrial Eve (Cann, 1987; Templeton, 2002)

By about 75,000 BC hunter gatherers of the Mousterian tradition were present in Arabia and settlements from this date are quite widespread (Masry, *op. cit.*). An interpretation of the field evidence indicates that the main factors accounting for the distribution of Palaeolithic settlements are the fossil ecological settings of ancient lake beds, extinct *wadi* drainage systems and raised terraces. Such locations would presumably be rich in game and have reasonable supplies of fresh water. Nomadic hunter-gatherers continued to roam the Arabian Peninsula throughout the Late Pleistocene and there is evidence for still later hunters in areas of former lakes at the north-western edge of the Rub' al Khali. The oldest set of lakes for which there is still evidence date from about 34,000 BC but about 17,000 BC the numerous lakes dried up during a long period of aridity and the hunters were force to migrate (McClure, 1976).

From 7,000 to 5,000 BC the climate over the Arabian Peninsula improved and lakes re-formed in the Rub' al Khali. By this time Neolithic hunter-pastoralist had entered the region from the north and established partially settled communities. Concomitant with these emerging Neolithic communities was the establishment of long-range migratory herding of domesticated cattle, goats and sheep on the savannah grasslands. Cleuziou and Tosi (1998) discuss the subtle relationship between the evolution of these hunter-gatherer populations and various climatic and environmental changes in the Holocene. These researchers argue that the evolution of human communities was not the result of a direct response to environmental change, nor was it an attenuated echo of the Neolithic economy in the Fertile Crescent or even later, the products of direct influences from the urban societies of the ancient Orient. They suggest that there were elements of a local evolution of the Holocene hunter-gatherer societies of Arabia, grounded in their own adaptive strategies to highly constraining environments in which they lived.

It is thought that at about 5,000 BC the hunter-pastoralists came into contact with settled village peoples of the Tigris-Euphrates Valley. Coastal traders of the 'Ubayd period in Mesopotamia made contact with the hunter-pastoralists living along the shores of the Arabian Gulf and the first villages were established – the so-called pre-pottery Neolithic sites. It was about this time that agriculture was introduced into the area as knowledge of Mesopotamian farming methods diffused. In the Jabal Mokhyat area (N20°12', E43°28') in the south-western Nadj charcoal from campfire sites used by hunters has been dated at 5,930 ± 300 years BP (USGS Carbon Laboratory, W-2949, January 9, 1974). Charcoal from a similar fire sites in the Nafud Hanjaran area, 125 km north-west of Jabal Mokhyat, gives a similar date of 6,350 ± 350 years BP (USGS Carbon Laboratory, W-3282, December 3, 1975).

The cultural legacy of the Neolithic settlements in Arabia, apart from the chipped-stone industries, is the widespread rock art (petroglyphs in low relief) which is found particularly in the northern Nadj and the Asir (Anati, 1972, 1974; Courtney-Thompson, 1975). The best known rock art site is at Jubbah, 80 km north of Ha'il in the central Nafud desert. Here, on the faces of red sandstone cliffs a wide variety of finely carved relief figures depicting adorned humans, long-horned bovids and equids are found in profusion (Figure 2.5). Nearby traces of pre-pottery settlements are found along the edges of a fossil lake bed. Masry (1977) suggests that the rock art is related to these settlements and puts a date on the occupation at between 6,000 and 3,000 BC.

2.3.2   The first permanent settlements (3,000 BC to 200 AD)

Coastal villages of mud and reed houses were first established in spring-fed areas between Dhahran and Jubail on the Gulf coast and then spread inland to Hufuf along *wadi* and former lake systems (Masry, 1974). About 3,000 BC these settlements gave way to more permanent villages with houses made of mud bricks. The permanency was accompanied by the development of trading centres near Qatif and Tarut Island about 30 km north of present-day Dhahran (Larsen, 1983). From about 3,000 BC to 2,500 BC the coastal settlements of eastern Arabia probably made up the ancient Bronze Age Kingdom of Dilmun (today's Bahrain and neighbouring mainland inland to Hofuf) which supplied wood, pearls and dates to the growing urban population in Mesopotamia. Masry (1977) suggests that it is not an exaggeration to describe the commercial activities of the coastal traders of the Gulf at this period as the busiest 'international' trading system that ever developed in the history of ancient civilisations with trade networks stretching east as far as the Indus Valley.

A further phase of aridity set in towards the end of this period and the coastal trading centres at Qatif and those on Bahrain declined along with the collapse of major cultural centres in Mesopotamia, the Indus Valley and south-west Iran. For much of this period hunter-pastoralists still occupied vast inland areas of the Peninsula and in the north-west of Arabia along the Red Sea some trade with Eygpt was conducted from newly established coastal settlements. Both Zeuner (1954) and Field (1960) describe Neolithic arrowheads from sites in the Rub' al Khali during this period. At a camp site (N18°46', E50°16') some arrowheads where collected along side a hearth the charcoal from which was dated at 3,131 ± 200 BP.

2.3.3   Camel pastoralism and the ancient kingdoms (1,800 BC to 570 AD)

The desiccation of the Arabian Peninsula between about 1,800 BC and 1,500 BC coincides with the influx from the north of camel-mounted, Semitic nomads referred to for the first time as Arabs. The domestication of the camel with its wonderful adaptation to desert life gave the incomers distinct advantages over the hunter-pastoralists and over the course of about 1,500 years, these proto-*badu*, as they are some time known, developed into a unique cultural group of pastoralists in the heart of Arabia. The material conditions under which the Arabs lived began to improve around 3,000 BP when a method of saddling camels to transport large loads had been developed and caravan routes were established across the harsh interior of the Peninsula.

By 600 BC increased trade in spices and incense brought wealth, particularly for the small Kingdoms, such as Saba, Main, Qataban and Haramawt, which flourished in the south-west of Arabia (present-day Yemen) and grew rich by trading frankincense and other aromatic gums. One caravan route from the south headed north through Makkah and via the oases of the Hijaz to Egypt, Syria and the Mediterranean. Another passed along the western edge of the Rub' al Khali through Qaryat al Faw, now just an archaeological site, and onward to central Arabia and the Gulf. The famed Kingdom of Kinda, centred perhaps at the trading settlement of Qaryat al Faw is widely regarded as the earliest identifiable politco-cultural administration of the Arabian heartland.

In the northern part of the Peninsula the Nabataean Kingdom was established about 400 BC by nomads from the northern Hejaz who settled in Jordan, the Negev and northern Arabia. The Nabataeans grew very rich by controlling the caravan routes from the Kingdoms of the south and established Petra (in Jordan) as their capital. Their southernmost stronghold was Madain Salih in north-western Saudi Arabia. This is the most famous archaeological site in the Kingdom and has wonderful remains of rock tombs and dwellings. Foreigners need to get a permission to enter the fenced site but this can be obtained in Al Ula – be prepared for a wait.

With the rise of Rome, the Peninsula's commercial wealth came under military threat but an invasion launched by Aelius Gallus in 24 BC ended in failure. The Kingdoms of the Peninsula were, however, doomed. The Nabatean Kingdom was eventually incorporated into the Roman Empire as Arabia Petraea in 106 AD, and once the Greeks and Romans had discovered the Arab methods of sailing to India on the Monsoon winds the people of the once-prosperous south migrated northwards as trade withered. In the fourth and sixth centuries AD south-western Arabia fell under Abyssinian rule and it was during the year 570 AD, when Makkah successfully fought off an attack by the Abyssinian Abraha, that the Prophet Muhammad was born.

2.3.4 The early Islamic period (570 AD–700 AD)

The Prophet Muhammad was born in Makkah, the son of Abd Allah and Aminah who belonged to the wealthy and powerful Quraysh tribe who controlled pilgrimage water rights in the city. Abd Allah had died before Muhammad was born but the young Muhammad was fortunate to have the protection of his uncle Abu Talib who was one of the leaders of the Hashimite clan of the Quraysh. Muhammad gained financial security when, at about the age of 25, he married a rich widow, Khadija. Muhammad would often go into the hills that surround Makkah where he saw visions and Muslim belief holds that the angel Gabriel recited versus to him which later became part of the Qur'an. In 613 AD he proclaimed the word of God, as revealed to him by the angel Gabriel, in the streets of Makkah. Fearing the economic repercussions of Muhammad's preaching against the deities worshiped by the pilgrims at Makkah's central shrine, the Kaaba, the leading families of the city persecuted Muhammad and his followers. In about 620 AD Muhammad's guardian and protector, Abu Talib died and an enemy uncle nicknamed Abu Lahab (Father of the Blaze) became head of the clan. Muhammad's life was no longer safe in Makkah and in 622 AD he and his followers secretly left the city and moved 445 km north to the small agricultural settlement of Yathrib which was renamed Madinah (*madinat al-nabi*, the 'city of the prophet'). The Muslim era dates from the *hijra* – Muhammad's migration to Madinah (A.H. 0).

Over the next ten years or so Muhammad rose to become the political leader of virtually all of central and southern Arabia. A Muslim victory over a larger force of polytheists from Makkah at a caravan watering site called Badr in 624 AD became the turning point in Muhammad's rise to power. In 628 AD the Makkah authorities signed the Treaty of al-Hudaybiya which led to the peaceful surrender of the city to Muhammad and his followers in 630 AD. By the time of his death in 632 AD Muhammad enjoyed the loyalty of almost all the tribes of Arabia though not all had become Muslim. The Prophet Muhammad had no spiritual successor in the sense that the word of God had only been revealed to him. There were, however, successors to the Prophet's temporal authority called caliphs.

2.3.5   The caliphate and the Ottomans (700 AD–1700 AD)

The caliphs ruled the Islamic world until 1258 AD when the last caliph and all his heirs were killed by the Mongols. Under the caliphs the Arab armies, united under Islam, established a vast empire from what is now Spain to Pakistan. After the third caliph, Uthman, was assassinated in 656 AD, the Muslim world was split and the fourth caliph, Ali, spent most of his time in Iraq. When Ali was murdered in 660 AD the Umayyads established a hereditary line of caliphs in Damascus but in 750 AD they were overthrown by the Abbasids who ruled from Baghdad. The effect of all this murder and intrigue was to shift the centre of political power northward and Islamic civilisation was no longer centred on Makkah and Madinah. From the 8th to the early 10th century Arabia was merely a province of an empire ruled from Baghdad. Makkah remained the religious centre of the Islamic world and pilgrims from all over the Islamic world passed through the Hijaz to perform *hajj*. By far the most important pilgimage route was was Darb Zuybaydah (the Road of Zuybaydah or Pilgrims' Road) stretching 1,600 km from Baghdad to Makkah. Even in its ruined state the Darb Zuybaydah is regarded one of the most significant architectural monuments in the Kingdom.

The expansion south of the Seljuk Turks in the eleventh century and the beginning of the First Crusades in 1095 AD had little real impact on the Peninsula. The Ayyubids of Egypt, who had invaded south-western Arabia in 1173 to control the India trade route help protect coastal region and in 1181 AD Reynard de Châtillon's raids on the Red Sea ports as far south as the strait of Bab al Mandab were foiled when his vessels were destroyed by the Ayyubid Saladin. A more important threat to the Muslim world came with the first expansion of the Mongols in the thirteenth century when, in 1258 under Hulaga Khan, they invaded Baghdad and put an end to the Abbasid caliphate. The Mongols attempted to move against Palestine and Egypt where they were met tough adversaries in the Mamluks (Arabic: taken into possession), fearsome soldier slaves from the Turkish steppe who had been recruited by the caliphs of Egypt. During the 14th and 15th centuries AD the Egyptian Mamluks became an important power maintaining political control of the Hijaz and a body of cavalry in Makkah. At the beginning of the 16th century the economy of the Red Sea ports, and indeed of Arabia as a whole, was severely damaged by the Portuguese penetration of the Indian ocean where they blockaded the Indian trade routes to Europe via the Arabian Gulf and Red Sea.

In 1517 AD the Ottoman sultan Selim I conquered Egypt and took control of the Hijaz, the two Holy cities and Jiddah, and exercised authority southward into what is present-day Yemen. By 1550 Ottoman authority had reached the Arabian Gulf but by 1635, the Zaydis of Yemen, supported by northern tribes, expelled the Ottomans whose power base lay in the Hijaz where they controlled the Holy cities particularly for prestigious reasons. The effects of these invasions can still be detected in the Hijaz and Asir where literally thousands of fortified towers remain testament to this period of Ottoman rule.

2.3.6   The rise of the Wahhabis

Until this time the tribes of vast interior of the Kingdom were largely unaffected by the Ottoman presence and little had changed over hundreds of years. In the mid-eighteenth century that was all to change with the rise of the puritan Wahhabis in Nadj. The Wahhabi movement, founded by Muhammad ibn 'Abd al-Wahhab (1703–1792), was regarded as dangerous and heretical by the Turkish rulers in the Hijaz who felt threatened by this pre-modernist, revivalist movement who set out to purify Islam from the heresy of 'innovation'. Moreover Wahhabism aroused a national spirit against Turkish domination. 'Abd al-Wahhab was born into a family of religious scholars at Ayayna in Najd. In 1745 he moved to the small village of Ad Dir'iyah about 20 km to the west of Riyadh where he won the protection and adherence of its chief, Muhammad ibn Saud.

Muhammad ibn Saud and his son propagated the doctrines of the movement widely and soon had followers all over Nadj. The Wahhabis took Makkah from the Turks in 1802 and Madinah in 1804. The Turks were horrified at the loss of the two Holy cities and won the help of the viceroy of Egypt to regain the two cities in 1812. Repulsed from Hijaz the Wahhabis strengthened their grip in the Nadj and made Riyadh their capital in 1818.

There was much warfare between the various Wahhabi factions, and rivalry over succession weakened their control of the Nadj allowing Turkish incursions. At about the same time there was also a challenge from a rival dynasty in north-central Arabia, the al-Rashids, who moved on Riyadh and defeated the Wahhabi forces in 1891. The Wahhabis were forced to retreat south to the fringes of the Rub' al Khali where they were provided shelter by the Al-Murri tribe – forging a bond of loyalty that exists up to the present day.

In 1901 the young Abdul Aziz ibn Saud, Sultan of Najd, gathered a band of about 40 men and moved on the Musmak fort in Riyadh. In a minor skirmish Abdul Aziz ibn Saud, seized the fort, thereby returning Riyadh into Saud family hands. By 1906 the various faction of the Wahhabi movement had become united under the leadership of the resourceful, intelligent and imposing Abdul Aziz ibn Saud, under whose inspired leadership the Wahhabi state expanded to the Gulf and south to the Yemen.

2.4   THE MODERN KINGDOM – GROWING ENVIRONMENTAL PRESSURES

At the start of World War I the Ottomans held all of western Arabia and were supported in central northern Arabia by the Rashids of Ha'il. Ibn Saud, however, cooperated with Great Britain in fighting the Turks and victory resulted in their expulsion from the Hijaz, 'Asir and Yemen. Six years of civil war followed between followers of ibn Saud and other tribal leaders, particularly the ibn Rashid who was captured in 1921. In 1924 ibn Saud conquered the Hijaz becoming its king on the 8th January 1926. A year later, in 1927, he proclaimed himself King of Hijaz and of Najd and on September 23rd 1932 he gave his two dominions the name of Saudi Arabia and declared it an Islamic state, with Arabic designated as the national language and the Holy Qur'an as its constitution. Ibn Saud's annexation of Asir to Hijaz in 1933 caused a short war in the spring of 1934 between Saudi Arabia and the imam Yahya of Yemen. The treaty signed by the two countries on May 20th, 1934 provided that Asir and the inland region of Najran were to remain within Saudi Arabia

Boundaries with Jordan, Iraq, and Kuwait were established by a series of treaties negotiated in the 1920s, with the establishment of two 'neutral zones' – one with Iraq the other with Kuwait. The Saudi-Kuwaiti neutral zone was administratively partitioned in 1971, with each state continuing to share the petroleum resources of the former zone equally. Tentative agreement on the partition of the Saudi-Iraqi neutral zone was reached in 1981, and partition was finalized by 1983. The country's southern boundary with Yemen, which was partially defined by the 1934 Treaty of Taif, was finally determined by agreement in 2000. The border between Saudi Arabia and the United Arab Emirates was agreed upon in 1974. Boundary differences with Qatar remained unresolved.

King Abdul Aziz "ibn Saud" died in 1953 and was succeeded by his eldest son, Saud, who reigned for 11 years. In 1964, King Saud abdicated in favour of his half-brother, Faisal, who had served as Foreign Minister. Because of fiscal difficulties, King Saud had been persuaded in 1958 to delegate direct conduct of Saudi Government affairs to Faisal as Prime Minister; Saud briefly regained control of the government in 1960–62. In October 1962, Faisal outlined a broad reform program, stressing economic development. Proclaimed King in 1964 by senior royal family members and religious leaders, Faisal also continued to serve as Prime Minister. This practice has been followed by subsequent kings.

The mid-1960s saw external pressures generated by Saudi-Egyptian differences over Yemen. When civil war broke out in 1962 between Yemeni royalists and republicans, Egyptian forces entered Yemen to support the new republican government, while Saudi Arabia backed the royalists. Tensions subsided only after 1967, when Egypt withdrew its troops from Yemen.

Saudi forces did not participate in the Six-Day (Arab-Israeli) War of June 1967, but the government later provided annual subsidies to Egypt, Jordan, and Syria to support their economies. During the 1973 Arab-Israeli war, Saudi Arabia participated in the Arab oil boycott of the United States and Netherlands. As a member of the Organization of Petroleum Exporting Countries (OPEC), Saudi Arabia had joined other member countries in moderate oil price increases beginning in 1971. After the 1973 war, the price of oil rose substantially, dramatically increasing Saudi Arabia's wealth and political influence.

The growth of oil revenue in the 1960s and 1970s had major impacts on the agricultural sector of the economy in which the majority of the population was still employed. Many poor peasant farmers and agricultural workers left the land and moved to the growing urban areas in search of higher incomes. Some sold their farms to rich merchants and others gave up their small rented plots to seek work in the booming oil industry. Even the *badu* did not go unaffected. Trucks replaced camels, and sheep breeding replaced camel breeding. Trucks not only moved the flock around but also brought water to remote encampments.

In 1975, King Faisal was assassinated by a nephew, who was executed after an extensive investigation concluded that he acted alone. Faisal was succeeded by his half-brother Khalid as King and Prime Minister; and his half-brother Prince Fahd was named Crown Prince and First Deputy Prime Minister. King Khalid empowered Crown Prince Fahd to oversee many aspects of the government's international and domestic affairs. Economic development continued rapidly under King Khalid, and the Kingdom assumed a more influential role in regional politics and international economic and financial matters.

King Khalid died in June 1982, and Fahd became King and Prime Minister in a smooth transition. Another half-brother, Prince Abdullah, Commander of the Saudi National Guard, was named Crown Prince and First Deputy Prime Minister. King Fahd's brother, Prince Sultan, the Minister of Defence and Aviation, became Second Deputy Prime Minister. Under King Fahd, the Saudi economy adjusted to sharply lower oil revenues resulting from declining global oil prices. King Fahd suffered a stroke in November 1995 and towards the end of 1997, Crown Prince Abdullah took on much of the day-to-day responsibilities of running the government. With the death of King Fahd in 2005 the crown passed to Abdullah who is pursuing a strategy to open up the Kingdom to tourists and foreign businesses spurred on by a desire to join the World Trade Organisation. The Kingdom finally joined the World Trade Organisation on December 11th, 2005.

On the 2nd March 1992 the Kingdom was divided into thirteen Provinces each with its own governor – several more are proposed (Figure 2.11). The greater Riyadh region was recently divided into three, Riyadh, Sudair and Dawadmi, two more regions are to be formed in the Northern Border Province and one in the southern Najran Province.

The government's awareness of the importance of the complementary nature of development and the environment is not new. Indeed, central to Islamic *shariah* principles is the direction of the individual to conserve and protect natural resources and to be wise and rational in developing the natural environment for the benefits of future generations (Khalid and O'Brien, 1992).

Arid environments, without a protective blanket of well-developed soil and vegetation are particularly susceptible to disturbance. Although we naturally think of disturbance by changes brought on my overgrazing, in a rapidly industrialising country such as Saudi Arabia, the dust and air pollution created by opencast mining and quarrying can also be a serious nuisance. Even pipelines, which may seem innocuous at first glance, have to be pressure tested with millions of litres water and when this is released it can contaminate the ground. The need to feed and water a burgeoning urban population and the exploitation of the Kingdom's vast mineral resources has put a great deal of pressure on the natural environment and in the following section I want to highlight the scale of these challenges.

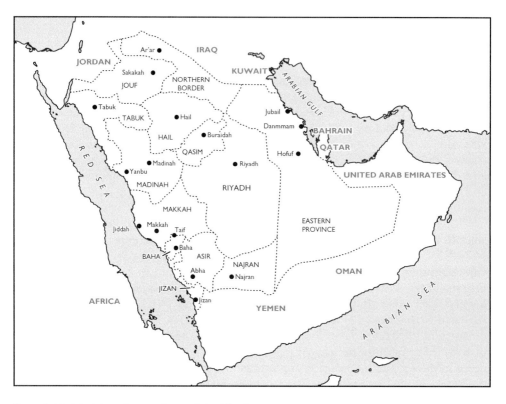

Figure 2.11 Administrative provinces of the Kingdom.

2.4.1   Population

The population of the Kingdom has grown rapidly over the last fifty years and will probably continue to do so into the immediate future. The most recent data from the 2004 census indicates a total population of around 22.67 million of which about 7 million are resident expatriates. These figures mask the large number of illegal immigrants, particularly in Jiddah and the hundreds of small towns and villages in the southern Asir near the Yemen border.

With the average Saudi woman giving birth to seven children it is no wonder that the Kingdom has one of the fastest growing populations in the world. On the one hand the burgeoning population is a wonderful human resource and it supports the process of Saudiization in the employment sector. On the other hand, the strains on the infrastructure are bound to be felt sooner rather than later. For example, demand for good quality drinking water has risen dramatically over the last twenty years and by 2002 Saudi Arabia had increased its desalinated water production capacity to 2.9 million m³ per day (Elhadj, 2004). This volume accounts for 30 percent of the world's total production and meets about seventy percent of the country's drinking water requirements. Some 35 plants are run by the Saline Water Conversion Corporation (SWCC) and during the current five year plan, seventeen additional desalination plants are to be built, twelve on the Red Sea coast and five on the Gulf coast. Some drinking water is provided by Kingdom's 190 dams which have total storage capacity of 775 million m^3 (MCM). Most of these dams are, however, in the south-west of the country and the future provision of good quality drinking water to the growing urban population in the Nadj remains an important issue. Even in Jiddah water to houses is rationed and cut off one day in four.

About 85 percent of the Kingdom's population now live in urban areas. Precise figures are not available but estimates of Jiddah's population at 2.4 million and Riyadh's at c.3 million are probably not too far from the truth. Much of the urban growth has been a matter of changing social values and rural poverty – towns provide jobs, education, good healthcare and some limited entertainment. Today less than three percent of the population of Saudi Arabia can be classed as nomadic and indeed it is the modern grazing practices of the rural populations and the demands of city dwellers for meat that has been instrumental in the severe decline in the quality of the Kingdom's rangelands. The decline of ecologically sound traditional nomadism has not just been a passive process. In response to the severe drought in the early 1950s, with its large losses in stock numbers, tribal rangelands were opened up by a Royal Decree in 1953. The intent was to free up grazing restrictions so as to combat the drought. It was also meant to encourage the *badu* to settle and in this regard was highly successful. In retrospect it is debatable whether this was actually a wise policy.

Urban waste management is now a major problem for cities in the Tihamah and the Gulf coast, made worse in these low lying regions by rising water tables. Much of the rise is probably due to the increase in domestic effluent entering the ground water system but general sea level rise is likely to cause increasing problems in cities such as Jiddah where the water table is only 30 cm or so below road level in some neighbourhoods. Proper sewerage systems and treatment works are essential in all the Kingdom's towns if public health is to be safeguarded.

2.4.2   Agriculture and land use

Agricultural in the Kingdom has become one of the largest and most successful sectors of the economy and has grown at an average annual rate of 8.7 percent since 1970. It now accounts for more than 9.4 percent of Saudi Arabia's GDP. The drive towards self-sufficiency has been a great success story (perhaps with a built in tragedy in terms of water demands) and in 1995 had reached 68 percent of the local consumption of chicken broilers, 85 percent of vegetables, 66 percent of fruit, and 46 per cent of red meat.

Land under cultivation in the Kingdom has risen from 149,696 hectares in 1975 to some 6 million hectares in 2000. Much of this expansion has been associated with the cultivation of crops such as wheat, alfalfa, dates, rice, corn, millet and palm seedlings. The expansion has been actively encouraged by the government. The Kingdom produced 1,787,542 tons of wheat, and some 2,170,794 tons of cereals in 2000. The excessive use of precious fossil groundwater for wheat production has not gone unnoticed and wheat productions and subsidies have recently been reduced. It is salutary to note that the water to make a 1 Riyal loaf of bread costs 2 Riyals to produce.

Date production reached 735 million tons in 2000, making the Kingdom the world's largest producer with some 18 million date palms under cultivation. These achievements are not without costs and there are immense demands made on fossil groundwater supplies by the arrays of sprinkler irrigation systems.

By 1998, the Kingdom was producing 816,000 tons of dairy products and 397,000 tons of poultry meat per annum. Saudi Arabia is also the home of the Al Safi Dairy Farm, the largest integrated dairy farm in the world with some 29,000 head of Holstein cattle and production of 125 million litres of high-quality milk annually. The farm is located south-east of Harad and can clearly seen on satellite images. There are many centre pivot irrigation sites in the surrounding *widyan* producing hay and sorghum silage for the herd. A moment's thought will temper these achievements when it is realised that vast amounts of water are needed to operate such large, open air, shed dairies – at present the water is free! For example, a typical adult dairy cow at the Al Safi dairy requires some 120 litres of water per day for drinking and the control of shed temperature via water coolers. It is too hot for the cows to go outside the sheds. Each adult cow also produces some forty kilograms of waste which has to be disposed of so as not to pollute the ground water.

2.4.3   Oil and gas

No account of the modern Kingdom would be complete without some reference to oil and gas. With it Saudi Arabia has become the world's 20th largest economy and has an important strategic rôle to play in the Middle East as a whole. Without it Saudi Arabia would be a desert backwater bypassed by the rest of the world apart from pilgrims visiting Makkah and Madinah.

In 1923 a New Zealander, Major Frank Holmes, acting on behalf of a British syndicate obtained the first Saudi Arabian oil concession from King Abdul Aziz ibn Saud to explore for oil and minerals in the Eastern Region. The concession, at a bargain price of £2,000 per annum to be paid in gold, came to nothing and the syndicate could not interest any oil company to invest sufficient funds for exploration. The King waited for three years before revoking the concession in 1928.

In 1930, there was a drop in pilgrims caused by the world wide recession and the King was faced with a substantial fall in revenues. He invited a wealthy American businessman, Charles R. Crane, to visit the Kingdom and as a result it was agreed that a mining engineer would be sent from the United States to assess the Kingdom's water, mineral and oil resources. In 1931, the engineer, Karl Twitchell arrived in Jiddah and after an extensive survey lasting many months he submitted a report indicating that the geological formations around Dhahran strongly indicated the presence of oil. While Twitchell's survey was proceeding, SOCAL, the Standard Oil Company of California which had two geologists in Bahrain, was becoming increasingly interested in the oil potential of Saudi Arabia. SOCAL made an astute move in signing on St. John Philby as a confidential adviser and paid him £1,000 pounds for six months; a sum which helped him develop his business interests in Jiddah and also enable him to pay the Cambridge University fees of his son, the spy Kim Philby. In 1933 Twitchell arrived in Jiddah with a SOCAL representative, Lloyd Hamilton and succeeded in negotiating a concession for the exclusive rights to oil in the Eastern region. The agreement which had a 60-year life received royal assent in July 1933. Initial explorations produced disappointing results but in 1935 a well drilled in Dhahran found indications of oil in commercial quantities. Later, in March 1938, the SOCAL subsidiary, CASOC (California Arabian Standard Oil Company) struck oil in Well No. 7 on the Damman Dome and it soon became evident that Saudi Arabia's eastern oil fields were among the largest in the world. By the end of 1938 a little less than half a million barrels had been produced. Some six years later, in 1944, the annual output had increased to eight million barrels and by 1946 had increased to 60 million barrels.

A vast increase in oil revenues came about in December 1950 when a new 50/50 contract was negotiated with Aramco and the economy now entered a period of rapid development which was to accelerated spectacularly in the 1970s, transforming the Kingdom. In 1976 Saudi Arabia assumed ownership of all Aramco assets and rights within the Kingdom and the company was renamed Saudi Aramco. Saudi oil reserves are the largest in the world, and Saudi Arabia is now the world's leading oil producer and exporter. The scale of the production is difficult to grasp but we can guestimate that a single day's production of Saudi crude oil is sufficient for a family car to make about 48 round trips between the Earth and Mars.

Oil now accounts for more than ninety percent of the country's exports and nearly seventy-five percent of government revenues. Proven reserves are estimated at more than 300 billion barrels, about one quarter of the world's oil reserves and possibly up to 1 trillion barrels of recoverable oil. Half of the reserves are contained in eight field including the world's largest offshore field, the Saffaniyah (estimated reserves 19 billion barrels mainly medium to heavy gravity), and the world's largest onshore field, the Ghawar (estimated reserves 70 billion barrels, mainly light gravity).

Most Saudi oil exports move by tanker from terminals at Ras Tanura and Ju'aymah. The remaining oil exports are transported via the east-west pipeline across the Kingdom to the Red Sea port of Yanbu. The northern Trans-Arabian (TAP) pipeline which runs parallel with the Iraqi border to Jordan has been moth-balled since the Gulf War.

In mid-1998, Saudi Aramco completed its $2 billion upstream project, the massive Shaybah oilfield (estimated reserves 7 billion barrels) in the eastern Rub' al Khali, which was designed to replace declining production from some of the Kingdom's older oilfields. The Shaybah oilfield began production in 1998 with an output of about 200,000 barrels per day of premium-grade extra-light crude. Oil is piped some 500 km north to a blending centre at Abqaiq. Shaybah is expected to help maintain the Kingdom's capacity to produce 10 million barrels per day and to boost the country's revenue potential because of the higher market prices likely to be gained from the high quality, lighter crude in its reserves. The Kingdom's gas reserves stood at c.204 trillion cubic feet in 2000 and 5 trillion cubic feet of recoverable gas is expected to be added to this figure each year. The South Rub' Al Khali Company Limited (SRAK), partly owned by Saudi Aramco, was formed in December 2003 to explore for gas.

2.4.4 Mining

The mining sector occupies a prominent position in the Saudi government's strategy to diversify the Kingdom's economic base. Saudi Arabia is home to the largest mineral deposits in the Middle East, estimated at some 20 million tons of gold ore, 60 million tons of copper, 10 billion tons of phosphates, and a large variety of other industrial minerals and rocks. The mining sector in the Kingdom is expected to become the second largest source of government revenue within the next decade.

A further catalyst for private sector investment in mining has been the creation of the Saudi Arabian Mining Company ("Ma'aden") in 1997 which is contributing to the development of water, electricity, and telecommunications in remote areas where some of the mineral ores are located. An integrated mining and transportation policy was unveiled by the government in 2000.

Commercial quantities of minerals and rocks available in Saudi Arabia include basalt, bauxite, bentonite, copper, dolomite, expandable clays, feldspar and nepheline syenite, garnet, gold, granite, graphite, gypsum, high grade silica sand, kaolinitic clays, limestone, magnesium, marble, olivine, phosphate, pozzolan, rock wool, silver, and zeolites.

It should be kept in mind that all mining has serious environmental implications in terms of ground disturbance, dust generation, soil contamination and water demands. This is particularly so in an arid country such as Saudi Arabia.

2.4.5   Transport and communications

The Kingdom's network of asphalted roads and highways has a total length of some 45,000 km. The agricultural road network is even more extensive, totalling 95,900 km. Saudi Arabia is also the home to the only rail system in the Arabian Peninsula. It consists of a 571 km single-track line between Dammam and Riyadh and a line in excess of 300 km between Riyadh and Hufuf. A 237 km link was recently built between Dammam and Jubail, which includes connections between the cities' industrial centres and port areas. In addition, the Saudi Government is building a 400 km link between Makkah, Mina and Arafat and there are well developed plans for a link between Madinah, Ha'il and Riyadh.

Saudi Arabia has eight major ports with 183 piers, making it the largest seaport network in the Middle East. The Kingdom receives 12,000 ships per year and is capable of receiving 183 ships simultaneously. Jiddah is by far and away the largest seaport and some eighty percent of the Kingdom's food imports pass through the port on their way to other parts of the country.

2.4.6   The way ahead

Through a number of five-year development plans, the government has sought to allocate its petroleum income to transform its relatively underdeveloped, oil-based economy into that of a modern industrial state while maintaining the Kingdom's traditional Islamic values and customs. The Seventh Economic Plan was approved by the Council of Ministers in August 2000. The plan stresses the provision of education and social and health services, as well as the training of Saudi manpower to tie in with the on going process of Saudiziation. Among the environmental details in the plan it confirmed that three desalination plants, now under construction with a total capacity of 826,000 m^3 of water daily, would be completed. Work on 12 new desalination plants is also planned. They will include six small plants with capacities ranging from 6,000 to 22,000 m^3 of water daily. The capacity of the remaining six plants will range between 106,000 and 420,000 m^3 daily. The total capacity of the new plants will reach 2.1 million m^3 per day. The Plan indicates that the producing sectors are expected to witness an average annual growth rates of 3.05 percent for agriculture, 8.34 percent for non-oil mining sectors, 5.14 percent for industrial sector, 4.62 percent for electricity, gas and water sector, and 6.17 percent for construction sector. The petrochemical industries and other manufacturing will grow at average annual rates of 8.29 percent and 7.16 percent respectively. The environmental impacts of such growth has not been assessed but will be large in a country with such fragile ecosystems.

Although economic planners have not achieved all their goals, the economy has progressed rapidly, and the standard of living of most Saudis has improved significantly. Dependence on petroleum revenue continues, but industry and agriculture account for a growing share of economic activity. There is still a shortage of skilled Saudi workers at all levels and this remains the principal obstacle to economic diversification and development. About 4.7 million non-Saudis are employed in the economy but there is now a firm policy of Saudiization. It is, however, totally, unrealistic to envisage Saudi nationals taking on many of the jobs done by workers from south and south-east Asia.

There is now genuine concern for the environment and its sensitivity to the changes brought about by industrial and urban development. To emphasise the point the Sixth Development Plan (1996–2000) included a number of ambitious goals focussed on sustainable development, environmental protection and the maintenance of biodiversity (Ministry of Planning, 1995). Similarly, the Seventh Plan (2001–2005) highlights the need for the conservation of wildlife. How these plans will work out in practice has yet to be seen. It is certainly the case that Government policies for conservation are not matched by funding.

As well as being blessed by enormous oil reserves, which will some day be depleted, the Kingdom also has the possibility of utilising, clean, renewable solar energy resources and at the Energy Research Institute (ERI) established by King Abdulaziz City for Science and Technology (KACST) in Riyadh, scientists have taken up the challenge of investigating the potential role of solar energy in the Saudi Arabian context. A *Solar Village* has been created some 50 km north-west of Riyadh. As well as investigating the role of solar energy for electricity production in remote areas the ERI has also made some headway in the design of efficient solar powered seawater desalination plants. As the demand for drinking water rises year on year such environmentally friendly processes are a key tool in lowering atmospheric pollution which has become a problem at the Kingdom's major industrial complexes.

Until recently the Kingdom's attitude to domestic tourism has been extremely conservative. Tourism for many is driven by the need to escape the intense heat of summer. For the most part Saudi families have been content with picnics close a road on the edge of the desert in the cool of the evening but there. There is a growing demand for family holidays by the seaside. Wealthier Saudi families visit towns like Abha in the Asir which have a pleasant summer climate. So as not to be without transportation it is common for the family to fly in to the regional airport and pick up their car which has been sent ahead by a transported truck. A number of coastal recreation areas with accommodation are being developed on the Red Sea at towns such Al Qunfadah but in general the Red Sea coast is much less developed than the Gulf coast which is thick with villas and hotels from Dammam southward for nearly 100 km.

The formation of the Supreme Commission for Tourism in April 2000 signalled the beginning of a new era in tourism in Saudi Arabia, one that will see expansion of tourist facilities and services, as well as greater efforts to preserve historic sites and traditional crafts. The development of tourism is part of the on going Saudiization process and the Commission estimates the industry to provide about 30,000 jobs. It has to be said however that the development of specialist training courses is rather slow and language proficiency needs to be improved for the international market.

If handled correctly there are considerable economic gains to be had if the tourist industry were to be opened up – especially to foreign tourists, although the infrastructure needs to be much more fully developed (Ady and Walter, 1991). Many foreign tourists will want to see the unsurpassed dune landscapes, the Nabatean and petroglyph archaeological sites and the fascinating villages of the south-west of the Kingdom. Some will want to watch birds or dive in some of the most spectacular coral reefs in the world. To this end there are now plans being developed by the National Commission for Wildlife Conservation and Development for an ecotourism industry. As this type of development takes off over the next decade or so there is clearly a need to understand its possible environmental and ecological impacts which in some parts of the world have been disastrous.

Although it is easy to dwell on the environmental downside of rapid economic development it should not be forgotten that Saudi Arabia still has vast areas of spectacular landscape where the near wilderness experience, or something like it, can still be had.

Chapter 3

Geological framework

Contents

3.1 INTRODUCTION

Saudi Arabia has a very long geological history ranging from the Precambrian, with rocks more than 1.2 By years old, to young carbonate and saline sediments presently forming in the lagoons and on the beaches of the Arabian Gulf and Red Sea.

The geology is of fundamental importance to an understanding of the country's present-day environment and economic development. This is most readily understood in terms of the presence of vast oil fields in the sedimentary rocks of the Gulf

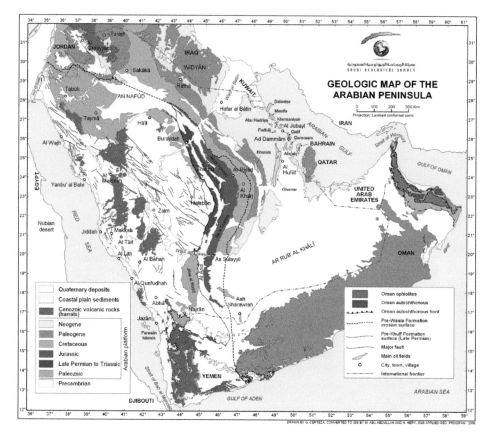

Figure 3.1 Geological Map of the Arabian Peninsula (Courtesy: Saudi Geological Survey).

coast region, but geology also has a major influence on the distribution of ground-water, the presence of hazards, and the distribution of landforms and soils in the Kingdom.

The purpose of this chapter is not to attempt a comprehensive discussion of this long geological history but to provide a general review which will help the reader construct a more meaningful picture of the Kingdom's environment. I have left a description of Quaternary sediments, such as dune sands and loessic silts to Chapter 6 – Geomorphology. In broad terms we can think of Saudi Arabia has having five distinct geological units (Figures 3.1 and 3.2):

1. The Arabian Shield
2. The Arabian Platform
3. The Tertiary and Quaternary *harraat*
4. The Red Sea Coastal Plain
5. The Major Sand Seas – Rub' al Khali and An Nafud

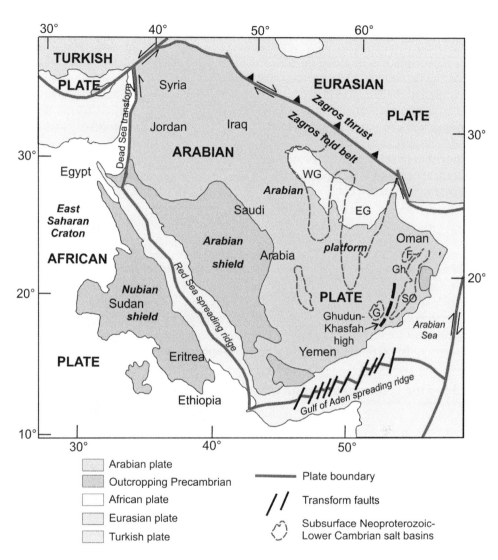

Figure 3.2 The Arabian Plate (Courtesy: Saudi Geological Survey).

3.1.1   The Arabian Shield

The Arabian Shield is the exposed region of the Precambrian Arabian Plate and crops out over some 777,000 km² in the western part of the Kingdom (Figure 3.2). It comprises metamorphosed sedimentary and volcanic rocks intruded by younger granites and gneisses. Exposure of the Shield from its blanket of Phanerozoic sedimentary and volcanic rocks is due entirely to Mesozoic and Cenozoic uplift and erosion. Since its creation, the Arabian Plate has moved north-eastward away from Egypt and Sudan, north away from Somalia, and has rotated counter-clockwise about a point in the vicinity of the Gulf of Suez. Such movement has been accommodated by compression and strike-slip faulting along the Bitlis and Zagros zones, where the Arabian Plate collides with, and subducts beneath, the Eurasian Plate, and by strike-slip displacement along the Dead Sea transform.

At the present time, the northern part of the Arabian Plate is moving north-west, with respect to the Eurasian Plate, at a rate of about 20 mm yr^{-1}. Because they are regions of extension, the southern, south-western, and south-eastern margins of the Arabian Plate have weak to moderate earthquake activity; the compressive northerly and north-easterly margins, conversely, are regions of strong earthquake activity. Overall, the plate has moved as much as 350 km away from Africa, depending on where the margins of the initial rift are placed with respect to present-day exposure of Precambrian basement, and on how much stretched continental crust is believed to remain in the Red Sea basin.

3.1.2   The Arabian Platform

The sediments of the Arabian Platform, also called the Arabian Sedimentary Basin, lie to the east of the Shield and rest on the buried Arabian Plate (Figure 3.2). The strata dip very gently away from the edge of the Shield and form zones of cuesta topography with scarps more or less following the curvature of eastern edge of the Shield. The Palaeozoic strata are arenaceous but are replaced by carbonate sequences in the Mesozoic and Cenozoic. The main cuesta is just to the west of Riyadh – the Tuwaiq Escarpment. Towards the Gulf the cuestas become barely perceptible.

3.1.3   The Cenozoic *harraat*

The Cenozoic basaltic lava fields, or *harraat*, crop out in a more or less north-south belt on the central Arabian Shield. Their study has revealed a fascinating picture of volcanism associated with mantle plume hot spots under the Shield.

3.1.4   The Red Sea Coastal Plain

This geological unit comprises the narrow Red Sea Coastal Plain which formed as the rifted escarpment of the Red Sea graben was uplifted and eroded. The rocks mainly comprise Late Mesozoic and Cenozoic sedimentary rocks, raised coral reefs and small patches of dune sand and sabkhah.

3.1.5   The Rub' al Khali and An Nafud

These two large sand seas (ergs) for the most part blanket the sedimentary rocks of the Arabian Platform. Both the Nafud and the Rub' al Khali are developed in shallow sags developed at depth in the basement.

3.2   BRIEF GEOLOGICAL HISTORY

Much of the early geological history of Saudi Arabia has been pieced together by an appreciation of the mechanisms of plate tectonics on a primitive earth with the repeated accretion of plates, the growth of ancient supercontinents and their subsequent splitting apart at constructive boundaries (Camp, 1984). These so-called Wilson Cycles repeat themselves every 500 Ma or so.

Two supercontinents are of relevance to the history outlined here. They are Rodinia – a Russian word meaning 'homeland', and Gondwana – derived from the Sanskrit *gondavana*, from *vana* "forest" + *gonda*, name of a Dravidian people living in north central India. It is also helpful to recall some terminology regarding the Precambrian Eon timescale which is conventionally divided into an Archean Eon (3,800–2,500 Ma) when the first unicellular life appeared, and a Proterozoic Eon (2,500–570 Ma) when the first mulitcelluar life appeared. The Proterozoic Eon is further divided into a Palaeoproterozoic Era (2,500–1,600 Ma) a Mesoproterozic Era (1,600–1,000 Ma) and a Neoproterozic Era (1,000–570 Ma). The supercontinent Rodinia is thought to have accreted during the Mesoproterozoic. Globally, the Neoproterozic is associated with the so-called 'Snowball Earth' a period formerly called the Varangian glaciation which lasted some 240 My. We now know that there were several glacial episodes during the Snowball Earth period.

3.2.1   The Proterozoic growth of the Arabian-Nubian Shield

Although the Arabian-Nubian Shield contains small outcrops of Archaean and Palaeoproterozoic rock most strontium isotope measurements show a preponderance of Neoproterozoic rocks. The Shield is one of the best exposed and largest assemblages of rocks of this age in the world. (Figure 3.3).

Between about 1,350 Ma and 900 Ma the supercontinent Rodinia formed from the collision of two other supercontinents called Atlantica and Nena. By 800 Ma Rondinia itself began to spit up into rifted cratons (Unrug, 1997). At this time the Mozambique Ocean was formed on the margins of the East Sahara craton (west Gondwana) and formed an accretionary basin into which a conveyor belt like sequences of island arc formations were subducted and squeezed under the approaching East Gondwana – the Indian cratons (Johnson and Woldehaimanot, 2003). Subduction was apparently oriented in several directions.

The main period of subduction and island arc accretion took place about 870–620 Ma. From 620–540 Ma additional volcanic and sedimentary rocks were deposited in marine and continental environments, and a vast amount of plutonic rock was intruded into and beneath the deformed volcanic and sedimentary rocks mainly due to post-collision crustal extension. The general plate tectonic sequence as envisaged by Stern, (1994) is shown in Figure 3.4.

As Stern *et al.* (2006) remark, the Neoproterozoic accretion of the Shield took place during a period of very dramatic climatic changes. These have been expressed in the so-called Snowball Earth hypothesis when climates swung back and forth from glacial to tropical conditions. Four glacial episodes have been identified, two prior to c.700 Ma and two after c.630 Ma. Stern *et al.* (*op. cit.*) suggest that the older episodes (Kaigas c.735–770 Ma and Sturtian c.680–715 Ma) occurred whilst the island arc accretion was still active, with dropstone deposits and tills being deposited in a wide range of marine environments. They go on to note that deposits from the younger glacial events (Marinoan c.635–660 Ma and Gaskiers c.582–585 Ma) are less likely to have survived because of the collision events between East and West Gondwana at this time. The Marinoan and Gaskiers events were likely to have been terrestrial as the

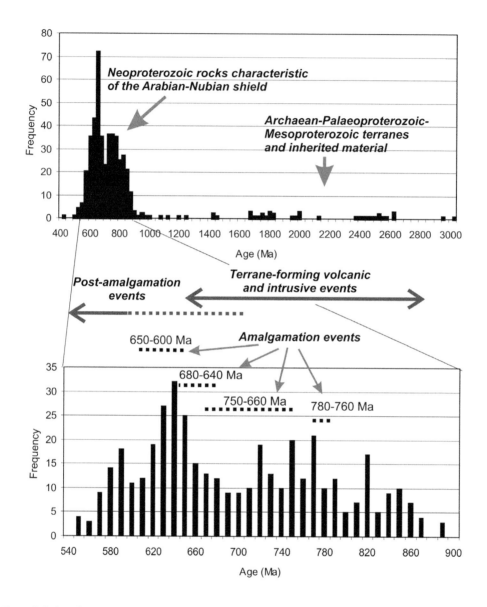

Figure 3.3 Age distribution of dated Precambrian rocks (Courtesy: Saudi Geological Survey).

Arabian-Nubian Shield was now probably above sea level. One interesting suggestion by Garfunkel (1999) is that the widespread erosion about c.600 Ma may have been due to the Marinoan glaciation.

Shield rocks crop out in the western third of Saudi Arabia and are composed of intra-oceanic arc/back arc basin complexes and terranes (microplates) amalgamated together along north- and east-trending sutures representing different subduction

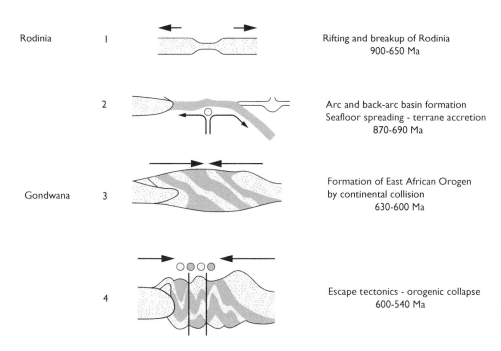

Rodinia — 1 — Rifting and breakup of Rodinia
900-650 Ma

2 — Arc and back-arc basin formation
Seafloor spreading - terrane accretion
870-690 Ma

Gondwana — 3 — Formation of East African Orogen
by continental collision
630-600 Ma

4 — Escape tectonics - orogenic collapse
600-540 Ma

Figure 3.4 General plate tectonic sequence. (modified from Stern, 1994).

events. The term terrane is used by geologists for a crustal block or fragment that preserves a distinctive geologic history and is bounded by faults or shear zones. The Shield's terrane sutures are often marked by ophiolitic rock sequences (Greenwood *et al.*, 1980; Gass,1981; Abdelsalam and Stern, 1996), with their characteristic pillow lavas and gabbros, which indicate that material at an oceanic spreading ridge has been squeezed and thrust up. Peter Johnson at the Saudi Geological Survey notes that terrane sutures are often stitched together by the intrusion of younger granite batholiths.

3.2.2 Palaeozoic tectonic stability

Voluminous Palaeozoic sandstone sequences were deposited in northern Africa and Arabia following an extended Neoproterozoic orogenic cycle that culminated in the consolidation of Gondwana and a long period of Palaeozoic stability. Throughout this time a major ocean, the Tethys Sea, lay to the northeast and erosion of the margin of Gondwana generated clastic deposition into the epicontinental environments over and beyond what is now the Shield.

During the Palaeozoic major unconformities developed which are probably related to epeirogenic movements of the Arabian-Nubian Shield and world-wide eustatic sea level changes. In the Late Proterozoic and Palaeozoic the Gondwana supercontinent lay astride the South Pole. It is thought that by the Late Ordovician the Shield was covered by an ice cap and was extensively scoured and up to 35 metres of glacial till were deposited.

3.2.3   Hercynian stripping

Although the details are sketchy it is thought that there was an east-west trending uplift of the Arabian-Nubian craton along with similar uplift in southern Egypt and northern Sudan caused by progressive convergence of Laurasia and Gondwana. (Ghebreab, 1998). The Late Carboniferous was also characterised by a significant fall of sea level in northern Gondwana which led to emergence and a significant increase in fluvial activity attuned to lower base levels. The Early Palaeozoic sediments are thought to have been completely stripped from the higher ground and much of the Shield together with basement rocks to the east of the were planed level. The evidence for this suggestion is the fact that the Permian Khuff Formation rests directly on the Precambrian on the eastern central edge of the Shield.

By the Late Permian the sedimentation pattern in Arabia has changed permanently and completely, and the Arabian Platform had essentially become a shallow carbonate shelf on the edge of the Tethys Sea. From time to time epeirogenic movements on the platform gave rise to clastic sequences and deltaic environments but for the most part the depositional environments were low energy and shallow enough on occasions to provide conditions for the development of gypsum, anhydrite and occasionally, halite.

3.2.4   Jurassic-Cretaceous

The Late to Early Jurassic period was marked by high sea levels that led to flooding of the Arabian Platform and the deposition of extensive carbonate ramp-type deposits. Throughout this period Arabia continued to drift northward and crossed the Equator in the Cretaceous. By this time some of the north-east margin of Gondwana had started to break up into microplates which collided with Laurasia. As the microplates drifted north the Neotethys Sea opened up along the line now marked by the Zagros Crush Zone. Eustatic sea level changes brought about marine invasions of the Arabian platform and the alternating deposition of deeper-water carbonates and shale with shallower-water, higher-energy carbonates. In the Late Cretaceous the Arabian Plate began to thrust into the Eurasian Plate and it may be about this time that the north-south alignment of the Gulf oil field structures were emplaced and the Arabian Gulf basin formed.

3.2.5   Cenozoic events

The present-day boundary between the Arabian and African plates is a series of troughs 2,000–4,500 m deep in the Red Sea, Gulf of Aden, and Gulf of Aqaba reflecting spreading along the Red Sea and Gulf of Aden. Seafloor spreading in the Red Sea began 5–6 Ma ago, although an earlier episode of spreading may have occurred 15–25 Ma ago. Intrusion of magma along the spreading axes created the oceanic crust of the southern Red Sea and formed pools of metal-laden brine and hot springs in particular deeps.

The floor of the northern Red Sea is probably a mixture of rifted continental crust and newly formed oceanic crust. Syn- and post-rift sedimentary rocks, including evaporites, flank the spreading axis in the Red Sea and underlie the Red Sea coastal plain. In most localities, the contact between Precambrian rocks of the Shield and Cenozoic rocks of the Red Sea basin is faulted. The Gulf of Aqaba pull-apart basins contain sedimentary rocks. Ocean crust in the Gulf of Aden began to form 11 Ma ago (Middle Miocene) by westward propagation from the Carlsbad ridge in the Indian Ocean. The ocean crust becomes progressively younger toward to west, and is about 2 Ma old near the Afar triple junction hot spot.

Processes related to spreading caused uplift of the south-western and south-eastern margin of the Arabian plate forming mountains in western Saudi Arabia and Hadramaut that peak at the erosional escarpment (2,500–3,300 m asl.) which now lies inland from the Red Sea coast itself. According to fission-track data, the Red Sea margin of southern Saudi Arabia has undergone 2.5–4 km uplift in the last 13.8 Ma. East of the Red Sea escarpment and north of the Hadramaut escarpment the land surface slopes to a broad plateau 900–1,000 m asl. and farther east slopes gently to the Gulf. Late Cretaceous-Tertiary events in the south-eastern part of the Arabian Plate include oblique obduction of the Masirah ophiolite (Palaeocene) onto the Arabian continent and rift-shoulder uplift and normal faulting of coastal southern Oman and eastern Yemen.

Flood basalts erupted in western Arabia as a result of the spreading process and formed the *harraat* which cover parts of the Shield and some sedimentary cover rocks rocks. The volcanic rocks are mainly alkali-olivine basalts extruded from numerous vents and volcanoes. The northerly alignment of cones and faults in the volcanic fields possibly reflects a northerly alignment of fractures in the underlying crystalline basement along which magma rose from great depth. The oldest flood basalts are in Yemen and southern Saudi Arabia (30 Ma); the youngest are in parts of Harrat Rahat and Harrat Khaybar which have erupted as recently as 700 AD.

Neotethys closed during the Cenozoic with the collision of the Afro-Arabian Plate and the Eurasian Plate. At this time the Shield was still part of an emergent Arabian- Nubian dome and the Arabian Platform was a marine carbonate environment the eastern edge of which was being subducted under the Eurasian Plate. Concomitant with the subduction and northerly movement of the Arabian Plate, was thinning, stretching and block faulting of the crust at the junction of the African and Arabian plates. These Early Tertiary events have to be seen alongside the fact that thick saprolites were widespread across the low-relief, low-altitude crystalline rocks of the Arabian Shield indicating some degree of cratonic stability (Brown *et al.*, 1989).

By the Late Oligocene the crust had rifted to initiate the development of the Red Sea graben – a process which spread from south to north. In the southern Red Sea, between N19° and N23°, oceanic crust was produced in the central trough or spreading centre (Girdler and Southern, 1987). The Red Sea trough flooded by mid-Miocene times and evaporites accumulated.

Over the last 2 million years it is estimated that the southern Red Sea has continued to rift open at between 5–10 cm yr^{-1}. Rifting has been accompanied by isostatic uplift of the Red Sea rift shoulders (Schmidt *et al.*, 1983) and the western edge of the Shield has been deeply eroded and stripped of its weathered mantle.

There is still no generally accepted tectonic explanation of the rift in spite of the fact that the area was one of the first in the world to have plate tectonic concepts applied to its geological development. Ghebreab (1998) provides an extremely useful summary of the main tectonic models.

The classic model of Lowle and Genik (1972) suggests that mantle doming and crustal stretching lead to the development of horst and graben structures along normal faults during widespread rifting. An alternative proposal by Cockran (1983)supposed that there was an initial period of 'diffuse crustal extension' lasting tens of millions of years before extension was concentrated along a single axis and gave way to sea floor spreading that took place mainly in the Miocene. This model contrasts markedly with those that suggest abrupt oceanic spreading along a simple fracture on the continental plate. More recently Voggenreiter and Hötzl (1989) have proposed a model based on lithospheric shear.

Several points of interest regarding the tectonics are worthy of note. In the western Yemen apatite fission track dating from basement amphibolites indicate cooling by erosional unroofing of at least 3–4 km of rock between the onset of volcanism at 24–29 Ma and 16 ± 2 Ma (Menzies *et al.*, 1992). And if uplift is caused by magmatic underplating from a plume, the timing of the uplift should be roughly coincident with the initiation of magmatic activity. Menzies *et al.* also point out that this is consistent with the observations of Overstreet *et al.* (1977) who observed that the As Sarat palaeosol, which are overlain by lavas dated from 29–25 Ma rests on a landscape of low relief and not much altered by tectonic disturbance.

Uplift of the Shield also led to the accumulation of vast quantities of coarse clastic debris on the shoulders of the developing Red Sea Basin. Seismic data suggests that 4–6 km of sediments exist under the Red Sea coastal plain and although about 3 km of this is evaporite there are thick wedges of Lower Miocene conglomerates and sandstones (Ahmed, 1972). This probably represents the main depositional sink for the several kilometres of post-Late Oligocene erosion implied by the fission track data. (Ghebreab, 1998).

Brown *et al.* (1989) argue that the uplift actually took place in two stages. After an initial Late Oligocene–Early Miocene phase there was a long non-tectonic interlude and a landscape of broad *widyan* developed on the early escarpment. Two such *widyan* crossing the escarpment west of Harrat Rahat north-east of Jiddah, are preserved beneath the Upper Miocene basalt flows from the *harra*. In contrast, present-day streams flow in steep *widyan* incised below the broad-valley level. During, and since, Pliocene times, the Red Sea escarpment was rejuvenated in a second-phase of scarp uplift which is probably still in progress. This uplift stage is marked by the deeply incised *widyan* and rugged topography of the modern scarp.

3.3 REGIONAL GEOLOGY

3.3.1 The Arabian Shield

Despite being referred to, where exposed, as a shield, the crystalline basement of Saudi Arabia has not been completely stable since the end of crust-forming events in the Precambrian. It has been affected, in response to plate movements during the ongoing history of Gondwana, by strike-slip faulting and rifting, graben formation, and also by uplift and subsidence, forming domes, basins, arches, and troughs. The effects of such deformation are considerable. The crest of the Ha'il arch, for example, is about 4 km above the trough of the An Nafud basin and the crystalline rocks in the easternmost part of the Arabian Plate are depressed beneath as much as 10 km of sedimentary rocks. The present-day Arabian Shield is exposed because of uplift along the Ha'il arch and Red Sea arch, and the Shield is partly concealed, between the arches, by Lower Palaeozoic and Upper Cretaceous-Palaeogene sedimentary rocks and Cenozoic volcanic rocks preserved in a north-south structural low.

3.3.1.1   Terrane

Over the last twenty years or so there has been a major re-think regarding the geological divisions of the Shield. A substantial compilation of research by geologists working for the United States Geological Survey (Brown *et al.*, 1989) divided the Shield rocks into Groups which they describe in considerable detail. Although this publication has some mention of plates and terranes, the real paradigm shift came through the research of geologists such as Peter Johnson at the Saudi Geological Survey and Douglas Stoeser at the United States Geological Survey, who have recast the whole story in terms of island arc-subduction-accretion events.

In their original paper, Stoeser and Camp (1985) identified five terranes but recent isotopic composition data for granites in the Shield (Stoeser and Frost, 2005) has led to the identification of at least eighteen microplates (Figure 3.5). These authors recognised three groups of terrane:

1. Western arc terranes – those microplates west of the Hulayfah – Ad Dafinah fault (Midyan, Naqrah, Hijaz, Jiddah, Bidah, An Nimas, Al Qarah, Tathlith-Malahah). These terranes have an ocean-arc affinity and are characterised by low radiogenic Pb and Sr and high Nd isotopic composition.
2. Eastern arc terranes – those microplates east of the Hulayfa – Ad Dafinah suture (south-eastern Hail, Siham, Nuqrah, Saqrah, Sawdah, Ad Dawadimi, Ar Rayn, Amlah). They have a similar affinity to the western arc terrane but have more elevated levels of Nd and Sr suggesting that their mantle source was less depleted than that supplying the western arc terranes (Stoeser and Frost, 2005).
3. The Khida terrane is a unique microplate comprising re-worked continental crust. It is thought to be the north-western edge of the Arabian craton which underlies the central and southern Arabian Peninsula.

The Hulayfa – Ad Dafinah fault which separates the eastern and western arc terranes probably represents a suture which developed as a result of the collision of East and West Gondwana from 750 to 650 Ma (Stern, 1994).

3.3.1.2   Nadj Fault system

The Nadj Fault system is a major belt of north-west trending wrench faults some 400 km wide that transect the central and north-eastern parts of Arabian Shield. It is the most prominent structural feature of the Shield. The displacement in the system has been estimated to be as much as 240 km (Brown, 1972). The system consists of at least four zones of intensely sheared rock subjected to considerable amount of vertical movement and accompanied by realignment of pre-existing structures. Schmidt *et al.* (1978) interpret the Nadj Fault system as a Late Precambrian oblique shear zone associated with the collision of East and West Gondwana. Late Precambrian and Cambrian

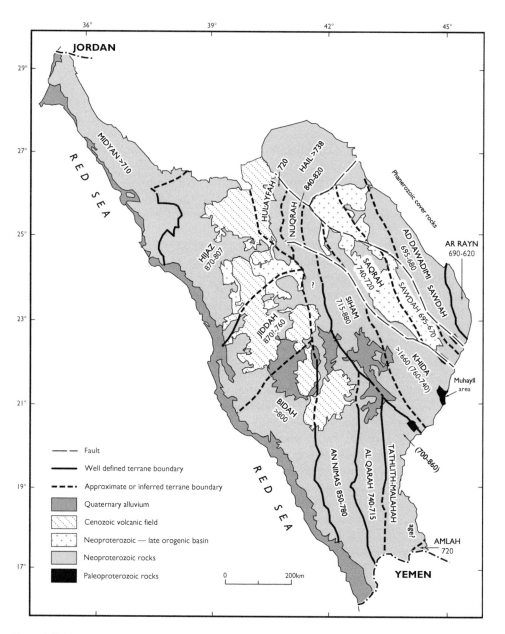

Figure 3.5 Main terrane of the Shield (Modified from Stoesser and Frost, 2005).

granitic plutons were emplaced into the Nadj system both during and after fault movement, thus providing a means of dating . A whole-rock K-Ar date of 530 ± 20 Ma suggests that movement had ceased by Cambrian times (Greenwood *et al.*, 1980).

Regionally, fault system is complex and its formation still poorly understood. It is
possibly correlated with similar NW-trending sinistral faults in Madagascar, Ethiopia and

Tanzania (Johnson and Kattan, 1999). The central problem seems to be whether the Nadj system of faults was formed under crustal shortening and compression or extension and rifting. Johnson and Woldehaimanot, (2003) strongly prefer a compressive origin based on several lines of evidence indicating sinistral wrenching.

3.3.1.3 Tertiary and Quaternary volcanics

A large area of the Arabian Plate – about 180,000 km^2 – is covered by extensive Tertiary and Quaternary lava fields related to the fracturing and faulting associated with the opening of the Red Sea which began at the end of the Oligocene (about 25 Ma). As the African and Arabian plates were dragged apart tensional stresses caused considerable subsidence and the development of linear vent fissures which became conduits for alkali-olivine basaltic lavas. The stress regime during the volcanism was clearly extensional but is poorly understood and does not generally run parallel to Red Sea normal faults. The plateaus are bleak, inhospitable places and their dark surfaces covered with coarse, angular, blocks. In addition to lava flows the *harraat* are strewn with a variety of volcano types. Their remoteness and the hostile terrain makes the *harraat* important refuges for wildlife

In western Arabia eruptions of lava have continued from the Oligocene to the present. According to Camp *et al.* (1987) within-plate volcanism has resulted in at least 21 eruptions in the past 1,500 years, the latest being at Dhamar in the Yemen in 1934. One of the most important eruptions in Saudi Arabia was that near the holy city of Madinah in 1256 AD when an eruption on the northern part of the Harrat Rahat resulted in a lava flow 23 km long that approached to within 8 km of the city. Camp *et al.* and Ambraseys *et al.* (1994) provide interesting historical detail of the flow based on eyewitness accounts of the 52 day eruption.

Camp and Roobol (1992) discuss in some detail the evolution of the west Arabian *harraat* and note that they differ from the contemporaneous lava fields in east Africa in that they are not associated with well-developed continental rifts. They suggest that because the lavas occur on uplifted eastern flank of the Red Sea depression, they could be considered as 'rift shoulder' lava fields similar to those in eastern Australia. Their presence in Jordan, suggests that they are not simply related to aesthenospheric upwelling adjacent to the Red Sea spreading axis.

The continental magmatic rocks of Saudi Arabia can be divided into older and younger groups which differ in their overall composition and structural setting. The first phase of magmatism, from about 30 to 15 Ma emplaced olivine poor, alkaline, tholeiitic lavas along north-west Precambrian trends reactivated during the Oligo-Miocene. This phase is attributed to the remobilisation of a fossil plume head accreted to the base of the Arabian lithosphere, during the extension of the Red Sea. These *harraat* are now deeply weathered and eroded so much so that no morphological volcanoes remain. The younger phase, dating from about 12 Ma to historical eruptions, produced mildly alkaline olivine basalts along north-south trends during a major period of crustal uplift associated with the West Arabian Swell which forms the northern part of the much larger Afro-Arabian Dome. It is suggested that the West Arabian Swell is underlain by a hot upwelling mantle. The central axis of the swell is referred to as the Makkah-Madinah-Nafud (MMN) volcanic line and represents a north-south fissure system more than 600 km long stretching from Makkah in the south, through

Madinah to the edge of An Nafud in the north (Figure 3.6). The MMN line continues northward as far as the Ha'il arch which was initiated in the Late Cretaceous by mantle processes prior to the formation of the Red Sea (Greenwood, 1973). Volcanism along the MMN volcanic line migrated northward with time, beginning on Harrat Rahat at about 10 Ma, on Harrat Khaybar at about 5 Ma and on Harrat Ithnayn at about 3 Ma (Camp *et al.*, 1992). These lava fields contain magnificent volcanic features, Harrat Khaybar together with the adjoining Harrat Ithnayn and Harrat Kura form the largest plateau-basalt field on the Shield, and in one or two places has been shown to have a

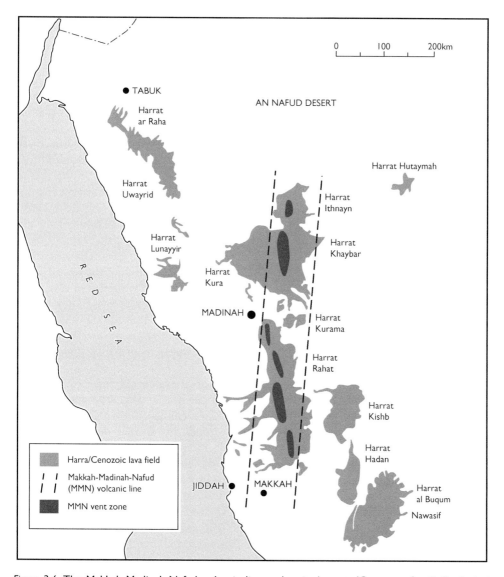

Figure 3.6 The Makkah-Madinah-Nafud volcanic line and main *harraat* (Courtesy: Saudi Geological Survey).

thickness in excess of 500 m, though in general the flows are very much thinner. The *harraat* have K-Ar dates range from 11.5 to 7.5 Ma. Doughty (1888) was informed by local *badu* that a warm, smoking vapour could be seen around the crest of Jabal Ithnayn – the largest volcanic cone on Harrat Ithnayn. Harrat Kishb is of interest because it is the site of the Wahbah explosion crater, some 2 km wide and 270 m deep. Recently a number of lava tubes have been discovered which contain animal bones, ancient throwing sticks and loessic silts. Further details of the *harraat* can be found in Brown *et al.* (1989).

3.3.2   The Arabian Platform

The blanket of Phanerozoic sedimentary rocks now found to the east and north of the Arabian Shield covers almost two-thirds of the Kingdom and was built up on the relatively stable Arabian Platform. This forms part of the vast Middle East Basin extending northward into Jordan, Iraq and Syria and eastwards into Iran. Gentle subsidence throughout the Phanerozoic has allowed some 5,500 m of continental and marine sediments to accumulate in the basin. During the Early Palaeozoic, central Arabia is thought to have been a stable, subsiding, passive margin flanking Gondwana onto which shallow-marine, littoral, and fluvial sandstone, siltstone, and shale were deposited. In the Late Ordovician-Early Silurian the Shield and Platform were covered by an ice-cap – Arabia at this time was within $30°$ of the South Pole. Sea level rise and fall caused regression and transgression of the ocean flanking Gondwana and there was a corresponding migration of sedimentary facies on the stable shelf.

Because of Hercynian orogenic activity that originated beyond the confines of Gondwana, its passive margin in Arabia became active in the Devonian, and central Arabia underwent uplift and tilting. A regional structural high that lacks Devonian sedimentary rocks marks the early development of the Central Arabian Arch.

Earlier deposits were depressed in fault basins or eroded across generally north-trending horst blocks resulting in an irregular topography preserved beneath the Unayzah-Khuff unconformity and resulting in the initiation of structures that eventually controlled the location of oil fields in central Arabia. The Unayzah Formation clastic rocks, which constitute major oil reservoirs where they overlie appropriate Hercynian structures, mark a resumption of sedimentation in the Late Carboniferous. Deposition of the Khuff Formation, representing the earliest major carbonate unit in Arabia, was concurrent with rifting and Gondwana breakup in the Zagros region. During this period, a second episode of glaciation deposited Permo-Carboniferous glaciogenic tillites in the south-western Rub' al Khali basin. Within the Arabian Platform three structural provinces are recognised: an Interior Homocline, an Interior Platform, a number of intra-shelf basins. (Figure 3.7).

3.3.2.1   The Interior Homocline

This is a sedimentary belt bordering the Shield and has a width of about 400 km. The older rocks dip about $1°30'$ maximum to less that $0°30'$ in the younger. Homoclinal strata have been affected by a number of arches and graben stuctures. The Central Arabian Arch was active from the Carboniferous through to the Tertiary and was responsible for a number of marine regressions and transgressions. The Arch has a profound effect on the surface distribution of sedimentary rocks and controls the curvature of the interior homocline. It stretches from the easternmost point of the Shield in an east-northeasterly direction towards Qatar. The axial region of the arch underwent Middle Jurassic and Early Cretaceous inversion and became a basin, followed by reformation of the arch during the Late Cretaceous as a consequence of uplift in southern Arabia and continued subsidence to the north.

A second Arch, the Ha'il-Rutbah Arch, is part of the longest Arch in the Middle East and extends from northern Saudi Arabia into northern Iraq (Greenwood, 1973). This north-plunging arch developed as a stable headland in the Precambrian Shield rocks around which early Paleozoic seas advanced. The north- south trend of the arch disregards the north-west, south-east Nadj wrench-fault directions in the Precambrian basement. Palaeozoic sediments were eroded from the arch by the mid-Cretaceous when sandstones were deposited without interruption across its crest. Further arching activity seems to have continued into the Cenozoic. The Ha'il-Rutbah Arch is clearly aligned with the Makkah-Madinah-Nafud (MMN) volcanic line and the whole lineation is thought to represent a zone of intraplate crustal thinning and

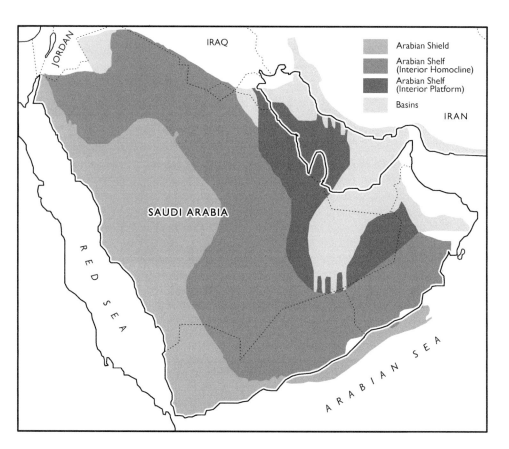

Figure 3.7 Location of Interior Homocline and Interior Platform.

warping and is not associated with compression related to Red Sea rifting as proposed by Burek (1969).

The Central Arabian Graben and Al Kharj Trough system is a narrow graben feature some 500 km long that resulted as a result of stresses associated with uplift of the southern Shield during the Late Cretaceous – Late Palaeogene.

3.3.2.2 The Interior Platform

The Interior Platform is a remarkably horizontal area bordering the homocline to the east. The width of the platform varies between 100 km along the southern and western side of the Rub' al Khali basin to 400 km or more across the Qatar Peninsula. Several major north-south anticlinal trends rise above the general level of the platform and the structures contain several of the the the great oil fields of the Kingdom such as the Ghawar (Figure 3.8).

3.3.2.3 Intra-shelf basins

Several depressions or basins have formed on the Arabian Shelf and are mainly recognized through sub-surface survey. The basins probably represent deep-seated basement faults but few details are available. The Tabuk and Widyan basins are found respectively on the west and east side of the Ha'il Arch. Both are floored with Lower Palaeozoic Saq sandstones which crop out in the north-west of the Kingdom. The largest basin is the Rub' al Khali basin, and is an elongate trench plunging gently towards the Arabian Gulf and extending almost as far as the Iranian coast. It is primarily a Tertiary feature with sediments of Palaeocene, Lower and Middle Eocene and Late Tertiary age thickening towards the centre.

Many oil fields in the western Gulf area are located on structural highs and gravity lows that may indicate salt flowage at depth. East of the Qatar peninsula diapiric salt structures occur frequently on islands. These salts are of Late Proterozoic

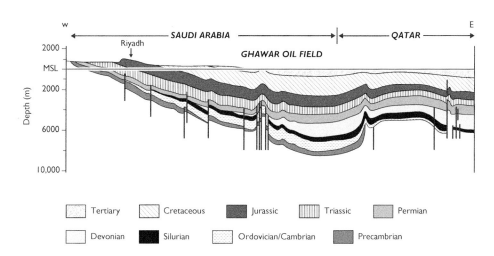

Figure 3.8 Geologic cross section of the sedimentary cover rocks.

age and their source is the infra-Cambrian Hormuz basin which underlies much of the Arabian Gulf.

3.3.3   The Red Sea Coastal Plain

The narrow Red Sea Coastal Plain marks the eastern edge of a large north-south graben bounded by escarpments and comprises thick sequences of Oligocene-Recent sedimentary rocks. In the Jiddah region, for example, the sedimentary succession is in excess of 1,500 m thick and rests on eastward-tilting basement blocks. Immediately underlying the first marine beds, directly related to the opening of the Red Sea in the mid-Miocene, are a series of now poorly preserved continental clastic rocks which have been affected by graben faulting. It is a matter of dispute whether these patches of Oligocene sediments are associated with pre-rifting tectonic activity and erosion or were once much more widespread and are only preserved in the later rift basins (Laurent, 1992). The continental clastic series are overlain by complex sequences of marine clastic-to-carbonate sediments of Upper Oligocene age.

From mid-Miocene times on evaporitic interactions of gypsum and anhydrite become common and can be found as far in land as the escarpment indicating a degree of erosional stability. In the post-evaporite period there were extensive changes in the sedimentation regime and the edges of the evaporite sequences are covered with a thick clastic sequence indicating pronounced erosion from the newly formed escarpment (Jado and Zötl, 1984). In addition to clastic sedimentation Pliocene reef rocks accumulated which are in turn covered by Quaternary reefs. The present coastal plain is a complex mosaic of reef terraces, marine benches, clastic accumulation terraces reaching back into *wadi* entrances (Braithwaite, 1987).

3.4   ECONOMIC GEOLOGY

3.4.1   Hydrocarbons

Alsharhan and Nairn (1997) have brought together much of the available information on the hydrocarbon habitats of Saudi Arabia. By far and away the most important hydrocarbon trap in the Kingdom is the Ghawar anticline which is draped over a basement horst structure (Figures 3.7 and 3.8). The anticline grew initially during the Carboniferous (Hercynian) deformation and was reactivated episodically, particularly during the Late Cretaceous.

The producing oil reservoir at Ghawar is the Late Jurassic Arab-D limestone, which is about 90 m thick and occurs 2,000–2,500 m beneath the surface (Figure 3.9). The oil originates from Jurassic organic-rich lime mudstones, which were laid down in intershelf basins. The integrity of the thick anhydrite top seal is helped by the general absence of faults. The field's production has been enhanced by water injection, which was initiated in 1965. Accurate information on the injection rates are not in the public domain but a rate of seven million barrels of seawater per day is probably a reasonable estimate.

As well as oil the Ghawar reservoir rocks yield about 2 billion cubic feet of associated gas per day. The Late Permian Khuff A, B and C stacked carbonate reservoirs are the main gas producing zones at depths of 3,000–4,000 m.

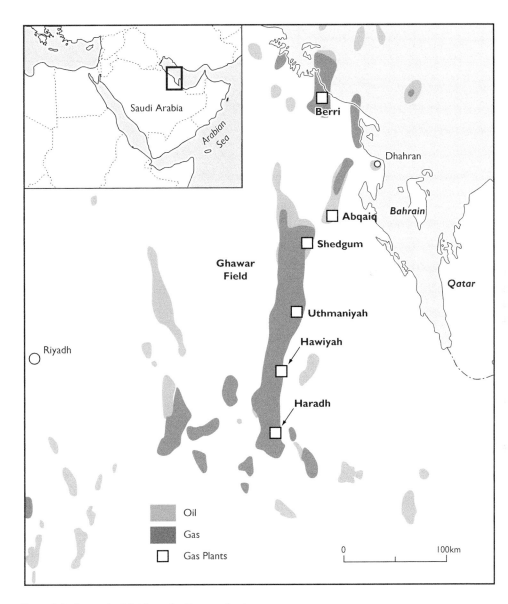

Figure 3.9 Gas and oil fields in the Eastern Region.

3.4.2   Metals and industrial minerals

The diversification of the economy has been a consistent objective of the Kingdom's development plans and in this regard the development and utilization of the country's extensive mineral wealth has an important rôle to play. Indeed, the Sixth Development Plan (1905–2000) explicitly defined the establishment of a substantial mining industry as a major development objective. To spearhead exploration and mining activities the state-owned Saudi Mining Company (Ma'aden) was created in April 1997. To some extent the development of the mining industry in the Kingdom is restricted by limited water supplies and most mining operations are accompanied by extensive groundwater surveys. In some cases, such as the development of the Wadi Sawawin iron ore mine (N27°56', E35°47') 700 km north of Jiddah, the available groundwater is heav ily charged with dissolved solids and a feasibility study recommended a desalination plant on the Red Sea coast some 60 km to the west. The Deputy Ministry for Mineral Resources has targeted the following for rapid development:

Bauxite

The only deposit of commercial interest is at Az Zabirah (N27°56', E43°43') in the Qasim region 180 km north of Buraydah. The bauxite is part of laterite profile exposed over a distance of 105 km in the scarp face of an Early Cretaceous sandstone-mudstone sequence. The estimated total reserves are 101.8 Mt.

Gold

Primary gold mineralisation has been found only in the Proterozic Arabian Shield. Two sites are presently mined. The Kingdom's most important mine is Mahd Adh Dhahab ('Cradle of Gold') is an underground mine 280 km northeast of Jiddah a few kilometres north of the small but growing town of Al Madh (N23°39', E40°51'). In 1997 Mahd Adh Dhahab processed nearly 200,000 metric tons of ore averaging 24.41 grams per ton of gold and 131.24 grams per ton of silver. Overall bullion production was 4.32 tons. The readily mined reserves at Mahd Adh Dhahab amount to 1,142 Mt with an average grade of 31.8 g/t of gold. A second mine is the Sukhaybirat open-pit site halfway between Madinah and Buraydah (N25°27', E 42°01') and some 750 km by road from Jiddah. The mine produced 1.94 tons of bullion in 1997.

Copper

Copper mineralisation is widespread in the Arabin Shield. The most important deposit is at Jabal Sayid (N23°51', E40°56') about 315 km north-east of Jiddah and 40 km north of the Mahd Adh Dhahab gold mine. The deposit includes about 80 Mt of copper. Little water is available for mining purposes at this site.

Magnesite

Petromin hold an exploratory license to investigate the small (4 Mt) but high grade magnesite deposit at Zarghat (N26°30', E43°35') about 150 km south-west of Ha'il.

Zinc

The Al Masane zinc sulfide deposit (N18°08', E43°51') is approximately 80 km north-west of Najran. The site was worked in ancient times during the Abbasid caliphate about 1,200 years ago. Proven reserves of 7.03 Mt, averaging 5.33 percent Zn encouraged the Arabian Shield Development Corporation of Texas, in partnership with the Al Mashreq Company for Mining Investments, to form a joint venture to mine and process the deposit (Metal Bulletin, 1998).

Phosphate

The Al Jalamid phosphate deposit (N31°32', E29°56') is located 120 km east-south-east of Turayf and about 25 km north of the Trans-Arabian Pipeline which has paved highway to the industrial centres of the Arabian Gulf, 1,150 km away to the east. The deposit has proven reserves of 213 Mt averaging 21 percent P_2O_5 and is to be developed by Ma'aden and a consortium of private suppliers. Transport to the coast is a major problem and one proposal is the construction of a slurry pipeline to Al Jubayl the water being supplied from the Tawil aquifer. An estimated 35,000 m³/day are required for the mine and the community. Phosphate production is particularly important for the production of fertilisers and the development of the Kingdom's agricultural sector.

Cement-grade Limestone

Saudi Arabia is richly endowed with limestone and dolomite which occur mainly in the Permian and Mesozoic rocks of the Arabian platform. Cement-grade limestones must have less than three percent MgO. Although cement as a manufacture product is sufficiently value enhanced to allow it to be transported long distances the raw materials (limestone and clay) are low-value commodities and must be extracted in the vicinity of the cement plant. Furthermore, the high capital costs involved in the construction of a cement plant necessitates sufficient reserves for a fifty-year life span.

Table 3.1 Major cement factories in the Kingdom

Name of company	Location	Capacity (Mt/yr)	Limestone formation
Red Sea coast			
Arabian Cement	Rabigh	1.85	Quaternary Coral
Yanbu Cement	Ra's Baridi	1.34	Quaternary Coral
Southern Province Cement	Umm 'Araj	1.56	Jurassic Amran
Southern Province Cement	Baysh	1.30	
Central Region			
Yamama Cement	Riyadh	1.85	Cretaceous Sulaly
Qasim Cement	Buraydah	1.20	Permian Khuff
Eastern Province			
Saudi Cement	Hufuf	1.60	Miocene Dam
Saudi-Bahraini Cement	'Ayn Dar	1.60	Miocene Dam
Saudi-Kuwait Cement	Kharsaniyah	2.20	Miocene Dam

Chapter 4

Climate and environmental change

Contents

4.1 INTRODUCTION

Early travellers crossing the Arabian Peninsula often made shrewd, if unscientific, observations about the weather and its seasonal patterns. Indeed, prior to the 1940s only a few meteorological stations existed on the Red Sea coast, Gulf of Aden and Arabian Gulf. In the immediate post-war period the need for a basic network of meteorological stations became apparent both to understand the reasons behind major breakouts of desert locusts from their spring breeding grounds in central and southern Arabia, and to provide basic environmental information for the developing oil industry in the Eastern Province. In fact, one or two of Saudi Aramco's meteorological stations, such as that maintained at Dhahran, have the longest continuous data series available for anywhere in the Peninsula. For many variables data stretch back to the early 1950s, and in the case of rainfall, to 1939 (Williams, 1979). By the end of 1960 there were still only ten government stations but with the development of off-shore oil fields in the Arabian Gulf in the 1960s a number of marine engineering companies such as Imcos Marine Ltd acted as consultants to the oil industry and collected good information on winds and tides (Imcos Marine Ltd, 1974a, 1974b, 1976).

Since the mid-1960s the network of government run stations has grown rapidly, spurred on in particular by the burgeoning agricultural sector, and by 1987 some 182 stations had been established though not all to the same standard. Government stations are maintained by the Ministry of Water and Electricity (since 2001 – was formerly part of the Ministry of Agriculture and Water – MAW), and the Meteorology and Environment Protection Administration (MEPA). Saudi Aramco also maintains eight stations mainly on the Gulf coast.

In addition to raw data supplied by the Ministries and Saudi Aramco there are a number of other sources of information. Several foreign companies have collected considerable amounts of environmental data notably Italconsult which throughout the 1960s and 70s collected climatological information for MAW as part of the company's survey work on water and agricultural development (Italconsult, 1967, 1969, 1979). Some government agencies, such as MEPA, also produce in-house reports of interest to climatologists. For example, Lee's overview of the climatology, produced as part of a series on the climate of Saudi Arabia by MEPA (Lee, 1984). More recently Al-Jerash (1989) has re-worked climatic water balance data for 67 stations. For each station ten tables have been constructed to show: mean monthly temperature, monthly potential evapotranspiration, monthly precipitation, (P-PE), monthly soil moisture storage, monthly change of soil moisture storage, monthly moisture deficit, monthly moisture surplus and monthly run-off. The secondary data were computed using a program written in 1981 by Robert Muller (Louisiana State University). These data are interesting and worthy of further investigation although one should caution users of the data set regarding the choice of the model's starting conditions and the lack of any meaningful vegetation or soil information.

Two major publication of immense use to those interested in the climatology of Saudi Arabia were published in the 1980s. In 1984 the Ministry of Agriculture and Water in cooperation with the Saudi Arabian – United States Joint Commission on Economic Cooperation published the *Water Atlas of Saudi Arabia*. This magnificently produced atlas has a whole chapter devoted to meteorological conditions in the Kingdom. Four years later in1988 the Joint Commission published their authoritative *Climate Atlas of Saudi Arabia*. This superb colour atlas not only contains excellent maps but also includes a substantial text and a large number of photographs.

4.2   SYNOPTIC CLIMATOLOGY AND SEASONS

Although factors such as continentality and altitude give rise to considerable variations in climate over the Arabian Peninsula, patterns in the general synoptic climatology and weather can be deduced from the invasion patterns of major air masses – widespread, homogenous bodies of air which develop their characteristics in their source region. The dominant air mass incursion over the Peninsula in winter is Polar Continental air which develops as an intense anticyclone over central Asia. High pressure conditions give rise to clear skies and dry, stable air (Figure 4.1).

In summer, the high temperatures over the Peninsula lead to the development of Tropical Continental air which form part of the Monsoon low circulation centred over north-west India. The air is very dry and the intense convection raises vast quantities of dust into the lower atmosphere and gives rise to poor visibility and generally hazy conditions. The summer heat over the Peninsula also draws in unstable, Tropical Maritime, Monsoon air from the Arabian Sea and the two tropical air masses merge at the Intertropical Convergence Zone (ITCZ). The northerly extent of the ITCZ varies from year to year but it usually penetrates the Red Sea as far north as Jizan Province sometimes generating intense downpours and flash floods as the air is forced to rise over the Red Sea escarpment. Further east the ITCZ gives rise to occasional downpours in the Empty Quarter. For example, in July 1997 three weeks of "freak" Monsoon-like rainfall occurred in the north-eastern Rub' al Khali.

In winter, Polar Maritime air sometimes invades the Kingdom from the Mediterranean. Depressions embedded in this air mass often track across the desert plains of northern Saudi Arabia and give rise to much needed rainfall.

Four seasons are generally recognised in the Kingdom although their distinctiveness varies from region to region. This is particularly true when comparing coastal locations such as Jiddah or Jizan with more continental locations such as Riyadh at 591 m asl.

4.2.1   Summer (June–August)

Over much of the Kingdom the summer season is period of stability characterised by high diurnal temperatures and hazy skies. The intense heat often gives rise to local low pressure systems over southern Iraq giving rise to north-westerly winds which blow southward across Kuwait and down the Gulf. These hot, dusty winds are known as *shamal* winds (Arabic: north). Strong *shamal* winds usually affect the northern Gulf in June and occasionally in early July when sustained winds blowing for 2 to 3 days at speeds of 40–50 km/hr transport vast quantities of dust into the lower atmosphere and severely reduce the visibility. By mid-July there is no longer an appreciable north-south pressure gradient in eastern Arabia and the still predominantly north-westerly winds weaken considerably.

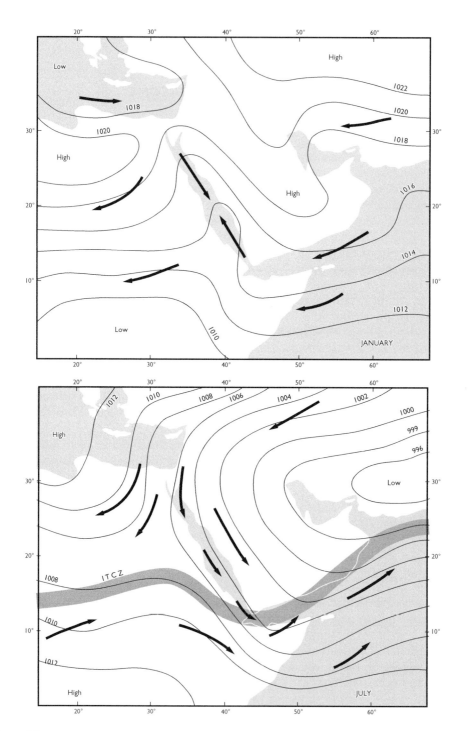

Figure 4.1 Wind direction and average sea level pressure (mb). (Modified from Edwards and Head, 1987).

4.2.2   Autumn (September–November)

This is a transitional season when pressure begins to build once more over the Peninsula and the ITCZ retreats south. Temperatures become more tolerable with cool nights and warm, sunny days.

4.2.3   Winter (December–February)

Away from the tempering effects of the sea, winters in Saudi Arabia are characterised by cold nights and bright, sunny days with good visibility. In Riyadh, temperatures fall below zero when cold Polar Continental air invades the region. In cities like Abha, in Azir Province, sleet and sometimes snow is not uncommon. Eighty kilometres away down on the Red Sea coast at Jizan, frost is unknown and temperatures rarely drop below 10°C. In the northern plains and the Nafud basin the winter season brings rains associated with depressions tracking in from the Mediterranean.

4.2.4   Spring (March–May)

The spring, like the autumn is a transitional period. The sun is higher in the sky and air pressure over the hot desert surfaces starts to fall. Cool nights and sunny warm days make this a very comfortable time of the year. In the northern part of Kingdom advective thunderstorms are common at this period when dense Polar Maritime air from the Mediterranean undercuts less dense air masses developed over the desert surfaces.

4.3   AIR TEMPERATURE

The general pattern of isotherms over the Kingdom is shown in Figures 4.2 and 4.3. Data for specific stations presented in Table 4.1. There is considerable variation in mean annual temperatures across the Kingdom but in broad terms there is a temperature gradient from the northern plateaus, which experience very cold winters, south-eastward towards the Rub' al Khali. Northern cities such as Tabuk, Ha'il and Jawf have winter averages barely reaching 10°C and northerly winds sometimes bring down temperatures to freezing on several days of the year.

  This general thermal gradient is interrupted by the mountains of the Asir and southern Hijaz which experience considerable adiabatic cooling. Compare, for example the mean annual temperature of Jizan on the Red Sea Coast with Khamis Mushait some 2,000 m higher and just inland from the escarpment. No wonder those who can, visit friends and relatives on the coast to escape the cold in winter.

  It is also interesting to compare stations on the Red Sea with those on the Arabian Gulf at the same latitude. Al Wadj, for example, on the Red Sea coast has less seasonality than Dhahran on the Gulf. Winters are warmer and summers less hot on the Red Sea coast. This can be partly explained by the protection against the cold winter winds given by the Hijaz Mountains.

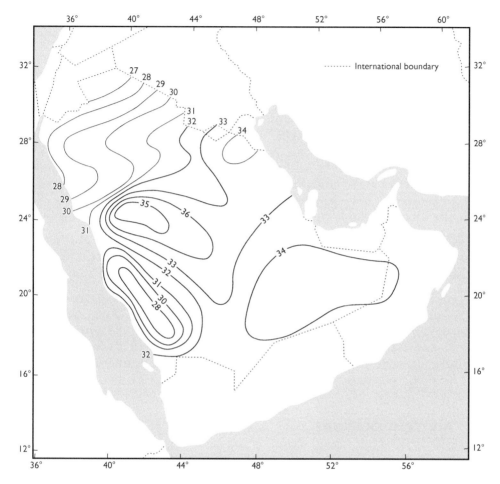

Figure 4.2 General pattern of summer isotherms over the Kingdom ($^\circ$C).

4.4   ATMOSPHERIC HUMIDITY AND EVAPORATION

Data for simple pan and reference crop pan evapotranspiration (ET$_r$) are presented in Table 4.2. In recent studies of evapotranspiration it is usual to investigate the depth of water loss per unit time evaporated and transpired from a specific reference crop. This approach replaces many of the ambiguities involved in the interpretation of potential ET measured by the simple pan method. This is particularly true if the pan is poorly sited. Well maintained pans can, however, provide a useful check on ET$_r$ estimates from meteorological data through the simple relationship

$$\mathrm{ET}_r = K_p \cdot E_p$$

where K_p is the pan coefficient, which is dependent on the type of pan involved and other factors, E_p is the evaporation from a class A pan. Pan coefficients for Riyadh and

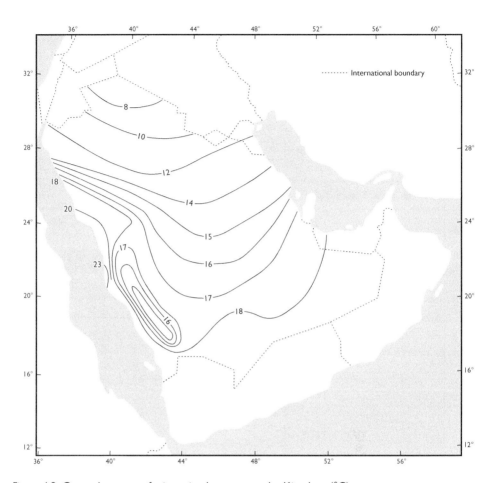

Figure 4.3 General pattern of winter isotherms over the Kingdom (°C).

Najran are 0.7 and Jizan and the Asir 0.8 (Al-Ghobari, 2000). Furthermore, the use of a reference crop ET_r permits a physically realistic characterisation of the effect of the microclimate. The reference crop most widely used in arid areas is alfalfa (Allen *et al.*, 1989) which will be usually greater than for a clipped grass surface by about 15 percent.

Several studies have been conducted to calculate ET_r in Saudi Arabia (Salih and Sendil, 1985; Saeed, 1986; Mustafa *et al.*, 1989; Abo-Ghobar and Mohammad, 1995; Al-Ghobari, 2000). This type of estimate has importance for irrigation schemes since it helps increase water use efficiency and irrigation scheduling. Al-Ghobari (*op. cit.*) examined a number of methods based on meteorological data, pan data and a alfalfa lysimeter at Riyadh. On the whole there are very good agreements between the various methods but of the meteorological methods the Penman method calibrated for Riyadh conditions is probably best for Saudi Arabian conditions (Abo-Ghobar and Mohammad, *op. cit.*). Some representative pan and crop ET_r data are shown in Table 4.2. Not surprisingly, evapotranspiration is highest in the dry summer air of the Riyadh area in the centre of the Kingdom. This is also one of the most

Table 4.1 Temperature data for selected stations

		Alt (m)	Jan	Feb	Mar	Apr	May	Jun	Jul	Aug	Sep	Oct	Nov	Dec	Year
Al Hanakiyah	24.50°N 40.31°E	849	13.6	17.3	20.8	25.0	30.3	33.5	33.7	33.4	31.8	26.4	19.5	15.1	25.0
Al Qatif	26.30°N 50.00°E	5	15.4	16.6	20.4	25.4	30.0	32.9	34.1	33.5	31.5	27.6	22.0	17.2	25.5
Al Qaysumah	28.19°N 46.07°E	360	11.0	13.0	18.0	25.0	31.0	35.0	36.0	35.0	33.0	27.0	18.0	13.0	24.0
Al Ula	26.37°N 37.51°E	681	14.3	16.5	20.5	25.7	29.7	32.5	32.6	32.2	31.3	27.3	20.4	14.9	24.8
Al Wadj	26.20°N 36.40°E	20	18.6	18.9	21.2	24.3	26.5	27.9	29.2	29.2	28.2	26.2	23.5	20.0	24.5
Ar'ar	30.55°N 41.08°E	554	8.0	11.0	15.0	21.0	26.0	31.0	33.0	32.0	30.0	23.0	16.0	11.0	22.0
As Sulayyil	20.47°N 45.60°E	615	17.1	19.1	23.5	27.8	32.3	34.5	37.3	35.5	32.6	27.2	22.2	17.8	27.4
Ash Sharawrah	17.28°N 47.07°E	721	19.3	22.3	25.5	29.5	33.2	34.4	35.7	35.7	32.7	28.1	23.6	20.6	28.8
Bishah	19.98°N 42.60°E	1167	17.2	18.9	22.7	24.4	28.1	30.6	31.6	31.5	28.5	23.8	20.3	17.0	24.8
Buraydah	26.18°N 43.46°E	649	12.0	15.0	19.0	24.0	30.0	32.0	33.0	33.0	31.0	26.0	19.0	15.0	24.0
Dhahran	26.27°N 50.10°E	26	15.8	17.1	21.0	25.6	31.2	34.4	35.8	35.3	32.6	28.4	23.0	17.5	26.5
Gassim	26.30°N 43.70°E	648	12.2	14.1	19.0	23.6	29.9	32.9	34.3	33.8	31.5	26.2	19.5	13.6	24.4
Ha'il	27.43°N 41.60°E	1115	10.8	12.8	16.6	21.8	26.6	29.7	30.9	30.7	29.4	24.0	17.0	12.4	21.9
Hufuf	25.17°N 49.29°E	171	14.3	16.1	20.2	25.3	29.9	32.9	33.9	33.6	31.0	26.4	20.9	16.0	25.0
Jawf	29.78°N 40.10°E	689	9.7	11.5	15.3	21.3	26.9	29.8	31.8	31.8	29.7	23.6	15.7	11.5	21.9
Jiddah	21.50°N 39.20°E	17	23.5	23.6	25.3	27.5	29.8	31.3	31.9	31.9	31.1	29.5	27.2	24.8	28.1
Jizan	16.89°N 42.50°E	6	25.8	26.4	27.9	30.2	32.1	33.4	33.4	33.1	32.5	31.2	28.7	26.6	30.1
Khamis Mushait	25.05°N 48.08°E	2066	13.5	14.8	17.0	18.3	21.1	23.5	23.5	23.1	21.9	18.7	16.5	14.0	18.9
Khurays	18.30°N 42.80°E	430	13.3	15.4	20.8	25.6	30.4	32.8	34.9	33.8	31.9	25.6	19.6	14.3	24.8
Madinah	24.55°N 39.70°E	654	17.7	20.0	23.3	27.3	31.3	34.3	34.1	34.3	33.5	28.9	23.3	19.3	27.3
Makkah	21.29°N 39.50°E	161	23.0	24.0	27.0	31.0	34.0	35.0	35.0	35.0	35.0	32.0	28.0	25.0	31.0
Rafha	29.38°N 43.29°E	446	10.0	12.0	17.0	23.0	28.0	32.0	33.0	33.0	31.0	26.0	17.0	12.0	23.0
Riyadh	24.70°N 46.70°E	609	14.2	16.2	20.4	25.0	29.9	32.6	33.7	33.1	30.5	25.5	20.0	15.2	24.7
Tabuk	28.22°N 36.38°E	769	10.3	13.3	16.8	22.2	25.9	29.4	30.7	30.3	28.6	23.5	16.4	11.9	21.6
Ta'if	21.48°N 40.50°E	1478	15.0	15.7	19.0	21.7	24.7	27.1	27.7	28.3	26.6	22.4	18.0	15.8	21.8
Tayma	27.38°N 38.29°E	820	9.9	12.6	15.8	20.0	23.9	26.7	27.5	27.5	26.8	22.0	15.8	11.6	20.0
Turaif	31.40°N 38.43°E	852	7.0	8.0	12.0	18.0	23.0	27.0	29.0	29.0	26.0	20.0	13.0	8.0	18.0
Unayzah	26.04°N 43.49°E	724	14.0	15.4	19.9	24.4	29.7	31.8	32.8	32.4	31.0	26.4	19.9	15.3	24.4
Yabrin	23.19°N 48.57°E	200	14.2	17.2	21.8	27.2	30.7	34.4	36.2	34.0	31.4	25.7	20.1	16.1	25.9
Yanbu	24.15°N 38.00°E	8	19.8	20.3	23.2	26.7	29.6	31.3	31.6	31.6	31.1	28.9	24.8	21.6	27.0

Source: Presidency of Meteorology and Environment.

Table 4.2 Average monthly pan and reference crop (ET$_r$) evapotranspiration (mm/day)

	Riyadh		Najran		Abha		Jizan	
	Pan	ET$_r$	Pan	ET$_r$	Pan	ET$_r$	Pan	ET$_r$
Jan	3.54	4.85	3.37	4.69	3.93	4.80	5.16	5.97
Feb	4.65	6.07	4.55	5.72	4.80	5.07	6.20	6.95
Mar	6.20	6.81	5.99	6.41	5.22	6.01	7.37	7.41
Apr	7.97	7.90	6.91	6.67	5.52	6.60	9.46	7.69
May	10.22	9.43	7.80	8.27	7.54	7.27	9.79	9.87
Jun	11.72	11.32	9.03	10.26	8.44	8.29	10.07	11.54
Jul	12.58	11.64	10.37	10.51	9.18	8.38	11.43	11.85
Aug	11.64	11.00	9.32	9.98	8.76	8.30	10.22	9.51
Sep	9.32	9.77	7.84	8.95	7.87	7.64	8.60	8.56
Oct	6.75	8.08	6.43	6.96	6.63	6.97	7.67	7.37
Nov	4.59	6.27	4.50	5.23	4.99	5.04	6.66	6.67
Dec	3.41	4.53	3.50	4.37	4.28	5.04	5.26	5.70

Source: Hussein Al-Ghobari, King Saud University.

important agricultural regions where huge quantities of fossil groundwater are used for irrigation. The lowest evapotranspiration rates are on the southern escarpment in the Abha region. This is due to a combination of altitude and cloudiness.

4.5   PRECIPITATION

Practically all precipitation statistics for the Kingdom are influence by extremes. This partly due to the patchiness of the events themselves and their failure or otherwise to be caught by the weather stations, and partly because of unpredictable nature of incursions of rain bearing winds. It should also be borne in mind that there are no long homogenous records of data for the Kingdom which would dampen down the influence of extreme events in the data.

4.5.1   The general pattern

Annual precipitation data for selected stations are shown in Table 4.3 and isohyets in Figure 4.4. Much of the Kingdom receives less that 100 mm precipitation per year but in the central Nadj this total rises a little in the Unaysah - Ha'il region and is due to winter depressions tracking in from the Mediterranean. For the rest of the year the core of the Kingdom is dry and little rain is recorded.

The wettest places in the Kingdom by far are the mountains of the southern Hijaz and the Asir. Abha, for example, located just inland of the Red Sea escarpment edge and not far from the Kingdom's highest mountain has c.430 mm per year some of which falls a sleet and even snow. Rain can fall during any month of the year in the mountains of the high Asir when moist air is forced up the escarpment from the Red Sea.

There is little data on rainfall in the deep Empty Quarter per se. Both Yabrin and Ash Sharawrah have rainfall in most months of the year but the amounts are tiny.

Table 4.3 Precipitation data for selected stations (1970–1986)

		Alt(m)	Jan	Feb	Mar	Apr	May	Jun	Jul	Aug	Sep	Oct	Nov	Dec	Year
Abha	18.12°N 42.29°E	2200	26.2	107.8	34.2	81.3	85.1	28.2	10.8	27.9	14.5	2.6	6.1	40.7	434.6
Al Hanakiyah	24.50°N 40.31°E	849	2.9	0.0	2.5	29.9	0.4	8.5	0.0	0.0	0.0	0.0	17.2	1.1	62.5
Al Ula	26.37°N 37.51°E	681	5.0	5.0	8.4	17.9	8.3	0.0	0.0	2.0	0.0	2.1	7.9	6.2	58.0
Al Wajd	26.20°N 36.40°E	20	2.1	1.3	2.8	0.0	0.0	0.0	0.0	0.0	0.0	8.1	4.4	6.8	24.6
As Sulayyil	20.47°N 45.60°E	615	3.6	3.3	21.6	12.4	1.3	0.0	0.0	0.3	0.0	0.0	0.0	1.5	44.2
Ash Sharawrah	17.28°N 47.07°E	721	0.3	5.0	2.8	13.7	0.3	0.2	2.8	2.0	0.9	0.0	0.0	1.5	29.5
Bishah	19.98°N 42.60°E	1167	10.0	6.0	17.9	33.8	17.3	3.0	4.0	3.2	0.0	5.5	6.4	2.2	109.4
Ha'il	27.43°N 41.60°E	1115	17.7	5.4	17.4	18.4	6.7	0.2	0.0	0.0	0.0	6.1	36.2	8.1	116.2
Hufuf	25.17°N 49.29°E	171	15.9	11.8	23.3	11.9	0.5	0.0	0.0	0.1	0.0	1.0	2.8	5.0	72.1
Jiddah	21.50°N 39.20°E.	17	14.0	11.4	0.5	1.5	1.5	0.0	0.3	0.0	0.0	0.8	13.0	8.6	51.3
Jizan	16.89°N 42.50°E	3	13.5	0.8	11.4	4.8	4.3	0.5	2.5	6.6	5.8	14.2	3.3	16.8	84.6
Khurays	18.30°N 42.80°E	430	6.6	12.6	18.9	12.0	1.1	0.0	0.3	0.3	0.0	2.0	2.1	6.3	62.2
Madinah	24.55°N 39.70°E	654	5.1	0.6	8.5	20.8	4.0	0.9	0.0	0.5	0.5	3.1	6.5	4.4	54.8
Makkah	21.29°N 39.50°E	161	16.4	1.8	7.0	18.2	0.3	0.0	0.0	7.7	2.1	4.2	10.7	29.8	98.2
Najran	17.37°N 44.26°E	1210	2.8	8.1	22.6	17.5	4.6	0.0	1.5	1.8	0.0	8.1	0.5	3.8	71.3
Qatif	26.30°N 50.00°E	5	14.7	20.9	18.7	10.1	0.5	0.0	0.0	0.0	0.0	2.0	8.5	10.2	85.5
Qurayyat	31.20°N 37.21°E	549	7.9	10.4	6.5	4.0	2.3	0.0	0.0	0.0	0.3	3.7	7.4	5.8	48.4
Rafha	29.38°N 43.29°E	446	15.7	13.2	13.2	17.3	4.1	0.3	0.0	0.0	0.0	14.5	15.2	12.4	105.9
Riyadh	24.70°N 46.70°E	609	14.4	8.8	17.8	19.9	7.7	0.0	0.2	0.1	0.0	1.2	3.3	10.0	83.1
Tabuk	28.22°N 36.38°E	769	5.4	4.2	4.1	3.9	1.2	0.0	0.0	0.0	0.0	2.9	3.7	1.2	26.7
Ta'if	21.48°N 40.50°E	1478	12.2	0.3	10.3	27.5	61.6	0.6	0.0	4.1	6.6	5.3	21.3	12.0	161.7
Tayma	27.38°N 38.29°E.	820	6.5	3.5	5.9	10.3	4.2	0.1	0.0	0.0	0.0	1.8	8.2	2.5	43.0
Unayzah	26.04°N 43.49°E	724	17.8	3.0	19.1	27.5	29.9	0.0	0.0	0.0	0.0	21.1	13.2	7.4	139.0
Yabrin	23.19°N 48.57°E	200	8.8	7.9	15.3	9.8	0.4	0.0	0.1	0.0	0.6	0.0	2.9	3.1	49.0
Yanbu	24.15°N 38.00°E	8	7.1	1.3	1.3	0.8	0.0	0.0	0.0	0.0	0.0	2.5	7.9	1.3	22.4

Source: Al-Jerash (1989).

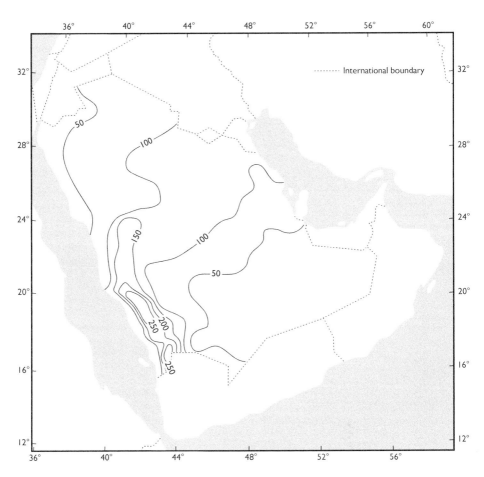

Figure 4.4 General pattern of isohyets over the Kingdom (mm).

4.5.2   Rainfall extremes

Extreme rainfall events are probably more common than is supposed, particularly in the south-west of the Kingdom. Both Wan (1976) and Wheater *et al.* (1989) have investigated rainfall intensities and recurrence intervals in escarpment drainage basins between Taif and Jizan. Wheater *et al.* based their studies on the comprehensive investigations by the consultant company Saudi Arabia Dames and Moore (1988) and their 100 autographic rain gauges in Wadi Yiba, Habawnah, Tabala, Liyyah and Lith. Wadi Yiba, north of Abha, is fairly typical of the five catchments and its estimated rainfall intensities for different recurrence intervals are shown in Table 4.4.

One well-documented storm event which caused a massive amount of damage to bridges and roads in Wadi Dellah was recorded at the meteorological station at Abha in 1982 (Table 4.5). This steep *wadi* drains westward from Abha at the top of the Red Sea escarpment down to Ad Darb on the Tihamah and is the route of the main road from the interior down to Jizan.

Table 4.4   Rainfall intensities (P) for different recurrence intervals – Wadi Yiba*

	T = 10 min	T = 30 min	T = 1 hr	T = 2 hrs	T = 6 hrs	T = 24 hrs
P (50 years)	34.7	67.0	91.7	112.3	125.6	128.8
P (75 years	38.0	73.3	100.3	122.8	137.4	141.0
P (100 years)	39.7	76.6	104.8	128.4	143.7	147.3
P (150 years)	41.6	80.3	109.8	134.6	150.6	154.4

* P is the precipitation in mm in the given time periods.
Source: Wheater et al. (1989).

Table 4.5   Rainfall intensities – Wadi Dellah flash floods – 1982

Date	Rainfall (mm)									Total duration in hours
	Maximum intensities for durations indicated									
	10 min	20 min	30 min	1 hr	2 hr	3 hr	6 hr	12 hr	Total	
Feb										
2	0.2								0.2	0.1
3	0.2								0.2	0.1
5	0.2					0.4			3.4	2.5
7	0.2								0.2	0.1
8	1.6	2.2	2.6		3.2	4.0	6.4	6.6	6.6	19.8
9	1.4		1.6	2.8	5.2	6.8	10.2	16.0	20.2	14.5
10	0.6				0.8		1.5		1.0	3.8
12	3.0	4.6	4.8	6.6	7.0	7.8	13.6	15.0	15.0	10.5
13	26.0	40.0	41.0	80.5	91.2	97.2	133.6	163.2	166.6	15.3
20	0.6	0.8		1.2		1.4	2.0		2.0	3.4
21	0.6	1.0	1.2	1.8	2.0		2.6	2.8	2.8	10.5
25	6.8	7.6			12.2		13.6	13.5	13.8	6.6
28	0.2								2.2	0.1
									229.2	

Source: Ministry of Communications (1992).

As one might expect rainfall extremes away from the mountainous south-west are much lower. Williams (1979), quotes the following 24 hour intensities for stations on the Arabian Gulf: Dhahran (1939–1977) 54 mm – 4th July 1969; Abqaiq (1951–1977) 80 mm – 12th December 1955; Ras Tanura (1951–1977) 66 mm – 4th May 1976.

4.5.3   Thunderstorms

Most rainfall extremes are associated with intense thunderstorms and Shwehdi (2005) has recently written on their occurrence and spatial distribution in the Kingdom. As might be expected most thunderstorms, and associated lightning strikes, occur in the south-west of the Kingdom on the escarpment mountain range, particularly in spring and summer (Table 4.6). Thunderstorms are the frequent cause of flash flooding in the escarpment *widyan* and deaths due to lightening strikes are not unknown.

Table 4.6 The annual average number of days with thunderstorms

Cities	Summer	Autumn	Winter	Spring	Annual	Coordinates	
						Longitude	Latitude
Abha	42.54	10.3	3.54	41.0	97.33	42° 40'	18° 14'
Al Baha	21.05	11.2	2.63	32.6	67.42	41° 39'	20° 18'
Al Hasa	0.26	0.8	6.26	9.7	17.05	49° 29'	25° 18'
Al Wadj	0.04	1.4	2.08	0.9	4.46	36° 28'	26° 14'
Arar	0.29	5.5	3.13	6.2	15.13	41° 01'	30° 59'
Bishah	3.89	4.1	2.11	25.2	35.21	42° 37'	20° 00'
Dhahran	0.04	1.7	5.88	8.6	16.17	50° 10'	26° 16'
Ha'il	0.58	8.9	4.25	15.5	29.21	41° 42'	27° 31'
Hafar al Batin	0.23	3.3	5.92	10.5	20.00	45° 58'	28° 26'
Jawf	0.16	4.3	2.04	5.5	11.96	40° 05'	29° 56'
Jiddah	1.17	4.0	1.92	1.8	9.00	39° 11'	21° 42'
Jizan	17.14	16.4	1.10	4.1	38.76	42° 33'	16° 54'
Khamis Mushait	29.83	7.3	3.42	35.5	76.13	42° 44'	18° 18'
Madinah	2.83	7.0	2.67	7.7	20.21	39° 38'	24° 27'
Makkah	2.89	9.9	2.95	3.7	19.47	39° 50'	21° 25'
Najran	3.74	0.5	0.79	11.11	12.75	44° 12'	17° 32'
Qassem	0.21	5.5	5.47	12.6	23.84	43° 46'	26° 18'
Qaysumah	0.37	5.4	8.11	11.6	25.47	46° 08'	28° 19'
Qurayyat	0.05	5.6	2.00	3.9	11.53	37° 20'	31° 20'
Rafha	0.13	4.4	2.67	6.0	13.29	43° 30'	29° 38'
Riyadh	0.25	1.0	4.46	10.0	15.75	46° 43'	24° 56'
Sharawrah	3.37	0.9	0.74	5.1	10.05	47° 07'	17° 29'
Tabuk	0.79	6.5	2.63	4.5	14.50	36° 36'	28° 23'
Taif	16.86	29.6	6.55	43.0	96.00	40° 33'	21° 29'
Turayf	0.32	6.9	1.68	6.9	15.89	44° 09'	26° 50'
Wadi ad Dawasir	0.78	0.50	0.89	5.0	7.17	44° 43'	20° 29'
Yanbu	0.47	3.8	1.77	1.1	7.14	38° 05'	24° 05

Source: Shwehdi (2005).

4.6   GLOBAL RADIATION AND SUNSHINE

The general pattern of radiation intensity over the Kingdom is fairly simple although clouds generated by the Red Sea escarpment and dust storms from the northern deserts distort the picture. During the winter months the radiation intensity is between 200 Langleys per day in the northern parts of the country and 450 Langleys in the central southern region. In the summer northern parts of the country e.g. Qurayyat receive about 550 Langleys per day and in the southern Jabal Tuwayq region values of 650 Langleys have been recorded at Sulayyil. Average daily intensities in the central Rub' al Khali probably exceed 600 Langleys though data are hard to come by.

Sunshine hours per day expressed as a percentage of the maximum possible for Abha, Al Qatif, Riyadh and Hail are shown in Table 4.7. In the summer months percentage sunshine is depressed in Abha. Although most days experience 7 to 8 hours of sunshine throughout the year the percentage figures clearly illustrate the impact of cloud cover in the Abha area which is located on the very western edge of the Red Sea escarpment. Al Qatif, on the Gulf coast has low winter and spring values as a result of dust in

Table 4.7 Sunshine hours per day as percentage of maximum

	Abha	Al Qatif	Ha'il	Riyadh
Jan	64.2	57.6	64.3	71.7
Feb	65.0	61.6	76.7	80.9
Mar	68.0	54.2	63.9	68.8
Apr	70.6	57.6	67.8	67.6
May	59.5	64.7	66.8	65.9
Jun	64.4	74.6	79.0	78.9
Jul	54.6	70.5	78.4	78.3
Aug	55.3	72.9	78.2	78.9
Sep	70.3	78.4	72.7	81.7
Oct	72.5	75.8	71.5	78.3
Nov	72.7	71.4	70.4	77.8
Dec	62.3	62.3	69.4	68.8

Source: Presidency of Meteorology and Environment.

the atmosphere. In the Nadj, Ha'il's percentages are depressed as compared with Riyadh due to cloudiness associated with weak fronts passing in from the Mediterranean.

4.7 WINDS

The Kingdom is not a windy place in spite of perceptions to the contrary. Atmospheric pressures are usually high and the seasonal pressure differences over much of the region relatively small. Table 4.8 illustrates the seasonal situation at representative stations. In spite of *shamal* winds the Gulf coast region is actually not that windy. Al Qatif has relatively low mean monthly wind speeds of around 2 m s^{-1} and even the monthly maximia are not particularly high. Tabuk on the other hand in the north-west of the Kingdom is much windier and monthly maxima are often twice as high as Gulf coast stations. Two desert stations, Yabrin on the northern edge of the Empty Quarter and Najran on its south-western edge illustrate the general decline in wind speed from east to west across this vast erg. And in spite of its altitude Abha is much less windy than Tabuk or Ha'il.

The synoptic situation for a *shamal* on the 9th June 1982 is shown in Figure 4.5. The intense low pressure over eastern Iran creates a large a anticlockwise wind system drawing northerly winds down the Arabian Gulf.

In addition to the *shamal* there are one or two other seasonal winds affecting the Kingdom. For example, the *aziab* is a strong, hot, dry, wind from the south-west which is most frequent during May–Sept. It brings dust to the Jiddah region and thunderstorms further south. It is generated by low-pressure disturbances associated with fronts or troughs passing across the region. The disturbances are either a *khamsin* low pressure centre coming from Egypt or a north-eastward shift of the Sudan low pressure zone.

4.8 CLIMATIC REGIONS

4.8.1 Climatic regions according to Al-Jerash (1985)

This scheme is the first published account to use modern statistical methods for the climatological subdivision of Saudi Arabia. Principal components analysis was used

Table 4.8 Mean monthly wind speed (m/s^{-1})

	Tabuk		Ha'il		Abha		Najran		Yabrin		Al Qatif	
	mean	max	mean	max	mean	max	mean	max	mean	max	mean	max
Jan	2.2	13.9	2.3	10.8	2.5	7.5	1.6	5.0	2.3	9.2	2.1	6.4
Feb	2.5	13.6	2.6	11.8	2.7	10.3	1.6	7.5	2.3	9.9	2.1	4.7
Mar	2.8	16.1	2.7	14.6	2.6	9.4	1.7	7.5	2.6	11.0	2.2	5.8
Apr	3.2	15.0	2.9	13.9	2.2	8.3	1.8	7.2	2.5	10.0	2.2	5.0
May	3.1	14.7	2.8	13.9	2.1	7.8	1.9	7.2	2.4	9.2	2.2	7.9
Jun	3.2	13.3	5.3	10.0	2.2	8.1	1.9	6.9	2.9	9.6	2.7	7.8
Jul	3.1	11.7	2.7	9.7	2.4	8.3	2.1	6.9	2.8	8.6	2.3	6.8
Aug	2.6	9.4	2.4	8.9	2.2	8.3	2.0	6.7	2.4	8.1	2.2	7.1
Sep	2.4	10.0	2.1	8.9	2.2	10.0	1.7	5.3	1.9	8.3	1.8	7.2
Oct	2.0	22.2	2.2	9.4	2.1	8.1	1.4	4.4	1.8	8.6	1.7	6.8
Nov	1.9	12.8	2.1	11.1	2.1	6.1	1.3	4.2	1.9	8.9	1.9	5.5
Dec	1.8	13.3	2.1	12.2	2.1	6.1	1.4	3.6	2.1	8.6	2.2	6.8

Source: Presidency of Meteorology and Environment.

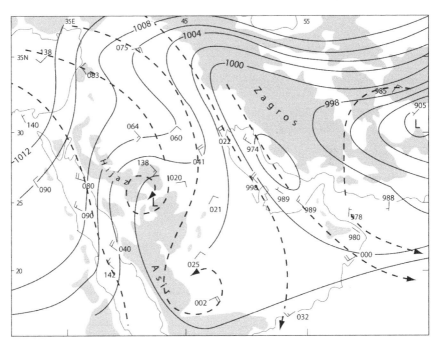

Figure 4.5 Synoptic conditions during the *shamal* 12:00 GMT, 9th June 1982. 850 mb streamlines (broken); land over 1000 m shaded (Modified from Membery, 1983).

to investigate a data matrix comprising 24 variables for 50 stations. The stations were selected on the basis of spatial distribution and the availability of complete records for the period 1970–1982. Potential evaporation and soil moisture information was calculated using software developed by Muller (1981). Five principal components were extracted from the data matrix. Al-Jerash interpreted these five components as: summer water balance; winter thermality, relative humidy, winter water balance and summer thermality. The first two of these accounted for 73.5 percent of the variance in the data and Al-Jerash suggests that variations in summer water balance and winter thermality are dominant features in shaping the climatological division of the country (Table 4.9 and Figure 4.6).

4.8.2   Agro-climatic zonation

An attempt at an agro-climatic zonation was undertaken by Al-Zeid (1988) for the Ministry of Agriculture and Water (MAW) in order to determine the irrigation requirements of major crops. In this zonation, 13 agro-climatological regions were defined using: temperature, relatively humidity, wind speed sunshine and evaporation data. There was little attempt to define regional boundaries based on meaningful interpolation and topographical features have been almost completely ignored.

Within the last decade the Land Management Department of the MAW has been assisted by experts from the Food and Agriculture Organisation of the United Nations and together they have concentrated on devloping climatic classifications for the Kingdom which have meaningful practical applications. In this regard any classification which puts too much emphasis on natural precipitation, when most crop production is based on irrigation, is of little practical use. Furthermore, an examination of local crop production information has amply demonstrated that relatively small variations in temperature regimes are major determinants of crop calendars and potential yields. With these factors in mind scientists at the Land Management Department, notably Kumul F.A. Fadil and M. Ashrif Ali, have sought to produce a classification which is based firmly on temperature regimes, condition by relative humidity. Their

Table 4.9 Characteristics of climatic regions as defined by Al-Jerash

	Central A	Tihamah B	Gulf coast northern Hijaz C	Highlands D	South-West highlands >2000 m E	Western slopes South-West highlands 200–500 m F
Average Annual Temp °C	24	30	23	17	17	30
Average January Temp °C	13	24	12	15	11	26
Average July Temp °C	33	32	32	26	22	34
Average Annual Rainfall mm	95	110	91	210	539	352
Average January Rainfall mm	13	20	17	14	97	19
Average July Rainfall mm	0.4	8	0.1	15	27	56

Source: Al-Jerash (1985).

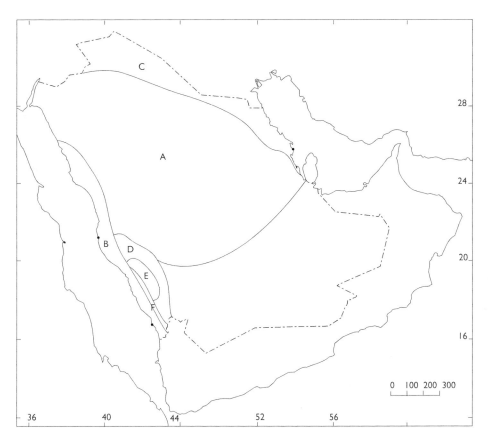

Figure 4.6 Climatic zones according to Al-Jerash (1985).

classification uses as its main differentiating criteria the mean maximum and mean minimum temperatures, especially of the coldest and hottest months. The annual mean temperatures do not in fact reflect the seasonality that coincides, approximately, with the growing periods of the summer of winter crops. Instead, it is argued, there is more impact on crop performance of mid-summer and mid-winter ranges through their impact on phenological stages. Only in the mountainous Asir region is precipitation taken as a classification determinant (MAW, 1995).

Twenty-one agro-climatic zones have been recognised in the Land Management Department's classification (Figure 4.7) The zones were initially delineated from isotherms given in the *Climatic Atlas of Saudi Arabia* (MAW, 1988). These isotherms were then adjusted to conform to major variations in topography so as to reflect local conditions more faithfully. The twenty-one agro-climatic zones are grouped into eight agro-climatic regions. Mid-summer and mid-winter temperature ranges of the eight agro-climatic regions are shown in Table 4.10.

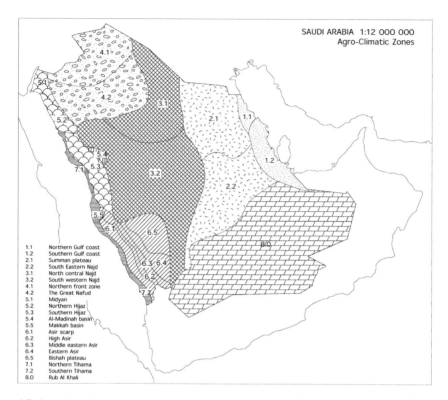

SAUDI ARABIA 1:12 000 000
Agro-Climatic Zones

1.1	Northern Gulf coast
1.2	Southern Gulf coast
2.1	Summan plateau
2.2	South Eastern Najd
3.1	North central Najd
3.2	South western Najd
4.1	Northern front zone
4.2	The Great Nafud
5.1	Midyan
5.2	Northern Hijaz
5.3	Southern Hijaz
5.4	Al-Madinah basin
5.5	Makkah basin
6.1	Asir scarp
6.2	High Asir
6.3	Middle eastern Asir
6.4	Eastern Asir
6.5	Bishah plateau
7.1	Northern Tihama
7.2	Southern Tihama
8.0	Rub Al Khali

Figure 4.7 Agro-climatic zones according to the Land Management Department, Ministry of Agriculture and Water, 1995.

4.9   QUATERNARY CLIMATIC AND ENVIRONMENTAL CHANGE

4.9.1   Global events and chronology

The probable nature of the relationship between Quaternary climatic changes in the Arabian Peninsula and global events is still rather sketchy. The area has not been as thoroughly investigated as northern Africa and environmental reconstruction is made difficult by the paucity of datable deposits. Computer models and the also the timing of events suggest that the most intensely arid and cool phases in Arabia approximately correspond to the maximum cover of ice in Fennoscandia and cold, ice-rafting phases in the North Atlantic (Heinrich events). In general, a greatly expanded glacial ice cover in the northern Hemisphere is thought to have weakened the influence of the summer Asian Monsoon zone, and thereby reducing precipitation in the southern parts of the Peninsula. In the winter, increased glacial ice cover enhanced winter trade wind circulation and might also have decreased precipitation by blocking the penetration of depressions tracking eastward along the Mediterranean

Table 4.10 Temperature ranges for agro-climatic zones as defined by Land Management
           Department – Ministry of Agriculture and Water

| Region | Zone | Temperature Range °C | | Representative | Zone in |
		Mean January	*Mean July*	*Station*	*Figure 4.7*
Gulf Coast	Northern Gulf Coast	10–14	>60	Khafji	1.1
	Southern Gulf Coast	14–16	34–36	Qatif	1.2
Eastern Najd	Summam plateau	10–14	34–36	Sarrar	2.1
	South eastern Najd	14–16	34–36	Kharj	2.2
Western and	North central Nadj	10–14	32–34	Unayzah	3.1
Central Najd	South western Nadj	14–18	30–34	Dawadimi	3.2
Northern Plains	Northern frontier zone	8–10	30	Qurayyat	4.1
	The Great Nafud	10–12	30–32	Ha'il	4.2
Hijaz Highlands	Middyan	10–14	<30	Bishah	5.1
	Northern Hijaz	14–20	30–32	Al Ula	5.2
	Southern Hijaz	20–22	30–32	Badr	5.3
	Madinah Basin	16–28	34–36	Madinah	5.4
	Makkah Basin	22–24	34–36	Makkah	5.5
Asir Mountains	Asir Scarp	16–22	26–32	Sayl al Kabir	6.1
	High Asir	14–16	24–26	–	6.2
	Middle eastern Asir	14–16	26–28	Abha	6.3
	Eastern Asir	14–16	28–30	Taif	6.4
	Bishah Plateau	>16	30–32	Bishah	6.5
Red Sea Coast	Northern Tihamah	22–24	32–24	Yanbu	7.1
	Southern Tihamah	20–22	32–34	Sabya	
Rub' al Khali	Rub' al Khali	14–16	34–36	Sharura	8

Source: MAW (1995).

(deMenocal, 1993). The enhanced glacial aridity of Arabia is also indicated by the
greatly increased accumulation of dust in Arabian Sea sediment cores. An analysis of
the periodicity of these dust accumulations shows a high correlation with glacial ice
volumes (Clemens and Prell, 1990).

The general chronology of climatic change events for Saudi Arabia is outlined in
Table 4.11. The most striking feature of this Table is the Late Pleistocene/Holocene climatic variability with relatively wet periods alternating with intensely arid periods.

4.9.2   Late Pliocene – Early Pleistocene climatic
           environments

There is some surrogate geomorphological evidence for the climatic regime during this long period of about three million years but the exact chronology is uncertain. The Late Pliocene-Early Pleistocene climates of the Kingdom are thought

Table 4.11 General chronology of climatic change in Saudi Arabia

Geological Epoch	Chronology BP	Climatic phase	Events in Saudi Arabia
Holocene	0–700	Hyperarid	
	700–1300	Slightly moist	Hufuf river noted by Yaqut and other geographers
	1300–1400	Arid	Dune movement
	1400–2100	Slightly moist	Sabatean Kingdom flourished and also Kingdom of Kinda at Qarayat al Fau
	2,100–5,000	Hyperarid	Dune movement
	5,000–5,500	Slightly moist	Neolithic camp site in Rub' al Khali
	5,500–6,000	Hyperarid	High crested dunes; and interdune corridors
	6,000–10,000	Wet (Pluvial)	Neolithic wet phase – lakes in south-west Rub' al Khali
Late Pleistocene	10,000–18,000	Hyperarid	Dune topography and extension of longitudinal dunes
	19,000–36,000	Wet (Pluvial)	Lakes in the south-west Rub al Khali and An Nafud
	36,000–70,000	Arid	Main movement of sand from from old *widyan* in shrunken Arabian Gulf
	70,000–270,000	Moist	Riss-Wurm interglacial and early phase of Wurm glacial
	270,000–325,000	Arid	Summan Plateau caves dry
Middle Pleistocene	325,000–560,000	Wet	Active karstification and cave formation in Summan Plateau
	560,000–700,000	Arid	Beginning of low dunes
Late Pliocene & Early Pleistocene	700,000–3,000,000	Wet humid (Pluvial)	Early Quaternary drainage systems and large alluvial fans formed.

to been humid and a considerable amount of runoff was generated. The evidence comes from the presence of widespread gravel trains and fans associated with palaeodrainage systems running eastward from the Shield into a dry Arabian Gulf. The Hufuf Formation on the Gulf coast was deposited by these systems.

The presence of lateritic soil profiles on dated lava flows also implies a tropical temperature regime for at least part of this time period. In Wadi Ranyah, for example, Hötzl *et al.* (1978) describe gravel deposits which are bedded between two K-Ar dated basalts. The upper basalts date from 1.1 Ma and the lower from 3.5 Ma. The same lower basalt also has a lateritic soil developed on it according to these authors. Chapman (1971, 1974) indicates that at least some part of the Early Pleistocene was semi-arid and he suggests that the thick duricrusts found in the Eastern Province started to form about this time although he provides no evidence for this suggestion.

One interesting approach for paleoenvironmental reconstruction is described by Thomas *et al.* (1998) who discuss the use of ^{13}C isotopic signatures in mammal enamel from early Pleistocene lacustrine deposits discovered in interdune deposits in the southwestern Nafud. Enamel tissue records the carbon isotopic signature of the diet which depend mainly on photosynthetic pathways of the plants consumed by herbivores. In tropical environments grasses are C4-plants whereas trees are C3-plants with different isotopic signatures. The authors suggest that an open environment will correspond to higher levels of ^{13}C indicative of C4-plants. These isotopic signatures are evidently preserved for millions of years in fossil enamel. Measured ^{13}C levels in early Pleistocene faunas in the western Nafud suggest to Thomas *et al.* that the vegetation was an open savannah. The absence of woodland and forest edge mammal species confirms this interpretation. This savannah landscape was dotted with lakes and was the home of giant buffalo and hippopotamus in the Nafud deposits. These lake bodies date from about 1.2 to 3.5 Ma and are not to be confused with the Late Pleistocene lakes in the Nafud described by Schultz and Whitney (1986).

The palaeodrainage systems are associated with vast, often thin, sheets of gravels and silty gravels forming very low angle fans. Both Powers (1966) and Edgell (1990) discuss the origin and nature of these fans. One of the largest forms the Ad Dibdibah plain, in north-eastern Saudi Arabia. This sheet of gravel was derived from the Precambrian Shield and was swept along the Wadi ar Rumah-Wadi al Batin drainage systems into Kuwait. The gravel train is now crossed by the sand dunes of the Dahna desert. The gravels associated with Wadi Sabha are thought by Hötzl *et al.* (1978) to have accumulated as a delta associated with a high Plio-Pleistocene sea level in the Arabian Gulf. Dabbagh *et al.* (1998) have reconstructed the palaeodrainage system for the Peninsula using Shuttle Imaging Radar (SIR-C/X-SAR).

4.9.3   Middle Pleistocene

700,000–560,000 BP

According to Edgell (1990) the beginning of the Middle Pleistocene in Saudi Arabia was arid and he suggests that low dunes began to accumulate in the Rub' al-Khali. Anton (1984) notes that basalts of the *harraat* dating from about this time are neither strongly weathered nor dissected by deep fluvial systems. These two lines of evidence suggested to Anton that during the rest of the Quaternary there was an absence of long humid periods.

560,000–325,000 BP

Edgell (*op. cit.*) suggests that the Middle Pleistocene ended with a wet phase when there was active karstification and cave formation of the Summan Plateau to the north-east of the Dahna desert.

325,000–270,000 BP

At this time the Summan Plateau caves were dry.

4.9.4   Late Pleistocene

270, 000–70,000 BP

This period was moist and U/Th dated carbonates accumulated in the As Summan Plateau caves (Rauert *et al.*, 1988).

70,000–36,000 BP

During periods of low eustatic sea level Saudi Arabian *wadi* floodwaters flowed in the dry Gulf to join the Tigris-Euphrates river system. It was during this arid phase that Arabian Gulf sands and silts from a dry Arabian Gulf were blown in land a short way where they mixed with deflated sands of the Hufuf Formation.

36,000–19,000 BP

Evidence from radiocarbon-dated lake deposits, duricrust and *wadi* alluvium indicates that between about 36,000 and 19,000 BP Saudi Arabia was in the grip of a major pluvial period. This pluvial may have started a little earlier but this depends on how one interprets isotopic dates for groundwater greater than 37,000 BP (Al-Sayari and Zötl, 1978). Looked at as a whole, isotopically dated deposits in the Kingdom show a large cluster of dates between 28,000 and 26,000 BP which demarcates the most intense and widespread part of the pluvial. Whitney (1983) remarks that deposits of this age exist both in the northern and southern regions of the Kingdom where-as deposits older than 30,000 BP are concentrated in the south (Table 4.12). The implication is that the Pleistocene pluvial began in the south. Monsoonal influences then expanded northwards with time and reached their maximum intensity at about 28,000 to 26,000 BP. A smaller cluster of deposits dating from 22,000 to 19,000 BP is found only in areas on the southern Shield. This distribution suggests to Whitney that the Monsoon-induced pluvial retreated from northern Arabia earlier than in the south. Pluvial activity in Saudi Arabia gradually decreased from about 24,000 BP and ceased altogether by the time of the last glacial maximum at about 18,000 BP. Glennie *et al.* (2002) suggest that the *shamal* during this time had a southern trajectory and did not curve south-westward into the Rub' al Khali basin.

Some of the best known lake deposits are those described by McClure (1976) in the Mundafan basin on the south-western edge of the Rub al Khali. Here, the lacus-trine deposits consist of calcareous, and often fossiliferous, marls, clays and silts. They characteristically form flat-topped benches and mounts often in the lee of longitudinal dunes an in the 'corridors' between the dunes where recent deflation has removed the superficial deposits. The lakes described by McClure were not enormous like the lakes in East Africa. They probably ranged in depth from 2–10 m but some were no more than ephemeral puddles. Radiocarbon dates indicate some lakes lasted for about 800 years. Despite the finds of hippopotamus teeth in the Mandafan lake beds McClure cautions against imagining that large herds of the beasts wallowed in lush lakes throughout the Rub' al Khali at this time. He envisaged that hippos, along with water buffalo, moved into the area from a lush bordering habitat but died out when the lakes dried up. Cattle, which are less demanding of water than water buffalos and

Table 4.12 Radiocarbon-dated lacustrine and interdune deposits

Locality	Location Lat/Long	Laboratory number	Material	Age BP
An Nafud	N27°36', E38°50'	B-3461	Organic material	5,280 ± 100
Nafud as Sirr	N26°09', E44°19'	B-2988	$CaCO_3$	5,480 ± 70
Nafud Urayk	N25°35', E42°40'	B-3466	$CaCO_3$	5,649 ± 90
Rub' al Khali	N20°20', E47°00'	W-4814	Organic material	6,230 ± 90
An Nafud	N28°00', E40°50'	Q-3118	Organic material	6,685 ± 50
Badr	N17°55', E43°41'	B-3425	$CaCO_3$	7,160 ± 100
Mundafan	N18°36', E45°21'	B-2989	$CaCO_3$	8,240 ± 90
Nafud Urayk	N25°37', E47°39'	B-3523	$CaCO_3$	8,440 ± 90
Rub' al Khali	N18°17', E45°36'	B-2981	Shells	19,390 ± 210
Rub' al Khali	N18°05', E46°00'	B-2980	$CaCO_3$	21,010 ± 210
An Nafud	N28°50', E40°15'	W-4835	$CaCO_3$	24,340 ± 300
An Nafud	N28°15', E41°15'	W-4847	$CaCO_3$	25,750 ± 600
An Nafud	N28°00', E40°50'	Q-3117	$CaCO_3$	25,630 ± 430
Rub al Khali	N19°03', E46°13'	B-2987	$CaCO_3$	26,050 ± 440
Rub al Khali	N19°02', E46°29'	W-4808	$CaCO_3$	26,400 ± 500
Rub al Khali	N18°05', E46°00'	B-2990	$CaCO_3$ (duricrust)	27,090 ± 320
An Nafud	N28°15', E39°45'	W-4864	$CaCO_3$	27,120 ± 420
An Nafud	N28°15', E41°15'	W-4855	$CaCO_3$	27,570 ± 500
An Nafud	N27°50', E41°04'	W-4838	$CaCO_3$	29,000 ± 600
Rub al Khali	N18°20', E45°29'	B-3467	$CaCO_3$ (duricrust)	30,500 ± 920
Nafud Urayk	N25°32', E42°36'	W-4972	$CaCO_3$	>31,000
An Nafud	N28°45', E40°45'	W-4959	$CaCO_3$	>32,000
Rub al Khali	N21°41', E50°50'	W-4797	$CaCO_3$	>32,000
Rub al Khali	N18°18', E45°37'	W-4922	$CaCO_3$	>33,000
Rub al Khali	N18°02', E46°28'	W-4811	$CaCO_3$	>35,000
Rub al Khali	N19°01', E46°19'	W-4815	$CaCO_3$	>37,000
An Nafud	N29°35', E40°13'	W-4978	$CaCO_3$	>38,000

Source: Whitney (1983); Garrard and Harvey (1981); McClure (1976).

hippos are also recorded in the deposits of the Mundafan basin and probably ranged widely both during this and the Holocene wet phase.

The presence of large numbers micron-sized opal phytoliths in the Mundafan deposits is of interest. These bodies are common in grass leaves and their presence in the lacustrine marls suggested to McClure that the lake surroundings were probably grasslands supporting herds of grazers. Fossil pollen suggests that the lake shores were lined with *Phragmites* and *Typha*.

Lake deposits are, in fact, much more widespread in the Rub al-Khali than the Mundafan basin and McClure charted them down the whole middle length of the desert – some 1,200 km. What remains of them are marl terraces and patches of hardened crust a few metres wide, distributed in thin, 'shoestring' forms, a kilometre or more long, between the dunes. Now, however, instead of lying below the level of the desert , the lake beds stand up in relief, the sand that once surrounded them having been deflated away. McClure, believed the lakes formed between the dunes and in the hollows on top of them when torrential rainfall spilled down dune slopes. The runoff brought with it clay and silt particles from the dune sides and pans formed where the water pooled. The pans impeded water penetration down into the sand and held any new precipitation. Lake beds, presumably of the same age as those in the Mundafan basin, are also present in the eastern Rub' al Khali in the south-eastern Uruq al Mutaridah (Pambour and Al Karrairy, 1991).

Schulz and Whitney (1986b) have investigated similar lake bodies in the Nafud. They discovered two major lake periods. In the Upper Pleistocene lakes occurred around the dune area and in the interior of the sand sea. The lakes were several km² in extent and about10 m deep and formed between 34,000 and 24,000 BP. The lacustrine deposits, which are up to 2.30 m thick, comprise calcareous crusts, marls and, rarely, diatomites. Schulz and Whitney observed that these deposits underlie the dunes and are not conformal with the present day dune configuration. A detailed investigation of the stratigraphy of these lacastrine deposits indicates that the lake phases must have been long – about 900 years.

18,000–10,000 BP (Arid)

Arid conditions prevailed once more and dune fields were reactivated. Some dune fields of this age actually spread into the dry Arabian Gulf only to be buried later by rising sea levels. Arid conditions are also confirmed by the complete absence of alluvial or lacustrine deposits during this period. Marine sediments in the Arabian Sea record this increase in aeolian activity by substantial amounts of terrigenous silt deposition. While the early lakes in the Rub' al Khali lie on white sand, the younger Holocene group all sit on foundations of red sand, indicating that the development of the iron staining both in the Empty Quarter and also probably in the Dahna, dates from this arid period.

4.9.5   Holocene

9,000–6,000 BP (Pluvial)

Radiocarbon-dated tufa, silty alluvium, soil carbonate and lake deposits in the Rub' al Khali and the Nafud indicate that pluvial conditions began again about 9,000–8,500 BP and lasted to about 5,000 BP (Sanlaville, 1992). This so-called Neolithic Wet Phase was not as intense as the Late Pleistocene wet phase and the lakes which formed were much smaller. Whitney (1983) suggests that his was a time of gentle and reasonably persistent rainfall over much of Arabia and during this period loessic alluvial silt accumulated in *widyan* on the Shield. In the Nafud shallow lake beds dating from this period are found at the bottom of interdune depressions, that is to say, conformal with the present dune configuration. Schulz and Whitney (1986b) note that it is not necessary to presume major climatic changes to account for the presence of these Holocene lakes and they suggest that more frequent invasions of cyclonic rainfall from the Mediterranean basin would have been sufficient to build up the shallow swamps in interdune sites. Pollen analysis of the lake sediments suggests that the vegetation was more or less the same as today.

A new lake body was recently found at Ranyah on the central Shield. Radio carbon measurements of shells from the lake marls indicate the deposits date from 8,000 BP (Figure 4.8).

Figure 4.8 Palaeolake Ranyah (Photo: author).

The Holocene chronology in Saudi Arabia is not closely synchronised with the Holocene chronologies of eastern Africa and the Sahara according to Whitney. He points out that pluvial conditions in northern and eastern Africa began about 11,000 to 10,000 BP and continued to 8,500 BP at which time climate suddenly became more arid. By contrast Holocene pluvial conditions in Saudi Arabia were not fully established until about 8,500 BP.

6,000 BP–present (Hyperarid)

A major hyperarid phase followed the so-called Neolithic Wet Phase. During this phase, which continues up to the present day, the system of large linear dunes in the Empty Quarter was established. The aridity has been broken by a number of wetter periods, notably between 5,500 and 5,000 BP, as indicated by Neolithic camps in the Rub' al Khali, and between 2,100–1,400 BP when the Sabatean Kingdom flourished at Madain Salih. There may also have been a brief wet interlude from 1,300 to 700 BP as noted by Arab geographers who describe the presence of a river in the Hufuf area.

The alternating sequence of lake formation and dune activity in the Upper Pleistocene is shown very clearly by Garrard and Harvey (1992) in their archaeological investigations near the village of Jubbah in the central Nafud. In one section they identify the following 7 horizons:

1. made ground
2. recent dune sand (1 m)

3. dark sandy silt (0.25 m) very rich in organic material and plant debris.Interpreted as a palaeosol – dated at 6,685 ± 50 year BP. An identical deposit of the same age has been observed some 35 km to the south-east near Qina.
4. iron-stained dune sand (25 cm) indicative of hyperarid conditions and contemporaneous with the main sand formations of An Nafud.
5. white, diatomaceous, lacustrine silts. Dated at 25,630 ± 430 years BP
6. hard, white, diatomaceous silts with high soluble salt content. Common over much of the Jubbah basin. Interpreted as shallow water sabhkah deposits (12 m)
7. brownish clay with some diatoms (12 m visible in well). Not dated.

The thick brownish clays at the bottom of the section are thought by Garrard and Harvey to represent a long humid phase. The deposits may well be synchronous with the speleothem development and karstification of the Summan Plateau mentioned previously.

4.9.6   Present-day aridity

While most observers would regard the Kingdom as arid, it is rather more difficult to quantify aridity because of the many different definitions of the term (Warner, 2004). Dzerdzeevskii (1958), for example, describes nineteen indices which purport to define aridity. One commonly used measure is the Budyko index, defined as the ratio of the annual depth of water which could be evaporated by the total net radiant energy at ground level, to the annual observed precipitation. Using this index, the Empty Quarter is the most arid part of the Kingdom with values in excess of fifty. This, however, is much less that most of the Sahara. It should also be noted that where climate stations are few and far between, as in the case of most of the Kingdom, isopleths maps of such indices are little more than informed guesses.

One weakness of the Budyko ratio, as with many other indices, is the use of annual data. As Warner (2004) notes, such indices cannot account for situations where there is sufficient precipitation relative to demand in certain months to allow the growth of non-arid vegetation.

4.9.7   Global warming

There are few studies regarding the possible impacts of possible global warming on Saudi Arabia. As the major international oil exporter the Kingdom maintains a healthy scepticism regarding doom-laden warnings from international agencies. This view finds some support in Soon and Baliunas (2003) whose major research review indicates that across the world, many records reveal that the 20th century is probably neither the warmest nor a uniquely extreme climatic period of the last millennium. On the other hand scenarios for the Middle East produced using General Circulation Models (GCM) suggest an increase in surface warming, a decrease in precipitation, and an increased temporal variability in both (DeMenocal, 1993). This has been recently confirmed by the Intergovernmental Panel on Climate Change (IPCC) whose Fourth Assessment Report was published in 2007. The report concludes that global temperatures will probably rise by the end of the century will be between 1.8C and 4C and that sea levels are likely to rise by 28–43 cm. The IPCC forecast that by the year 2050 temperatures may have warmed by as much as 1.6 °C over the Arabian Peninsula with interior regions warming more quickly than the coasts. Precipitation is forecast to be reduced by 10 percent over the same period. It is worth pointing out that current GCMs have relatively low predictive power both spatially and temporally. Models are only simplistic representations of the real world and are not the real world per se.

The implications for the Kingdom are, however, serious indeed. Notwithstanding the implications for dune mobilisation as vegetation dies off, the probable rise in sea level, both on the Tihamah and the Gulf coasts, has implications for both coastal aquifers, and infrastructure. Estimates from both tide gauges and satellite altimetry suggest a sea level rise rate of about 3 mm yr^{-1}.

Lioubimsteva (2004) has produced an interesting review of the possible impacts of climatic change in arid environments by examining past changes. For example, she notes that during the Last Glacial Maximum the 100 mm isohyet was about 13°–14°N as compared with 17°–18°N today. What this glimpse of climatic history shows is the remarkable latitudinal swings that have taken place across the Arabian Peninsula as a whole.

In one recent paper examining the influences of global warming on agriculture and water resources Alkolibi (2002) analysed temperature and precipitation records going back to 1961 for the cities of Jiddah, Riyadh, Dhahran and Madinah. Linear regression models were applied to smoothed data for July and January for temperature and annual totals for precipitation. As far as temperature trends are concerned only the July data for Riyadh showed any statistically significant increase but Alkolibi suggests that this is associated with a heat island effect due to rapid urban growth rather than climatic change. The results for precipitation trends indicate no significant change for Riyadh and Jiddah but statistically significant, but small, increases for Dhahran and Madinah – quite the opposite of the GCM model predictions. The one caveat worthy of mention is that the analyses were carried out on smoothed data rather than the actual data which certainly has the effect of reducing the statistical noise and increasing the regression R^2.

On the 21st December 2004 the Council of Ministers chaired by Crown Prince Abdullah (now King Abdullah) give its approval to the Kyoto protocol.

Chapter 5

Hydrogeology and hydrology

Contents

5.1 INTRODUCTION

Saudi Arabia is the largest country in the world without a natural, perennial, river running to the sea, and water has always been a scarce resource in Saudi Arabia. Urban life, industry, and above all agriculture, consume far more water than traditional life in the deserts and towns ever required (Beaumont, 1977). With continued population and industrial growth water management and mining have become important issues. A small reminder of the drastic state of affairs is the fact that south of the capital of Riyadh, in Wadi Hanifa, partially treated sewage water has actually created the longest permanent river in the Kingdom (c. 60 km). Only in the high Asir region are there small rivers reaching anything like this length (e.g. Wadi Batharah).

Good reliable data on water consumption on which to base any discussion of usage are difficult to come by and there are major discrepancies in the published literature. Table 5.1 compiled by Dabbagh and Abderrahman (1997) illustrates this point rather nicely. These authors obtained irrigation usage estimates for 1990–2 by summing the actual irrigation consumed in Saudi Arabia's eleven agricultural regions. This total, at the height of the development of the overproduction of wheat, greatly exceeds Ministry estimates. And even Dabbagh and Abderrahman's data are an underestimate of the total consumption of water through irrigation because it does not include the myriad of small private wells. Table 5.2, from the Sixth Development Plan data, provides a breakdown of water consumption by origin. The two tables, viewed together, make for gloomy reading. Agriculture accounts for about 84 percent of the total consumption and about 75 percent of the available water is from non-renewable ground water.

The surge in demand for drinking water by major urban and industrial centres has to a large extent been tackled by the development of 33 desalination plants making Saudi Arabia the world's largest producer of desalinated water. By the year 2000 desalinated water production met about eighty percent of the country's drinking water needs.

The water demands of agriculture, however, can only be met by tapping into extensive reservoirs of water stored in aquifers. Since the 1970s the government has provided funds for thousands of deep wells to provide water for irrigation, and particularly cereal production. In the last few decades the agricultural achievements have been remarkable, most notable is the Kingdom's rapid transformation from being an importer to exporter of wheat. In 1978 the country built its first grain silos and by 1984 it had become self-sufficient. Shortly thereafter, Saudi Arabia began exporting wheat, and by 1993 nearly production had reached 4.5 million tons – more than twice the domestic consumption. This success story, of course, has its downside and in 1993 the government, aware of the need to conserve precious groundwater supplies, announced plans to reduce the subsidy paid to farmers in order to hit a target production of 2.8 million tons by 1994. In 2000, wheat production stood at 1,787,542 tons.

Conservation of precious surface water supplies during seasonal floods which occur mainly in the south-west of the Kingdom is of major importance. By 1992 the Ministry of Agriculture and Water had built more than 200 dams with a total reservoir capacity of 774 million m³. Many of these dams were built for groundwater recharge and flood control purposes but the larger dams, such as those in Wadi Jizan, Wadi Fatima, Wadi Bishah and Najran, also supply irrigation water to thousands of hectares of cultivated land. A good deal of effort is also going into the recovery of urban waste water and it is estimated that by the year 2006 approximately 40 percent of the water used for domestic purposes in urban areas could be recycled.

Table 5.1 The growth of water use in Saudi Arabia (million m³)

Year	Domestic and industrial	Agricultural		Total
		MOP*	Authors**	
1980	502	1,850	–	2,352
1990	1,650	14,580	25,598	16,230
1992	1,710	15,308	29,876	31,586**
1995	1,800	16,406	–	18,206
2000	2,800	14,700	–	17,500

*Ministry of Planning Estimates; **Dabbagh and Abderrahman estimates.

Source: Dabbagh and Abderrahman, 1997.

Table 5.2 Annual water consumption estimates by origin (million m³ y⁻¹)

Resource type	1994–5	1999–2000	% Change
Renewable water	2,500	3,000	3.7
Non-renewable ground water	14,836	13,040	–2.6
Desalinated sea water	714	1,150	10.0
Reclaimed waste water	150	310	15.7
Total	18,200	17,500	–0.8

Source: Sixth Development Plan – Ministry of Planning.

5.2   THE DRAINAGE NET

Although the drainage net appears to be complex at first sight it can simply be thought of as comprising two basic units. By far and away the largest part of the net is that draining eastward across the Shield onto the cover rocks. The major *wadi* (Arabic: pl: *widyan*) systems can be thought of consequent *widyan* which originally flowed down dip on a surface of sedimentary strata. Once the sedimentary cover was eroded, the drainage network became superimposed on the Precambrian rocks below. There are countless examples of *widyan* breaching major structural barriers. Those breaching the Tuwaiq Escarpment are classic examples but there are many on the Shield itself such as the gorge created by Wadi Bishah west of Bishah. A similar gorge has been created west of Ranyah by Wadi Ranyah. On the cover rocks themselves the superimposed drainage has a broad trellised pattern with consequent *widyan* developed down dip and lateral, subsequent, *widyan* eroded along the strike. On the Shield there are also many examples of local drainage networks that are closely related to complex structures, and some of these are probably antecedent to neotectonic movements. Mapping the distribution of antecendant and superimposed drainage nets might be very revealing regarding the former distribution of cover rocks on the Shield.

A second unit of the drainage net comprises the short steep *wadi* systems that drain westward onto the Tihamah from the escarpment and coastal mountains. These *wadi* systems are neither superimposed nor antecedent and are closely adjusted to rock structures such as the faulting and shearing, and local lithological variations.

Two other aspects of the net are worthy of mention. First, some drainage nets are genuinely endoreic and developed around tectonic basins such as the faulted Sahl Rukkbah, a vast depression east of Taif. At a much smaller scale many *sibakh* and *kabra* on the Shield are sites of local internal drainage systems. Some *widyan* cannot be traced to the coast and disappear under a sea of sand, such as Wadi Najran which ends under the sands of the western Rub' al Khali. Second, and it might seem rather surprising, is the observation that *harraat* have well developed drainage nets obviously post-dating the lava flows themselves. It would be of some interest to examine the drainage development on dated harraat.

5.3 SURFACE RUNOFF

There are few permanent natural streams or rivers in Saudi Arabia due to the meagre rainfall. After heavy rain the upper reaches of *widyan* draining the escarpment towards the Red Sea may have a base flow lasting several months but none reach the sea. In spite of the scanty rainfall runoff occurs in most drainage systems in Saudi Arabia apart those areas with mobile sands. The pattern of runoff reflects the distribution of rainfall with low runoff in *widyan* in the Gulf region and higher runoff in the Hijaz and Asir. One man-made permanent river is the Ar Ha'ir river (N24°21', E4°57') which originated in 1976 and comprises runoff from the treated water sewage system of Riyadh. The river, which flows for about 60 km along Wadi Hanifa is now a highly favourable habitat providing large areas of woodland, marshland and open water and is frequented by many species of birds.

Most *wadi* systems are choked with sands and gravels and runoff rapidly sinks into the *wadi* bed after a few kilometres and forms a subflow which recharges the ground water. Major discharge events do occur, particularly in the Asir, but they are difficult to monitor and predict as they are generated by individual storm events. Storms usually generate high intensity rainfall and the associated runoff is short-lived. On the whole, runoff is greatest during the spring but can be very variable. In the Asir, storm runoff can occur at any time of the year but the greatest runoff is associated with unstable air at the northern margin of the summer Monsoon penetrating into the Red Sea.

Much of the runoff in Saudi Arabia is concentrated along the Asir escarpment where *widyan* draining to the coast carry as much as 60 percent of the Kingdom's total runoff. In the Red Sea coastal area as a whole, the mean annual runoff has been estimated at 39.8 m³/sec about 27 m³/sec of which occurs south of Jiddah (Ministry of Agricultural and Water, 1984). Estimates of mean annual runoff for individual *widyan* is shown in Table 5.3.

The total runoff draining inland from the escarpment is relatively small but some *widyan* such as Wadi Dawasir and Wadi Najran have higher runoff than *widyan* draining to the Red Sea because their headwaters are drained by number of very large tributaries (Table 5.4).

Since the 1980s a network of some 200 crest-stage gauges has been installed, mainly under bridges crossing major *widyan*. These gauges measure the peak flow which generally, but not always, coincides with the maximum stage or elevation of the flood. Very often peak flows occur a night. The intense heat of the land during the daytime triggers the development of convective storms, especially in the high escarpment mountains

Table 5.3 Mean annual runoff in the Red Sea coastal zone

Wadi system	Mean annual runoff m³/sec
Shafqah, Hali	3.49
Wasi, Shahdan, Qara, Akkas, Baysh	3.17
Jizan	2.38
Rabigh	2.06
Dhamad	1.90
Tayyah, Yiba	1.90
Khulays	1.74
Qanuanah	1.59
Iyar, Lith	1.43
Fatimah	1.43
Khaybar and Tubjah	1.11
Khulab	0.95
Itwad	0.95

Source: Ministry of Agricultural and Water (1984)

Table 5.4 Mean annual runoff – eastward draining wadi systems

	Mean annual runoff m³/sec
Wadi Dawasir Catchment	
Wadi Turabah	2.22
Wadi Ranyah	1.90
Wadi Bishah	4.76
Wadi Tathlith	0.95
Wadi Najran Catchment	
Wadi Idimah	0.16
Wadi Habaunah	0.95
Wadi Najran	3.17

Source: Ministry of Agriculture and Water (1984).

where the air is moist. These storms generate high intensity rainfall and as there is little vegetation on the slopes water levels in the *widyan* rise rapidly, usually a matter of minutes, after the start of the flow. Peak flows of large floods can be very destructive to both life and property. Wadi Bishah, Wadi Baysh and Wadi Rabigh have recorded peak flows in excess of 2,100 m³/sec during major floods.

Flood water in *widyan* is eventually lost from the channel either by evaporation or transmission into the sediments of the *wadi* floor. Transmission losses even over a short reach can be very great and although the water becomes part of the under-flow and will help recharge shallow aquifers. Knowledge of these losses is important, especially in the location of dams. Several studies of transmission loss have been conducted in the Asir. In one study cited by Abdulrazzak (1995) it was found that bed infiltration was as high as 95 percent of the total runoff in the Wadi Najran basin over a 50 km reach. Walters (1989) estimated transmission losses of 5,612, 27,174 and 186,007 m^3 km^{-1} for Wadi Tabalah, Habawnah and Yiba respectively.

One way to assess transmission losses in *widyan* is the mass balance approach and this was employed by Abdulrazzak (*op. cit.*) in his study of the Tabalah basin on the eastern side of the Asir escarpment. The upper reaches are largely on Tertiary basalt and the lower on Precambrian basement rocks of the Shield. This basin is important from a hydrological point of view in that is has been instrumented and monitored since 1983 through an extensive hydrological network comprising 14 rainfall recording stations, 12 observation wells, 3 runoff stations and 1 climate station. The main aquifer is located in the *wadi* alluvium which ranges from 50–300 m in width and from 0–30 m in depth. For the purposes of his study Abdulrazzak chose a channel reach some 23.8 km long in the centre of the basin.

In the mass balance approach the magnitude of transmission loss (TL) in a reach is estimated as follows:

$$TL = V_{UP} + V_{TR} - V_{DS} - E_P$$

where

V_{UP} = upstream flood volume
V_{TR} = downstream flood volume
V_{DS} = tributary runoff contributions to main channel
E_P = evaporation losses.

Some data from Abdulrazzak's investigations are shown in Table 5.5. Abdulrazzac also calculated an index (ANTEC) to represent antecedent moisture conditions in the *wadi* bed sands and gravels. ANTEC = $1 - 0.9^t$ where t represents the time in days since the last runoff event (Linsley *et al.*, 1975). Abdulrazzac makes the point that mass balance studies must take into account the input from tributaries, the output loss via evaporation and the antecedent moisture conditions in the *wadi* bed. The variations in magnitude of estimated transmission losses was associated with temporal and spatial runoff variability caused by random storm cell coverage over the basin. As the data in Table 5.5 illustrate, large floods are expected to induce higher transmission losses from the *wadi* channel and a greater depth of recharge. Runoff from small storms, such

Table 5.5 Mass balance characteristics – Tabalah Basin

Flood event date	V_{UP} 10^6 m^3	V_{TR} 10^6 m^3	V_{DS} 10^6 m^3	E_P 10^6 m^3	TL 10^6 m^3	ANTEC	Recharge depth metres
23/04/85	0.146	0.028	0.000	0.0037	0.166	0.72	0.08
28/04/85	0.120	0.052	0.068	0.0048	0.099	1.00	0.09
07/04/86	0.867	0.092	0.279	0.0079	0.672	0.19	0.16
09/04/87	2.950	0.213	1.970	0.0055	1.186	0.10	0.77

Source: Abdulrazzak (1995).

as that on the 23rd April 1985, is absorbed locally and may not be recorded at all by gauges placed a long way apart.

Within the Tabalah basin Abdulzarrak developed the following significant multiple regression model to predict transmission losses from the reach he investigated.

$$TL = 0.18 + 0.48V_{UP} + 0.019ANTEC$$

This model is not, of course, applicable in general because of different *wadi* characteristics and reach length. The mass balance approach is simple and effective but requires gauged *widyan*.

For ungauged *widyan* the geomorphological rainfall-runoff model (geomorphological instantaneous unit hydrograph – GIUH) is useful in predicting hydrograph characteristics and as been applied to *widyan* in the Asir by Sorman (1995) and Sorman and Abdulrazzak (1993). This GIUH approach has been most fully developed by Rodriques-Iturbe and Valdes (1979). Fundamentally, the approach seeks to derive the hydrograph through its connection with the general laws describing the architecture of the drainage network and the geomorphology. So, for example, the application of the GIUH approach requires geomorphological, soil, hydraulic, climate and hydrologic data (see Table 8 – Sorman and Abdulrazzak, 1993, for data applied to Wadi Midhnab). Not surprisingly in arid terrain the hydrograph features as predicted by the GIUH approach will be in error by 20 percent or more. This is primarily due to the inherent noise in the input data, the patchy spatial distribution of the rainfall input and the difficulties in estimating variables such as stream order and soil moisture in large basins.

Sen (2004) has recently developed a synthetic unit hydrograph model tailored to Saudi Arabian conditions. In developing his model he showed by a detailed study of *widyan* in the Baysh area that peak discharge was related to catchment area by the following power function:

$$\text{Discharge (m}^3\text{ s}^{-1}) = 43 * \text{Area}^{0.522}$$

5.4 DAMS

There is a long history of dam construction in the Kingdom, particularly in the Hijaz and Asir (Figure 5.1). The period of modern dam construction goes back to the 1950s and was triggered by urban growth and agricultural development. Before 1975, there were 16 dams and the number had increased to 180 in 1984. By 1993, there were 184 dams in total with a total storage capacity of 482,250,000 m^3. According to the Ministry of Agricultural and Water there are 58 dams in the central region, 23 in the Makkah region, 14 in Madinah region, 51 in the Asir, 14 in Ha'il region and 24 in the Al Bahah region. By 1998 more than 200 dams were in operation and there are many more planned.

The main reason for building dams is to trap runoff and to feed the underground water system although some dams also supply potable and direct irrigation water for agriculture. In one or two cases dams also protect settlements from the hazards of flash floods. In practice, however, dams are often ineffective. This is because they are subject to large amounts of evaporation and sedimentation. In addition farmers have noted

Figure 5.1 Ancient dam known as Sadd al Khanq south-east of Madinah (Photo: author).

that the water levels in their relatively shallow wells have gone down as the farmland no longer receives any flood water.

The King Fahd dam located in Wadi Bishah is largest dam in the Kingdom and the second largest concrete dam in the Middle East. It was opened in 1998 and has a storage capacity of 325 million m³. The dam is some 113 m high and 507 m long and lies across one of the largest *widyan* in the Kingdom. The dam is not only important for the supply of water to the Bishah region, which is considered one of the most important agricultural area in the Kingdom, but also controls dangerous flash floods.

Other major dams include: the Mudhiq dam near Najran with a storage capacity of 86 million m³; the Wadi Jizan dam with a storage capacity of 75 million m³; the Wadi Fatima dam on the outskirts of Makkah with a storage capacity of 20 million m³ and the Jizan dam with a capacity of 15 million m³. Abha has its own dam within the actual city limits to provide drinking water. The dam is also a small tourist attraction and has a cable car ride.

5.5   LAKES

There are few natural lakes in Saudi Arabia, the best known being the system of 17 small karst lakes found to the south of Layla (N22°15', E46°45') (Table 5.6). These lakes (also known as the Al-Aflaj lakes) are the product of the solution of the anhydrite and gypsum in the Hith Formation which has led to the collapse of the overlying rocks.

Table 5.6 Layla lakes before being drained

	Name	Maximum depth m	Specific conductance μmhos cm⁻¹ at 25 °C	Surface area m²
1	Not named	12.1	3,830	9,300
2	Arrwais	45.3	3,530	27,000
3	Not named	10.4	2,510	1,510
4	Not named	3.0	4,100	2,500
5	Arras	42.0	3,290	280,000
6	Not named	32.8	2,570	800
7	Not named	50.0	2,630	4,200
8	Not named	29.0	2,650	600
9	Um Daraj	23.0	2,810	3,900
10	Not named	3.0	8,600	700
11	Ashuqeeb	31.2	3,670	6,400
12	Alfodol	41.5	3,180	2,100
13	Um Burj	34.3	2,770	7,800
14	Um Heib	22.0	2,620	28,000
15	Baten	20.5	2,670	9,500
16	Not named	8.9	3,730	500
17	Molaiha	8.1	8,450	3,400

Source: Ministry of Agricultural and Water (1984).

It is thought that during the last pluvial episode from about 9000–6000 BP one large lake some 35 km long and about 5 km wide existed. During the present arid phase the lake body has gradually shrunk. The lake was clearly an important site for early agriculture and a series of ancient *qanats* drained water from the lake to irrigate farms. From the position of these old underground aqueducts the lake was about 2.7 km² at that time. By the early 1990s the total area of the lakes was just 385,200 m² and recent direct pumping and lowering of the water table for irrigation has more or less dried the lakes up completely. Perhaps at some time in the future when water is no longer wasted for expensive irrigation systems these interesting habitats might be restored, although the whole areas is heavily overgrazed by domestic livestock. A tourist resort has been constructed but never opened owing to the demise of the lakes.

In 1990 the National Commission for Wildlife Conservation and Development supported a small expedition into the Uruq al Mu'Taridah in the eastern Rub' al Khali near the borders of Oman and the United Arab Emirates. This part of the Rub' al Khali has some of the most spectacular barchan and star dunes found anywhere on the globe and it must have come as rather a large shock when the expedition members found an apparently spring-fed wetland not previously known. Details of the site at N20°41', E45°42' are sketchy but the wetland covers some 40 hectares and comprises large pools and reed beds surrounded by dunes. The pools are thought to be permanent. A word of caution is necessary as it is also know that Saudi Aramco have explored for oil in the same area and the lakes might well be the result of dynamite explosions! The 1:2,000,000 maps of the Arabian Peninsula published by the Ministry of Mineral Resources (1984) marks the Khawr Hamidam in the same general area (Arabic: *khawr* – a salty place or brackish well) and Wilfred Thesiger, in his first journey across the Empty Quarter, passed through the area on his way north-north-east from Muqshin.

5.6   SPRINGS

In a land where there are no permanent rivers running to the sea, and where rainfall is so meagre it is not surprising that permanent springs, and their life-giving waters, have long been regarded as special places. None have captured this essence better than the English explorer Charles M. Doughty in describing his long and perilous journey through the basalt wastes of Harrat Khaybar north of Medina:

> The few weeks of winter had passed by, and the teeming Spring heat was come, in which all things renew themselves. I had nearly outworn the spite of fortune at Khaybar, and might now spend the sunny hours without fear, sitting by the spring of 'Ayn er-Reyih, a pleasant place little beyond the palms, and the only place were the eye has any comfort in all the blackness of the Khaybar.
>
> Arabia Deserta, 1888

In modern Saudi Arabia, springs are still special places to be visited, to picnic by and rest awhile, but with the massive development of groundwater resources and particularly the multiplicity of modern wells tapping shallow aquifers, they are of little significance in terms of water supply.

The most famous spring in the Kingdom is that which feeds the Zamzam well in Makkah. It is thought that the spring has been running for around 4000 years. According to Islamic belief the well marks the site of a spring that, miraculously, issued forth from a barren and desolate *wadi* where the Prophet Abraham had left his wife Hajar and their infant son Ismail. In her desperate search for water, Hajar ran seven times back and forth in the scorching heat between the two hills of Safa and Marwa to provide water for Ismail who was dying of thirst. The Angel Gabriel, scraped the ground, causing the spring to appear. Fearing that it might run out of water, Hajar enclosed it in sand and stones. The area around the spring, which was later converted to a well, became a resting place for caravans, and eventually grew into the trading city of Makkah, birthplace of the Prophet Muhammad. The Zamzam well, is located within the precinct of the Holy Mosque in Makkah and its water is sacred to Muslims because of its miraculous origin. The rituals of *hajj* are in memory of Hajar and the word *hijrah*, meaning migration, has its root in her name.

From a hydrogeological point of view springs in Saudi Arabia are, however, of some interest in terms of their location and the information they provide on water quality. A thorough study of the springs of Saudi Arabia has been conducted by Bazuhair and Hussein (1990) who provide a map of the major spring locations. They recognise five type of spring: alluvial, sub-basaltic, fracture, solution opening, and interstratified. Alluvial springs are particularly associated with Quaternary *wadi* deposits on the Arabian Shield. Such springs have variable discharge according to fluctuations in the piezometric surface in the *wadi* fill. They are common in Wadi Fatima, Wadi Al Fara, Wadi Naaman and in the Taif and Al-Kamil regions. Sub-basaltic springs are associated with the lava sheets (*harraat*), particulary Harrat Rahat and Harrat Khayber. Large amounts of rainwater seep into the fractured, blocky surfaces of the lava flows several of which have flooded across the floors of pre-existing *widyan* buring the fill of sands and gravels. Fracture springs issue from joints and faults in bedrock.

The major fracture springs in the Kingdom are found in the Jizan and Al Lith regions of the Tihama. These springs are all thermal with temperatures up to 86 °C. Solution Opening springs are typically found in limestones and anhydrites and are associated with karstification. Springs of this type are associated with the Layla lakes and anhydrite karst of the Al Aflaj district, some 200 km south of Riyadh. Interstratified springs occur in the Al Hassa and Qasim oases. They are associated with thin, discontinuous, interbedded, aquicludes. Many springs in the central and eastern part of the country have now ceased to flow as a result of excessive groundwater pumping.

5.7   GROUNDWATER

Almost all the groundwater resources in Saudi Arabia are found in eight large tectonic basins of sedimentary rocks bordering the Shield. Very much smaller amounts of shallow surface underflow is found in the *widyan* of the Shield, particularly those originating in the high escarpment mountains. Total groundwater reserves are very large, being estimated at 1919×10^9 m^3 by Al-Alawi and Abdulrazzak (1994) with a further 160×10^9 m^3 stored in deeper secondary reserves. Even so, at current rates of consumption groundwater sources will be depleted in about 120 years – a worrying prospect. Bazuhair (1989) has constructed an interesting set groundwater depletion charts for selected aquifers relating number if wells, years to deplete the aquifer and pumping volume. The charts are useful planning tools.

Hydrogeological investigations in the Kingdom show that groundwater is stored in more than 20 primary and secondary aquifers. Edgell (1990c) classifies the aquifers into two groups, (i) primary aquifers, including quartz sandstones, conglomerates and some limestones and Quaternary alluvium; (ii) secondary aquifers, primarily in limestones where later diagenetic changes and sometimes karstification, have increased the original porosity. Apart from shallow aquifers associated with the *harraat* and *wadi* fills of the Shield, all the major aquifers are associated with the sedimentary cover rocks (Table 5.7).

Edgell provides a thorough geological description all the aquifers of Saudi Arabia and this will not be repeated here. It is, however, useful to comment on some of the more important aquifers listed in Table 5.7.

5.7.1   Wasia-Biyadh aquifer

This vast aquifer of Lower and Middle Cretaceous sandstones is located in the Northern Interior Homocline and the Widyan Basin margin. It is named after Biyadh (N22°00', E47°00') an extensive gravel plain in the outcrop area, and Ayn Wasia near Khasham Wasia. The Wasia Formation component is the single most important lithostratigraphic unit for water resources in Saudi Arabia. The aquifer supplies water to Riyadh and at Khurais has good quality yields with 555 mg/L total dissolved solids (TDS). At depth in the Eastern Province the water quality is very poor. Runoff from Jabal Tuwaiq and some *wadi* underflow provides 480×10^6 m^3 recharge per year. Near Kharj, the water has been isotopically dated at 8,000 BP and 10,000 BP down dip at Khurais. In the deeper parts of the aquifer near the Gulf, the water is over 20,000 years old.

Table 5.7 Estimated deep groundwater reserves

Aquifer	Reserves $10^6 m^3$	Recharge $10^6 m^3$	Salt content ppm
Wasia-Biyadh	590,000	480	900–10,000
Saq	290,000	310	300–3,000
Wajid	220,000	104	500–1,200
Tabuk	205,000	455	250–2,500
Umm er Radhuma	190,000	406	2,500–15,000
Minjur-Dhruma	180,000	80	1,100–20,000
Neogene	130,000	360	3,700–4,000
Jilh	115,000	60	3,800–5,000
Jauf & Sakaka	100,000	95	400–500
Aruma	85,000	80	1,600–3,000
Dammam	45,000	200	2,600–6,000
Khuff	30,000	132	3,800–6,000
Total	**2,185,000**	**2,762**	

5.7.2   Saq Sandstone aquifer

The Saq Sandstone is of Early Ordovician age and forms the major aquifer in northern Saudi Arabia and especially in the Tabuk basin which is almost 300,000 km² in area. The aquifer is named after Jabal Saq (N26°16', E43°18'). Water from the aquifer is of good quality and is often initially artesian with recorded pressures of up to 12.3 atmospheres. In the Qasim area the aquifer has been overexploited and the groundwater table lowered by as much as 100 m. Isotopic dating indicates that the water in the aquifer is from 22,000 to 28,000 years old and there is very little present-day recharge.

5.7.3   Wajid Sandstone aquifer

The Lower Ordovician Wajid sandstone extends over and area of 196,000 km² on the south and southern eastern edge of the Arabian Shield. The aquifer is named after Jabal Wajid (N19°16', E44°27'). It dips gently to the north-east under the Permo-Carboniferous and Mesozoic rocks of the Rub' al Khali where it forms a confined aquifer with artesian pressures up to 9.1 atmospheres. Water quality is not ideal for drinking but generally good enough for irrigation and livestock. Isotopic dating indicates that the water in the aquifer is more than 30,000 years old and is therefore 'fossil'.

5.7.4   Tabuk aquifers

The Lower, Middle and Upper Tabuk Ordovician quartz sandstone aquifers of the Tabuk Basin and Interior Homocline contains about 205 × 10⁶ m³ of groundwater. The Lower Tabuk aquifer is a significant, confined, aquifer in the Tabuk Basin. There has been considerable over-exploitation of this aquifer and a large cone of depression has developed.

5.7.5 Umm er Radhuma aquifers

The Paleocene-Lower Eocene Umm Er Radhuma aquifer system is one of the most extensive and important in the Eastern Region. It extends from the northern Haudraumat in the Yemen to the Saudi-Iraq border and underlies almost all of the Rub' al Khali. Where the aquifer crops out, such as on the As Summan plateau, it is extensively karstified. Water quality declines towards the Gulf due to the presence of anhydrite in adjacent formations. The important springs of the Al Hasa area are fed by water which originates in the Umm er Radhuma aquifer as are some *sibakh*. Indeed, Bakiewicz *et al.* (1982) indicate that *sabkhah* discharge prior to the advent of large scale abstraction by wells was as much as 855×10^6 m^3 per year as compared with 285×10^6 m^3 per year via springs. Isotopic dating indicates that the water in the Umm er Radhuma aquifer is fossil and accumulated 10,000 to 28,000 years BP.

5.7.6 Minjur-Dhruma aquifer

The Upper Triassic-Jurassic Minjur-Dhruma sandstones extend from Wadi Birk south to the Rub' al Khali. In the Riyadh region the Minjur Formation is 400m thick and at a depth of between 1,200 and 1,500 m. It is of considerable importance since it yields about 90 percent of the city's groundwater supply. Isotopic measurements indicate that the water is about 25,000 years old. Heavy pumping from the aquifer has considerably lowered the potentiometric surface. In 1956 this was estimated to be 45 m below the land surface and by 1980 it had dropped to 170 metres below.

5.7.7 Neogene aquifer

The Neogene aquifer is the collective name for aquifers developed in Mio-Pliocene formations in the Eastern Region. From oldest to youngest the Formations are: Hadrukh, Dam and Hufuf. Individually, the Formations range in thickness from 20 m to 120 m. The Neogene aquifer is an important source of water near the Gulf. It is heavily pumped in the oases of Al Hasa where it also feeds some of the springs. In the Yabrin area, south of Riyadh, evaporation from the Dam Formation is thought to be associated with the development of an extensive inland sabkah.

5.7.8 Alluvial aquifers

Unconfined Quaternary alluvial aquifers in *wadi* basins are important local sources of renewable, shallow groundwater, particularly where fed by runoff from the mountains of the Hijaz and Asir. Dabbagh and Abderrahman (1997) note that these aquifers store about $84,000 \times 10^6$ m^3 with an average annual recharge of $1,196 \times 10^6$ m^3. Along the Red Sea coast, the Ministry of Agriculture and Water have estimated that *widyan* such as Jizan, Dhamad. Fatimah, Lith, Khulays, Rabigh, Baysh and Hali have a storage of $14,250 \times 10^6$ m^3 and a potential annual yield of 105×10^6 m^3 of poor to good quality, shallow ground water. The summer capital of Taif is supplied with its water from groundwater in the nearby Wadi Wajj and Wadi Liyyah as well as from wells sunk into Wadi Turabah (El-Khatib, 1980).

In the Wadi ad Dawasir, the alluvial fill is up to 100 m thick and supplies relatively good quality ground water for centre pivot irrigation. Ministry of Agriculture and Water estimates indicate that about $17,000 \times 10^6$ m^3 of groundwater are available. In the fertile oasis of Najran, in the southern Asir, relatively high rainfalls from the mountains to the west have charged the Wadi Najran aquifer with up to $33,350 \times 10^6$ m^3 of good quality water. In the Riyadh region, Wadi Birk, Wadi Nisah, and the upper reaches of Wadi Sahba store about $14,100 \times 10^6$ m^3 which is extracted for local irrigation purposes.

5.7.9   Basalt aquifers

Basalt lavas provide some of the most productive aquifers in the world. Basalts are often highly porous and have high hydraulic conductivities. Potentially, therefore, the *harraat* of the Shield, which cover some 180,000 km^2, have great potential as aquifers. Al-Shaibani (2002, 2005) has made a particular study of Harrat Rahat, south of Madinah. This *harra* has a volume of about 1,999 km^3 and lies over a saprolite developed on the Precambrian basement and Ordovician Saq sandstones. Al-Shaibani notes that wells yield as much as 29,000 m^3 per day. Natural leakage from the Harrat Rahat aquifer also feeds a number of springs the runoff from which feed inland *sibakh* on its eastern margin.

5.8   AQUIFER WATER QUALITY

The water quality of the main aquifers varies from place to place and partly depends on the depth of extraction. Water suitable for domestic consumption is stored in the the Saq, Tabuk, Wajid and Neogene aquifers. Most other aquifers, such as the Minjur, Wasia, Dammam, Umm er Radhuma, produce water which usually requires treatment for temperature and high concentrations of ions such as Ca^{2+}, Mg^{2+}, SO_4^{2-} and Cl^{2-}. Some of this water is use directly in agriculture but this incurs the cost of soil salinization.

Table 5.8 Aquifer water quality – averages

Aquifer	Conductance $\mu mhos\ cm^{-1}$	Ca	Mg	Na	Cl	SO$_4$	NO$_2$	TDS
		mg/L						
Wasia	8,240	535	102	1,210	3,130	858	31	4,860
Biyadh	7,440	839	208	404	6,640	1,270	15	2,820
Saq	2,820	232	49	442	983	168	–	2,680
Wajid	2,210	128	28	149	269	294	–	1,200
Tabuk	1,680	104	40	196	305	613	–	1,360
Minjur	1,560	162	60	204	165	338	409	1,390
Dhruma	2,740	353	140	358	606	953	105	2,740
Umm er Radhuma	3,860	355	123	547	1,750	703	–	2,950
Neogene	3,450	256	157	495	1,630	865	38	2,690
Alluvial	2,600	210	100	437	757	649	15	2,320

Source: Ministry of Agriculture and Water (1984).

An important consequence of water extraction from aquifers which receive little or no recharge is a marked deterioration in water quality. Two factors contribute to this process. First, there is often a geochemical sequence in groundwaters with bicarbonate waters near the ground surface and saline waters in deeper, less mobile zones. Second, evaporation and ineffective leaching by rainwater increases the concentration of salts in water percolating back down to the aquifer (Todd, 1980).

Al-Faruq et al. (1996) investigated the chemical characteristics of the groundwater in the agricultural region of Qasim. They found that dissolved salts and pH levels of the drinking water had increased between 1978 and 1987, presumably as water was being withdrawn from deeper parts of the aquifer. And El Din et al. (1993) observed increasing levels of some nitrates and ammonia in their study of groundwater in the Riyadh region during the period 1984–1989; they attribute this increase to the use of fertilisers. Perhaps of more concern was the fact that fecal coliforms were present in twenty-one percent of their samples, indicating that human and animal wastes were finding their way into the groundwater. This is not surprising as a considerable volume of waste water is emptied in *widyan* and is not treated adequately. As Al-Rehaili (1997) notes, there is no enforceable quality criteria for effluent disposal and re-use in Saudi Arabia and the tentative wastewater re-use criteria published by the Ministry of Agriculture and Water were never approved and never enforced. Abderrahman and Rasheeduddin (1994), in their study of the Umm er Radhuma aquifer in the Greater Dhahran region note that the TDS have increased from 2,750 to 3,545 mg/L between 1967 and 1990 and they project that it will continue to rise to 4361 by 2010. These trends mirror the extraction from the aquifer. In 1976 the extraction rate in the Greater Dhahran area was 15 million m^3 per year and is forecast to rise to nearly 130 million m^3 per year by 2010. The serious deterioration of water quality is thought also to threaten the whole water supply network.

5.9   AQUIFER RECHARGE

Since most deep aquifers in Saudi Arabia contain fossil groundwater up to 40,000 years old there is little evidence of modern recharge. The aquifers were evidently last recharged during Quaternary pluvial periods but for the last 6,000 years the climate has been arid and recharge minimal as compared with extraction (Table 5.9). An interesting approach to the age of the fossil water has been published by Subyani and Sen (1991). They estimated the groundwater velocity through the Wasia aquifer to be 2.1 m/year and from this calculated that water entering the unconfined aquifer would take 38,000 years to reach a well site in the vicinity of Khurais. This compares with an isotopic age of 35,000 years.

Bazuhair and Wood (1996) have made a study of ground water recharge in the Hijaz and Asir using the chloride mass-balance method. The method only requires knowledge of annual precipitation, chloride concentration in the precipitation and the chloride concentration of the groundwater of the aquifer of interest. It assumes that no surface runoff leaves the aquifer area.

The method is particularly appropriate in remote arid areas. The method does however assume that all chloride in groundwater is derived from precipitation and hence is probably unsuitable for studies in aquifers away from the Shield.

Table 5.9 Ground water recharge rates for *widyan* in western Saudi Arabia

Wadi	Annual rainfall (mm)	Chloride (mg/L)	Recharge rate (mm year⁻¹)	Recharge as a % of precipitation
Abha	300	50	60	20
Al-Fara	75	195	3	4
Al-Yamaniyah	200	215	8	4
As-Safra	75	215	3	4
As-Shamiyah	250	285	8	3
Fayidah	75	435	2	3
Chulah	50	1,700	<1	2
Jizan	25	70	3	12
Khulais	75	885	<1	<1
Nahdah	75	710	1	1
Uoranah	225	220	9	4
Usfan	75	435	2	3
Wadj	250	220	10	4

Source: Bazuhair and Wood (1996).

The basic equation is:

$$q = (P)\,(Cl_{wap})/Cl_{gw}$$

where q is the recharge flux, P is the average annual precipitation, Cl_{wap} is the weighted-average chloride concentration in the precipitation and Cl_{gw} is the average chloride concentration in the ground water. P is area weighted over individual drainage basins using a method such as Thiessen polygons.

Bazuhair and Wood used an average chloride concentration in the precipitation of 9 mg/L. The results of these studies are shown in Table 5.9 where it can be seen that recharge is between 3 and 4 percent of precipitation. The high recharge in Abha is thought to be due to a combination of altitude, high rainfall and very loose, permeable material in the *wadi* floor.

A study of the recharge of the Umm Ar Radhuma Formation in north central Saudi Arabia has been conducted by Al-Saafin *et al.* (1990). The outcrop is intensely karstified and the presence of many *dahls* gives runoff water fast access to caves and deeper horizons. Runoff in this region is closely related to high intensity, short duration, thunderstorms. The authors conclude that surface runoff into the cave system only takes place if the precipitation rates exceeds 4.5 mm per 30 minutes. Such events occur, on average, at least 4 times per year with a mean rate per thunderstorm of 12.55 mm. As much as 47.7 percent of the precipitation ran into the cave systems during the three year period of the investigation. Although the authors conclude that considerable amounts of recharge can be expected in open karst areas it should not be forgotten that overall rainfall amounts are very low in the area.

Groundwater levels are monitored by a network of observations wells in all the major aquifers. Some shallow aquifers show a seasonality in water level due to recharge after winter rain but even so the general downward trend for all aquifers has been remarkably consistent and often spectacular. For example, an inspection well in the Saq aquifer the water level fell 9 metres between 1980–83. In the Umm Er Radhuma aquifer levels fell 7 metres between 1977–83. The decline has continued to this day.

An interesting study by Dincer et al. (1974) examined the impact of dunes on groundwater recharge in central Saudi Arabia. They showed that on dunes composed of medium sand, a significant portion (c.25 percent) of the 80 mm rainfall moves through the active dunes and recharges groundwater. In fine-grained dunes, however, no recharge takes place because the rate of infiltration is slower and evaporation is enhanced. Flow within the unsaturated zone is strongly influenced by slope-parallel heterogeneities, particularly when infiltrated water is redistributed by unsaturated flow after infiltration ceases.

There are, at present, no schemes to manage waste water for recharge purposes. As Ishaq and Khan (1997) succinctly note, treated wastewater is being wasted. The main sources of wastewater include: agricultural drainage water, run-off losses from irrigation systems and fields, sewage from cities, wastewater from water treatment plants and leakage from water supply systems.

The municipal wastewater disposal in urban areas is accomplished by a mixture of sewer and cesspit. In small towns and rural areas the disposal is limited to cesspits. By 1989, thirty wastewater treatment plants had been built in urban areas and some of the effluent used for irrigation or public parks and roadside verges. Where reuse is not practiced the treated, partially treated, or untreated wastewater is discharged into *widyan* or the sea (Abu-Rizaiza, 1999). Ishaq and Khan (1997) estimate that about 1,500 million m³ is discharged annually to the sea or is wasted and they indicated that using waste water for aquifer recharge has several advantages. They suggest that it is very economical and when used for agriculture may lead to a reduced need for fertilisers due to the presence of nitrogen and phosphorus.

The aquifers can be recharged using waste water either through injection wells or through spreading basins and contaminants are removed as the water percolates through the soil. Apparently, the reuse of such water drawn from a recharged aquifer is lawful from an Islamic point of view though it is likely that few people would wish to use such water for drinking purposes.

5.10   WATER EXTRACTION FOR IRRIGATION

Dabbagh and Abderrahman (1997) have thoroughly researched the effect of irrigation water use on groundwater conditions. They examined the distribution of agriculture and irrigation water consumption, and the extreme water stresses on shallow aquifers, particularly in heavily irrigated regions such as Jizan and Makkah. The twenty-seven percent of the total cultivatable area located on Shield utilises some thirty-five percent of the total irrigation water supply (Table 5.10).

There is also an excessive demand for groundwater in agricultural regions such as Al-Hassa, Al-Kharj, Qasim and Wadi ad Dawasir where there has been remarkable growth in the number of drilled wells from about 26,000 in 1982 to 52,500 in 1990. The whole problem of irrigation water supply is exacerbated by the fact that there is a spatial mismatch between demand and supply. For example, in the northern agricultural regions Dabbagh and Abderrahman indicate that there should be little deterioration in groundwater conditions apart from small areas around Tabuk. On the other hand there is excessive pumping

Table 5.10 Annual irrigation consumption for agricultural regions, 1992

Region	Aquifer	Water consumption million m³
Al Baha	Shallow alluvial aquifers	62
Asir	Shallow alluvial aquifers	239
Najran	Wajid, shallow alluvial aquifers	250
Madinah	Basaltic, shallow aquifers	769
Eastern Region	Umm er Radhuma, Dammam, Neogene	1,215
Northern Region	Saq, Tabuk	1,288
Makkah	Shallow alluvial aquifers, Basaltic	2,470
Jizan	Shallow alluvial aquifers	6,736
Qasim & Ha'il	Saq, Tabuk	6,970
Riyadh	Riyadh, Wasia, Minjur, Wajid	9,825

Source: Dabbagh and Abderrahman, 1997.

from all aquifer sources in Al-Kharj, Wadi ad Dawasir, Jizan, Makkah, Al-Hassa, Qasim and Ha'il resulting in unacceptable changes in groundwater conditions.

In their study of the Wadi ad Dawasir basin Al-Ahamadi *et al.* (1994) provide a classic description of the conflict between the need to conserve important groundwater resources and the need to provide intensive irrigation for Saudi Arabia's increasingly successful agricultural sector. Productivity in cereals and fodder was particularly spectacular and between 1986 and 1992, Saudi Arabia exported approximately 12 million tons of wheat. But to turn some of the most inhospitable terrain in the world into highly productive agricultural land is not without environmental costs. In the absence of controlled groundwater abstraction there is the danger of the rapid depletion of aquifer storage and the consequent lowering of water levels until pumping becomes either uneconomic or technically impossible. Not only does this threaten the viability of irrigation schemes themselves, but also creates acute difficulties for domestic water supplies in surrounding settlements burdened with the cost of deepening wells.

The upper Wadi ad Dawasir basin in the south-west of the country is of the largest cereal growing areas in Saudi Arabia (Figure 5.2).

Here, the main source of groundwater is the Wajid Aquifer. This thick, coarse grained quartz sandstone rests unconformably on the Precambrian Shield and dips gently eastward towards the south-western Rub' al Khali where it is overlain unconformably by Permian Khuff Limestone. Transmissivity varies between 5.7×10^{-4} and 2.1×10^{-2} m² s⁻¹ with storativity in the confined part of the aquifer in the range 2×10^{-4} to 2×10^{-1} (Authman, 1983). In 1965 a number of wells were bored in the aquifer. Estimated water levels ranged from 90 m above ground level in the confined zone to 40 m below ground level in the unconfined outcrop (Italconsult, 1969). Abstraction from the Wajid aquifer increased rapidly after 1984 and by 1990 more than 1,600 irrigation boreholes had been drilled. These boreholes, typically between 400 and 650 m deep, are pumped at rates between 6,500 and 10,000 m³ per day for 180 days per year. Al-Ahamadi *et al.* (1994) estimate that the total amount of water extracted in 1984 was 400 million m³ rising to 4,000 million m³ in 1988. The effect of all this extraction has been a lowering of the piezometric levels in the confined aquifer by as much as 150 m below the surface.

Figure 5.2 Centre pivot irrigation – Wadi ad Dawasir (Photo: author).

Much of the agriculture in the Wadi ad Dawasir area is based on large farm enterprises using centre pivots usually spaced 1 km apart. The sheer intensity of the irrigation is such that it can be seen clearly on Landsat imagery. Each well supplies two centre pivot systems, 500 m long, which are used to irrigate a circular field 1,000 m in diameter with crops such as wheat and alfalfa.

By the early 1990s wheat production consumed about thirty-three percent of the national irrigation water supply and was clearly was not desirable. In 1993 the Government cut the price support for wheat to one quarter of the previous wheat production of each farm in order to bring wheat production down to the annual national consumption level, encourage farmers to diversify and to reduce water consumption. Dabbagh and Abderrahman (*op. cit.*) conclude that if twenty-five percent of the wheat area is cropped and the rest is left without cultivation, the national irrigation water use will be reduced by some 7,422 million m^3 per year or twenty-five percent. A nineteen percent water saving can still be achieved by reducing the wheat area cropped to twenty-five percent and doubling the area under vegetables. As a result of the government measures there is some evidence that watertables are recovering where drawdown has been severe.

5.11   URBAN DEVELOPMENT AND GROUNDWATER

Agricultural development is not the only stress on ground water resources in Saudi Arabia. Over the last twenty years or so there has been a staggering amount of urban and industrial development placing huge demands on precious water supplies. A good example of such demand is discussed by Abderrahman and Rasheeduddin (1994) who have monitored groundwater depletion rates in the urban area of Greater Dhahran. The total population of Greater Dhahran, including the cities of Dhahran and Al-Khobar was around 400,000 at the time of publication and the whole urban areas is intensively developed due to the presence of the Saudi Arabian Oil Company headquarters, the King Fahd University of Petroleum and Minerals and an international airport. The main source of groundwater in the Dhahran region for industrial, domestic and landscape irrigation is the Umm er Radhuma (UER) aquifer. This aquifer is composed of fine grained and highly porous limestones and dolomitic limestones (effective porosity $\geq 30\%$) which are more than 300 m thick. The UER is highly productive and localised secondary openings allow transmissivity to reach average levels of 6.4×10^{-1} m^2 sec^{-1}. Estimated storativity is around 1.5×10^{-2} at the Dammam dome indicating unconfined to semi-confined aquifer behaviour and around 5×10^{-4} elsewhere, suggesting a strictly confined condition.

Extraction rates in the UER aquifer grew steadily from 15.38 to 29.31 million m^3 yr^{-1} during the period 1967–1979 but after 1979 rates increased dramatically as a result of large scale industrial growth and urban construction projects. By 1990 extraction had reached 65.24 million m^3. Abderrahman and Rasheeduddin (1994) estimated that by the year 2000 annual extraction rates reached about 91 million m^3. The increase in water extraction rate between 1967 and 1990 has resulted in a decline of the piezometric surface by about 4 m in the Dhahran region and numerical simulations indicate a further drop of 8 m by the year 2100 if present rates continue. By that time water levels will have fallen below mean sea level. In order to avert supply problems it is suggested that water extraction should be maintained at present levels if not reduced. This might be achieved by effective water conservation practice at the domestic level by reducing water pressure and by the introduction of water saving devices in kitchens and bathrooms. As far as landscape irrigation is concerned, the authors suggest the introduction of more drought tolerant plants and more efficient irrigation schedules. Away from built-up areas there is also the possibility of using treated sewage effluent for municipal landscape irrigation. Some further water saving might be made by more closely linking daily extraction rates and daily water demand.

Estimates indicate that a significant portion of water produced and piped is lost in major Saudi cities and Dharan is probably no exception. It is reported, for example, that fifty percent of the water supplied to Madinah is lost to leaks and the figure is even higher in Jiddah and Riyadh. Al-Dhowelia and Shammas (1991) suggest that the gross water leakage in the Riyadh water distribution network is, on occasion, as high as sixty-seven percent of the total network water supply.

Presently, Saudi Arabia consumes some 230 litres per capita daily, as compared with 150 litres in Europe. According to predictions from the Central Department of Statistics, the Kingdom's total population will exceed 29 million by 2010 and rise to 36.4 million by 2020. Assuming a baseline consumption of 300 litres per person per day, the resulting demand for water will increase to over 3,000 million m^3 per year by 2010.

5.12   WATER HARVESTING IN URBAN AREAS

Water harvesting in urban areas is potentially of importance in the battle against water wastage and the high demands of irrigation. Fast developing urban catchments create impervious substrates and storm runoff need not necessarily go to waste if it meets certain criteria. In this regard Ishaq and Alassar (1999) have made a study of urban storm runoff in a small catchment in Dhahran. They investigated the Assalamah catchment which is about 181.7 ha and is partly situated on the campus of King Fahd University of Petroleum and Minerals and the north-eastern section of the Arabian Oil Company (Saudi Aramco) compound. Of the catchment area a total of 52.1 ha was judged as contributing directly to the storm hydrograph generated from a drainage system composed of storm sewers in the shape of box culverts. Data were available for storm events during the period 1983–6. During this time 17 storm events produced 176.7 mm of rain and a total sewer runoff of 267.1 m^3 ha^{-1} and taking 81 mm as the average annual rainfall for Dhahran the authors estimated a total volume of 22,170 m^3 from the Assalamah watershed alone.

Of course, the potential usage of storm runoff depends very much on its quality and Ishaq and Alassar (*ibid.*) investigated a number characteristics including; suspended sediment (SS), total dissolved solids (TDS), total hardness, chemical oxygen demand (COD) and the coliform count. Water quality characteristics varied from storm event to storm event. The maximum levels found were: 6,169 mg/L SS, 2,652 mg/L TDS, 753 mg/L total hardness, 489 mg/L COD, 3.6×10^4 coliform count per 100 ml. The authors indicate that the runoff quality appears to be suitable for various re-use schemes excluding potable use. Concentrations of suspended solids, COD and coliforms were of major concern. It was suggest that if all the runoff were collected in a single reservoir, where it could be treated, it would be suitable, depending on the methods of treatment, for irrigation, car washing, aggregate washing and in some industrial processes. The potential volume of storm runoff for Dhahran as a whole is obviously considerable but to a certain extent unpredictable. Furthermore, treatment costs are not likely to negligible and the economic feasibility of such schemes needs to be considered before implementation.

Chapter 6

Geomorphology

Contents

6.1 INTRODUCTION

Although we now have a fairly good general picture regarding the geomorphology of Saudi Arabia there are still large gaps in our knowledge. The few geomorphologists who have worked in the Kingdom have, for the most part, been involved in recon-naissance surveys, and much of the early work was actually done by oil geologists for whom the geomorphology was of secondary interest (Fourniguet *et al.*, 1985). Thus, there is a good deal known about the oil-rich Eastern Region, particularly the karst, and much less about the Shield. Sand and sand dune research has obvious attrac-tions but in a way this had led to a neglect of plate tectonics, basic geomorphological processes, particularly rates of processes, the volcanic landforms associated with the *harraat* and the steep fluvial systems draining the Asir escarpment. Benchmark proc-ess studies are, however, extremely problematic. The terrain is harsh, off-road travel sometimes impossible, the mapping base, until recently, poor, and it is sometimes unwise to leave equipment unattended.

Another important gap concerns the distribution of landforms although this is not surprising given the sheer size of the Kingdom and the difficult terrain. Of course, satellite imagery can help but fieldwork is indispensable. Figure 6.1 is a slightly modi-fied version of a general landform map developed at the Saudi Geological Survey. At the time of the map's development the concepts of etchplanation, deep weathering and stripping were not on the agenda and it is evident in the field that the vast major-ity of pediments are in fact cut across older weathered surfaces so Figure 6.1 should be regarded as work in progress.

6.2 WEATHERING

Weathering in deserts, and the deserts of Saudi Arabia are no exception, is distinctive for a number of reasons: i) there are both daily and seasonal temperature/humidity changes; ii) contrary to what one might expect moisture is available either as dew, or the occasional rainstorm, and the relative humidity is often high at night as any one who has camped in the desert will know. It is also worth noting that many weathering products are inherited and not related to present-day climate.

6.2.1 Salt weathering

In the coastal zones, particularly on the Gulf, salt weathering is very potent and especially so on soft limestones. High saline water tables intersecting the topography i.e. within the so-called capillary fringe zone, are potent sites for the crystallisation

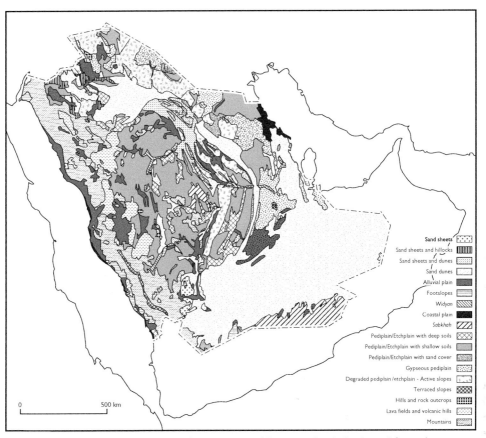

Figure 6.1 Generalised landform map of the Kingdom (Courtesy: Saudi Geological Survey).

and expansion of salts in rock pores and fissures. The three main pressure-producing processes are the growth of salt crystals themselves, changes in volume due to hydration and thermal expansion of the crystals (Figure 6.2). These processes lead to mealy weathering products, and the undermining of small cliffs and isolated outcrops. The main salts often present are:

calcium carbonate	$CaCO_3$
calcium sulphate (gypsum)	$CaSO_4 \cdot 2H_2O$
anhydrite	$CaSO_4$
sodium chloride	$NaCl$
sodium sulphate	Na_2SO_4
sodium carbonate	Na_2CO_3
magnesium sulphate	$MgSO_4$

The chlorides and sulphates are particularly aggressive, especially to man-made structures. Goudie and Viles (1997) provide a very thorough review of salt weathering.

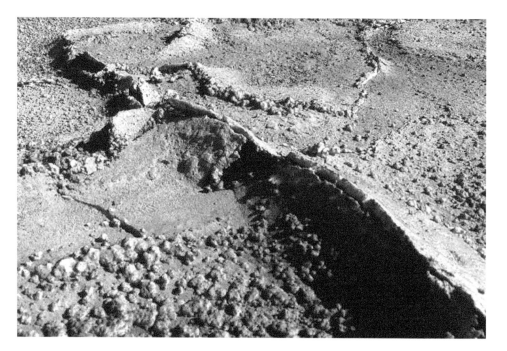

Figure 6.2 Lateral heave due to crystal growth. *Sabkhah* near Dhahran (Photo: author).

This type of weathering is not thought to be important on the Shield as evidenced by low conductivity measurements on weathered silts that choke many *widyan*. On the other hand, cavernous tafoni (singular: tafone) are widespread, and if salt weathering is factor in their formation then this type of weathering is more widely distributed than supposed (Figure 6.3).

6.2.2   Mechanical weathering

The comminution of rocks by mechanical weathering is a universal process in the Kingdom's deserts and granular disintegration, exfoliation, sheeting, splitting and flaking are widespread. Much has been written regarding the efficacy of insolation weathering in arid environments and whilst rock surfaces do undergo large ranges of temperature this process does not seem to be an effective agent. It is certainly not the case that desert sands are primarily the product of insolation weathering. On the Shield most rock surfaces are covered with desert varnish suggesting stability. Many granite plutons exhibit sheeting and this is probably due to thermal expansion and contraction, though there exists the possibility that it could be due to the reduction in loading as the mantle of weathered saprolites, and indeed other cover rocks, were stripped off in the Tertiary.

Surprising at it may seem, the most widespread weathering product, particularly on the Precambrian and Palaeozoic rocks, is that due to deep chemical weathering. Although this is now fossil, and often wonderfully preserved under the flood

Figure 6.3 Tafoni developed on granites near Taif (Photo: author).

basalts, some chemical weathering may have been active during former pluvial episodes (Figure 6.4). Good sections showing chemical weathering are visible along many of the highways in the mountainous parts of the Hijaz and Asir and it is not uncommon to find c.10 m of rotten rock and isolated core stones in a matrix of floury kaolinitic clays. Indeed, chemical weathering is probably responsible for the weathering of young *harra* surfaces though no studies have been undertaken.

6.3 DENUDATION HISTORY

Although undoubtedly complex it is obvious when travelling in the Kingdom that there are multiple denudation surfaces truncating complex rock structures giving rise to accordant summits, rolling upland plateaus, superimposed drainage systems and so on. In the past these surfaces have been called, rather naively, peneplains and this label has persisted in much of the literature.

Any understanding of denudation surfaces must, however, take into account plate tectonic movements and the fact that in the geological past the Arabian Peninsula has been several times both in equatorial and polar positions. Remarkably, evidence for these vast changes in geomorphic regime can still be detected in the modern landscape. Many intriguing details still need to be worked out but the ideas are potentially fascinating. For example, Peter Johnson at the Saudi Geological Survey has noted that there is some evidence that the erosion of the Shield began about 620 Ma, before its

Figure 6.4 Harrat as Sarat – lava overlying mid-Tertiary saprolites blanketing Precambrian basement (Photo: author).

final amalgamation around 550 Ma. He notes that 620 Ma old granite plutons are weathered down to their middle parts, whereas the 580–560 Ma granites are still exposed within their apical, topmost parts, intact together with rhyolites that probably represent volcanic eruptions from the granite magma chambers. This means, the younger granites are less eroded than the older granites. It might be possible to map this in the field but so far has not been attempted.

6.3.1   Pre-Wajid/Saq surface(s)

The pre-Tertiary evolution of the Shield's geomorphology is only known in outline and there is much to dispute. Practically all the evidence comes from the careful mapping of Palaeozoic outcrops and their underlying unconformities which have then been extrapolated on to 'surfaces' in the adjacent terrain. Most of the Palaeozoic rocks which may have once covered the Shield, and there is some debate about the extent of this cover, have been stripped away and are now only found in marginal positions. Exposed surfaces, therefore, are mostly exhumed but not a great deal is known about the dates of exhumation or how synchronous the events were. Obviously the Mid-Tertiary uplift of the Shield must have been important in this respect but earlier uplift episodes and Mesozoic stripping were possibly significant. For example, the Permian Khuff Formation rests on Precambrian rocks along the eastern edge of the central Shield implying erosion of Lower Palaeozoic cover rocks in that region. Also it is not always possible to differentiate between exhumed pre-Wajid/Saq surfaces and exhumed Tertiary surfaces – or indeed overprinted surfaces.

It is probably safest to think about these old surfaces as polygenetic denudation surfaces, rather than fluvial peneplains, arid pediplanes or tropical etchplains. Just to clarify matters it is generally thought that during the Protoerozoic 'Arabia' drifted southward from a position near the North Pole. It crossed the Equator about 600 My and was well south in the southern Hemisphere by the time of the Late-Ordovician glaciation. Throughout the Mesozoic Arabia drifted northwards and by the Upper Cretaceous was astride the Equator.

The primary denudation of the Shield is thought to have occurred near the end of the Proterozoic and continued into the Cambrian. The base of the Cambro-Ordovician Saq sandstone which now crops out on the north-eastern edge of the Shield is marked by a weathered boulder conglomerate and thin saprolite (Powers *et al.*, 1966; Whitney, 1983). Extrapolation beyond the outcrop shows the stripped exhumed surface to be characterized by a gently rolling etchplain dotted here and there by inselbergs which are probably sited on granite plutons emplaced between 660 and 580 Ma during the conversion of the Arabian Shield to a craton. The thinking here is that since the deep weathering and inselberg formation are part and parcel of the same geomorphological episode, the pluton emplacement provides an upper date for their formation since the stripping to reveal the plutons must postdate their emplacement.

On the south-eastern margin of the Shield the Cambro-Ordovician Wajid sandstone laps on to the Shield and overlies several metres of saprolite which Whitney (*op. cit.*) and Brown (1970) suggests is part of the same weathering episode as that below the Saq sandstones (Figure 6.5).

Weijermars and Asif Khan (2000), working in the Asir, note the presence of Wajid outliers away from the main outcrop suggesting that the saprolite and the late Precambrian surface are probably extensive. They suggest that the entire Phanerozoic cover sequence, which once must have overlain the Wajid sandstone, had largely disappeared by mid-Tertiary times. This can be concluded unequivocally, because of the emplacement of extensive flood basalts directly on to weathered Precambrian basement, now exposed in the As Sarat mountains as early as 29 Ma ago (Overstreet *et al.*, 1977). The absence of rocks between the basalt and the basement might also be due to their never having been there in the first place and there are still many unresolved questions as to how much of the Shield was ever covered by younger rocks. However, if most of the Shield has remained uncovered since the Late Proterozoic it difficult to see how such a surface has remained fundamentally unaltered and undissected. Indeed, because of periods of uplift the Shield remained the source area for vast accumulations of marginal Palaeozoic and Mesozoic formations and hence must have been subject to continued erosion. Given such a complex history is seems highly likely that any surfaces initiated in the Precambrian have been greatly altered.

Of particular interest regarding the pre-Wajid/Saq surfaces is the notion that the Neoproterozoic-Cambrian-Early Ordovician atmospheric pCO_2 was raised by as much as twenty times present atmospheric levels due to widespread volcanism. Precipitation would have been very acid and deep chemical weathering greatly enhanced (Avigad *et al.*, 2005). Stern *et al.* (2005) suggest that the pre-Wajid/Saq surfaces are part of a vast surface stretching from North Africa to Arabia which formed over 100 million

Figure 6.5 Eroded edge of the Saq Sandstone outcrop on the northern Shield revealing the sub-Saq denudation surface. View looking north-east (Photo: author).

years following the final collision of East and West Gondwana. It is relevant here to note that the Wajid/Saq sandstones were laid down on a surface of low relief the final development of which probably occurred in the early Cambrian since the surface truncates dikes as young as 532 Ma in southern Israel (Beyth and Heimann, 1999).

In Late Ordovician times, Arabia was subject to intense glacial and periglacial conditions and it is entirely reasonable to think of the Shield as being covered by an ice cap (McClure, 1978). The evidence for the glaciation and periglaciation is spectacular when seen from the air (Figure 6.6 and Figure 6.7). Vaslet *et al.* (1994) mapped the Sarah Formation in the Tayma quadrangle, just off the northern edge of the Shield, and discovered tillites up to 50 metres thick, glaciated pavements, large scale glaciotectonic structures, fossil eskers and periglacial features such as pingos, all of which are clearly visible on aerial photographs. More recently, the present author has discovered periglacial patterned ground in the same area. The glacial features in particular are often preserved as inversed relief due to later cementation and are concentrated in a series of paleovalleys up to 60 km long and 4 km wide. The presence of eskers indicates warm-based ice and it is therefore very likely that the Shield underwent extensive glacial erosion.

At the time when the pingos and patterned ground were forming, permafrost must have been widespread and seasonal frost shattering effective on weathered rock surfaces.

Figure 6.6 Fossil periglacial patterned ground exhumed in the Sarah Formation south-west of Tayma (Photo: author).

In Permo-Carboniferous times the Shield was glaciated for a second time and striated pavements, tillites and erratics have been found. Saudi Aramco geologists, using borehole information, suggest that the glacial landscape stretched far to the east under what is now the Empty Quarter.

Erosion of the Shield continued throughout the Mesozoic but events are sketchy. During the Upper Trias, for example, deeply weathered acid igneous rocks were stripped and re-deposited to form the Minjur Sandstones now easily identified in the field by their black, weathered, surfaces. Fluvial erosion and stripping of the Shield continued throughout the Cretaceous and into early the early Tertiary. Some small patches of Cretaceous rocks have been preserved in graben structures in the Umm Ladj region on the northern Red Sea coast. These are the only fossil dinosaur sites in the Kingdom so far discovered.

Unconformities in adjacent cover rocks indicate three separate periods of uplift probably related to epeirogenic movements of the Ha'il arch. During each period of uplift there was widespread erosion and especially of the marginal cover rocks. Some indication of environmental conditions can be deduced from the development of bauxite at Az Zabirah (N27°56', E43°43') some 200 km from the north-eastern edge of the Shield. This deposit, which averages 8.5 m in thickness over an area of 250 km², developed on a mid-Cretaceous (Albian-Aptian) unconformity. According to Whitney (1983) the deep weathering at this site followed a massive erosional episode which

Figure 6.7 Fossil glacial esker in the Sarah Formation south of Tayma, implying a wet-based glacier (Photo: author).

removed most of the earlier Mesozoic formations in the area. Environmental conditions probably included high rainfall, dense vegetation and a tropical or subtropical climate. Further evidence of Mesozoic stripping is found south-east of Taif where Oligocene marine beds rest directly on the Precambrian.

6.3.2   Late-Cretaceous – Early Tertiary Najd 'pediplain'

The Najd or central Arabian pediplain (Chapman, 1978; Brown *et al.*, 1989; Salpeteur and Sabir, 1989) is a mature surface (or surfaces?) of low to moderate relief. It is well developed on the Precambrian Shield east of the Red Sea Escarpment and it extends imperceptibly onto the sedimentary cover rocks on either side of the Tuwaiq Escarpment. It corresponds to the Tertiary peneplain described by Brown *et al.* (1989) though none of these authors seemed to have paid attention to plate movements, climatic changes or the effects of Tertiary uplift. It is also incorrect to think of this vast surface as a simple pediplain formed by the coalescence of pediments because the surface is in fact highly weathered and it has already been noted that Arabia was in the humid tropics during the Late Cretaceous. A truer description of the surface is one as having been formed by the stripping of a weathering mantle during the uplift of the Shield and subsequently pedimented as arid geomorphic processes took hold in the Late Tertiary and/or Quaternary. The reader will note that I have queried the fact that there is just one surface. My impression, having flown over the whole Shield in the last five years or so is that there are many surfaces, some better developed than others. Most have been disturbed by neotectonics, and piecing together their lateral extent is a complex task still to be undertaken.

Johnson and Jastaniah (1993) describe the western region of the Nadj surface as being characterized by pediments surrounding granite inselbergs, undulating hills of volcanic rock and anastomosing, weakly integrated drainage systems which are barely incised in spite of some 900 m of uplift since the Oligocene. To the west, the surface is strongly dissected because of base-level lowering. Accordant summit levels, incised dendritic drainage systems, rounded hill forms, intensely desert-varnished lag deposits and weathered bedrock permit the surface to be traced as far west as the edge of the present-day Red Sea Escarpment.

In the Early Tertiary deep weathering was widespread on the Shield and by the Oligocene etchplains had developed covered by a thick saprolite cover. This phase of deep weathering was possibly overprinted on to the remains of the exhumed sub-Wajid/Saq surface. Evidence for the mid-Tertiary deep weathering is now preserved beneath Tertiary *harraat* but elsewhere has mostly been stripped away by later events. Spectacular sections of saprolite crop out over an area of about 1,000 km² in the As Sarat mountains to the north-west of Najran (Overstreet *et al.*, 1977). The overlying *harra* dates from about 29 Ma and is now deeply dissected and eroded (Figure 6.4).

The Oligocene phase of deep weathering was the last such episode to effect the Shield and it most likely eliminated the divide between the eastward and westward-flowing drainage which had developed as a result of the uplift of the Ha'il Arch in the pre-Permian (Greenwood, 1973). Later Tertiary and Quaternary uplift and tilt of the Shield induced an eastward-flowing drainage system across the Oligocene surface. The effect of the uplift on the denudation of the Shield has probably been slight since there has not been much surface lowering postdating the Oligocene surface preserved under the older *harraat*. However, there are interesting exceptions. For example, Jabal at Tinn, east of Bishah, is an isolated inselberg developed in Precambrian rocks, and topped by a dense sandstone a few metres thick and then by a metre or so of basalt (Figure 6.8). The total height of the inselberg is 200 metres above the surrounding desert plain indicating that at least 200 metres of surface lowering has taken place since the outpouring of the lava. The lava is undated and the nearest *harra* is some 60 km away. There is also a hiatus between the sandstone and the basalt and one can speculate that this must represent some Mesozoic stripping phase.

Barth (1976) mapped flight of pediments in sedimentary Nadj abutting against the Tuwaiq escarpment (Figure 6.9) though the exact age of the proposed pedimentation is not known it is clear from the way the pediments truncate the Triassic Jihl Sandstone that they are young and probably post-date the Late Cretaceous- Early Tertiary denudation phase. Pediments are usually associated with sheet floods - perhaps across weathered surfaces, and it was not until the Miocene that arid conditions and this type of geomorphic activity became widespread. What is problematic about Barth's work is why the pediments are developed as a flight. Changes in base-level, climate or even lithology might all bring about pediment incision. What is clear, however, is that Barth suggests that pedimentation has brought about a considerable retreat of the escarpment down dip in an easterly direction.

Recently, the author has visited the Tuwaiq Escarpment in the Wadi ad Dawasir region and questions Barth's interpretation. It is evident on the ground that what

Figure 6.8 Jabal at Tinn east of Bishah, an isolated inselberg topped with lava and thin Palaeozoic sandstone – evidence of extensive stripping (Photo: author).

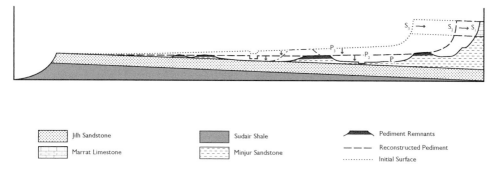

Jilh Sandstone	Sudair Shale	Pediment Remnants
Marrat Limestone	Minjur Sandstone	— — — Reconstructed Pediment
		.... Initial Surface

Figure 6.9 Flights of pediments on the Tuwaiq Escarpment (Modified from Barth, 1976).

Barth has identified as pedimented surfaces may in fact series of dissected gravel fans and outcrops associated with scarp retreat due to extensive spring sapping on the escarpment face (Figure 6.10). The Tuwaiq Mountain Limestone, which forms the main cliff of the escarpment, is underlain by the Minjur Sandstone. Water draining through the limestone is perched on the sandstone and emerges along the foot of the escarpment in lines of spring-sapped hollows. The rocks above the hollows

Figure 6.10 Spring-sapped cliffs – Tuwaiq Escarpment, Wadi ad Dawasir (Photo: author).

collapse and are rapidly comminuted. It is worth noting that the dip of the strata forming the escarpment is less than 1° for the most part.

6.3.3   Yanbu-Muwaylih surface

This surface is developed between the present-day Red Sea coastal plain and the escarpment foothills for a distance of more than 500 km from Yanbu northwards to Al Muwaylih. The essential nature of the surface was first described by Pellaton (1979) in the Yanbu area and by Davies and Grainger (1985) in the Al Muwaylih area. According to Johnson and Jastaniah (1993), the surface is characterized by conspicuous summit accordance defining a gently inclined slope seaward from about 300 m to 100 m asl. at its contact with the coastal plain. The junction is marked by a small cliff, dropping about 50–75 m to the coastal plain. The present-day drainage dissects the surface and isolated mountain massifs rise above it. The eastern limit of the surface is an abrupt change of slope at its contact with the mountains of the escarpment foothills (Figure 6.11).

The Yanbu-Muwaylih surface is thought to be of Late Miocene-Pliocene age as it bevels Miocene sedimentary rocks of the Raghama Formation. The surface is overlain by Tertiary-Quaternary basalts and Quaternary gravels. Johnson and Jastaniah (1993) note that the origin of the surface is currently unknown but it may represent a major wave-cut platform associated with high sea level during the Late Tertiary.

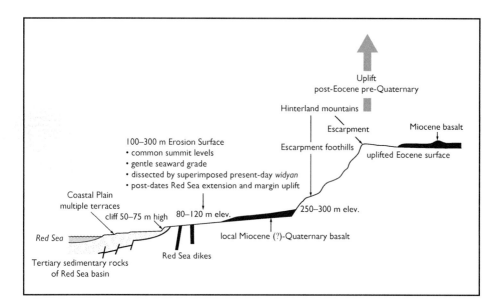

Figure 6.11 Main characteristics of the Yanbu-Muwaylih denudation surface (Modified from Johnson and Jastania (1993)).

This might be the case but it should be noted that no marine fossils are associated with the surface. At Al Wadj the surface is about 40 km wide and it is difficult to see how such as surface could have formed during a period of falling sea level since most marine surfaces are usually thought to have formed during transgressive phases. Furthermore, the low wave energies in the Red Sea were probably not that much different than to day. Probably the most serious criticism of the marine surface origin of the Yanbu-Muwaylih surface is the fact that it appears also to be present on the landward side of some of the mountains which project from its surface. There would have been little wave energy at these sites to create an abrasion platform. There is no reason why the surface cannot be interpreted as a pediment or a pedimented etchplain.

One of the most conspicuous denudation surfaces of Late Tertiary age is found in the Najran region. Here, the Wajid sandstone has been heavily dissected down to a very dense, resistant, ferruginous sandstone – an ironstone. The surface forms a vast plateau over several thousand square kilometres stretching southward into the Yemen. Here and there it is interrupted by sandstone inselbergs. Where broken through, inliers of the weathered sub-Wajid surface are exposed. Thus two denudation surfaces lie in close altitudinal juxtaposition – a rather unique situation (Figure 6.12).

6.3.4   Pediment surfaces in the Eastern Region

The Late Tertiary and Quaternary rocks of the Eastern Region dip almost imperceptibly towards the Arabian Gulf. Over thousands of square kilometres there is a marked unconformity between the topographical surface and the regional dip indicating that

Figure 6.12 Najran region – Late Tertiary denudation surface held up on dense Wajid ironstone overlying the bevelled top of the weathered Cambrian/sub-Wajid surface (Photo: author).

for the most part the surfaces are erosional and join the coastal plan almost without a break. In some places it is possible to recognize clear pedimented surfaces. Two such sites are described by Chapman (1971) who investigated the landforms of the Dammam Dome and the Shedgum Plateau.

6.4   TERTIARY DRAINAGE DEVELOPMENT

As has already been mentioned, for the most part the drainage pattern both on the Shield and the cover rocks is fairly simple. On the Shield, most of the *widyan* drain eastward and are well adjusted to the complex rock structures. On the cover rocks the drainage is essentially that of consequent *widyan* draining down dip, joined laterally by obsequent and subsequent tributaries.

In detail, there are, however, many puzzling features. For example, Wadi Bishah flows eastward towards Bishah and about 10 km west of the town it breaches a north-south topographic barrier via a wide rock-cut gorge. Is this drainage antecedent or superimposed? If superimposed, it can only be from a drainage system on former cover rocks and the system might be very old. If antecedent, the drainage system must at least pre-date the Shield's mid-Tertiary uplift.

Equally enigmatic are the breaches of cover rock escarpments in the cover rocks by easterly flowing *widyan*. These are almost certainly superimposed drainage. A good example is that of Wadi ad Dawasir in the southern section of the Tuwaiq

Escarpment. As the *wadi* gravels contain Precambrian clasts and no Precambrian gravels are found on the top of the escarpment the *wadi* must have been flowing eastward on cover rocks when the initial incision took place.

Brown (1970) makes the interesting observation that steep, active, *widyan* draining the escarpment westward to the coast have, in places, captured older *wadi* systems. A good example is that of Wadi Hamd which once flowed from the Madinah area north-westward towards the Gulf of Aqaba but south of Al Ula has been captured by headward erosion and now flows down to the coast near Umm Ladj.

Miller (1937) describes an interesting 'fossil' drainage system in the north-eastern Nadj that he first observed from a plane flight. According to his report the system occurs over more than 10,000 square kilometres some 350 km from the Gulf coast, south of the Bagdad-Makkah pilgrim road (Darb Zubiada) and east of the Dhana dune belt. This region is part of the vast Al Batin dendritic *wadi* system running north eastward towards Basra in Iraq and the As Summam plateau. The whole area is one of low relief and the low ridges are all capped by grey and tan calcretes. The fascinating observation made by Miller is that the stream courses when viewed on airphotos are actually the tops of ridges and are frequently more than 12 metres above their surroundings. Miller suggests that the original stream courses were wet zones and the intense evaporation precipitated calcretes. The interfluves were thus relatively less protected from erosion and were lowered differentially to form inverted relief. Miller uses the term 'suspended dendritic drainage' and 'suspendritic' drainage to describe the feature but the term has not caught on. Maizels (1988) describes similar features in Oman.

6.5   DEEP WEATHERING AND STRIPPING

Of central importance in understanding the recent geomorphological development of the Shield is the relationship between plate tectonic movements and former climatic weathering and erosional regimes. In particular, even though Arabia is now perceived as an arid environment seventy million years ago it lay astride the equator and the landscape must have been similar to that now found in central Africa (Figure 6.13).

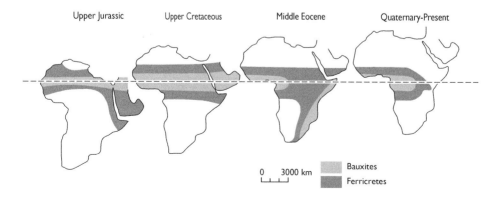

Figure 6.13 Possible deep weathering zones in relation to plate movements of Africa and Arabia (Modified from Tardy *et al.*, 1991).

Presumably the whole landscape was covered with dense vegetation and rivers flowed across a landscape covered with thick saprolites beneath which was a zone of intense chemical etching.

6.6 INSELBERGS

The stripping of the saprolitic cover during the mid-Tertiary uplift exposed a landscape comprising residual inselbergs (German: island mountains) and smaller kopje rising from surrounding etchplains. Whilst most inselbergs are composed of coherent rock covered with debris of weathered core stones, kopjes often look like a pile of core stones and probably represent the final stage in the destruction of inselbergs. Stripping of the saprolitic cover and deep weathering must have been comcomitant since there are many inselbergs more than 100 metres above the surrounding plains and the saprolite is unlikely to have been this thick.

Some inselbergs are domed (these are sometimes called bornhardts) and have sheeted slopes, while others are developed in jointed bedrock. Particularly nice granite inselbergs are to be seen on the edge of the Shield near Taif. Most inselbergs on the Shield are developed on resistant granites and probably represent the remnants of buoyant plutons injected into the Neoproterozoic host rock.

6.7 PEDIMENT DEVELOPMENT AND GRAVEL PLAINS

Pediments fringing inselbergs are common on the crystalline Nadj where they truncate weathered etchplain surfaces. Vincent and Sadah (1996) have described in detail pediment morphologies in the Al Aqiq area and have shown how their slopes decline as their lengths increase. The age of the pediments is difficult to determine but they were probably initiated towards the end of the Tertiary when a semi-arid regime replaced one of deep tropical weathering.

Pediments seem particularly well developed on granites. This may well have something to do with the effectiveness of sheet floods as erosive agents on granite slopes already weathered with a surface veneer of grus (Vincent and Sadah, 1995).

Gravel plains are common on the crystalline Nadj are probably result from the deflation of *widyan*. Deflation is also a possible process for the formation of extensive desert pavements where the gravel lag is underlain by poorly sorted gravels and sands. Other reasons why pavements form include wetting and drying and even fluvial sand and silt transport into the gravel lag from surrounding slopes. There are also extensive gravel plains south of Harad. These are, in fact, the deflated surfaces of the Hufuf Formation which was deposited by a palaeodrainage system draining from the Shield. The surfaces abound with ventifacts (Figure 6.14).

6.8 ALLUVIAL FANS

Alluvial fans are very common the Shield and practically every large *wadi* has drapes of alluvial fans along its side slopes. These fans have not been studied but are worthy

Figure 6.14 Ventifacts formed from the deflated Hufuf Formation south of Harad (Photo: author).

of closer inspection – especially those which are clearly segmented. Whether the segmentation is due to climatic change or perhaps neotectonics is the sort of question that might generate intriguing answers.

Several vast fans, presumably related to the uplift of the Shield, pass through the Tuwaiq Escarpment eastward onto the younger cover rocks. These fans can be traced easily since they have clasts of Precambrian igneous and metamorphic rocks. The Wadi ad Dawasir fan can be followed as far as the Gulf coastline, whilst the Wadi Al Batin fan has produced a vast spread of gravels, the Ad Dibdibah plain, which reaches north-eastward into Kuwait. These low angle fans of regional importance give rise to extensive desert pavement where the fines have been winnowed away.

6.9   SAND DESERTS AND DUNES

At this moment in time it is only possible to hypothesis the sequence of events which brought about the development of Saudi Arabia's sand deserts and the early history is particularly vague. Circumstantial evidence suggests that the Neogene was probably a dry tropical climate, but there is no direct evidence available. Certainly by Late Miocene times the Mediterranean had undergone severe desiccation and thick evaporites accumulated in the Red Sea basin. In addition, the beginning of the uplift of the Shield would have cut the centre of the country off from rain-bearing south-westerly winds. Whybrow and McClure (1981) examined botanical and fauna remains from the Miocene Dam Formation along the Arabian Gulf and concluded that the climate was dry and tropical but they found no evidence of dune formation. One possible pre-Pleistocene dune system has been hinted at by Thomas *et al.* (1998) who described sections in Early Pleistocene lacustrine deposits in the western Nafud which overly a partly consolidated dune system. The white, fine to medium-grained sands show giant trough cross-stratification. By the Mio-Pliocene, and indeed in the Early Quaternary, fluvial systems draining the Red Sea Escarpment flowed eastward to the Gulf. Large alluvial fans developed and vast quantities of sandy alluvium were deposited in the *wadi* floors.

The first low dunes and sand sheets may have formed in the Middle Pleistocene as early as 700,000 BP and certainly before the wet phase lasting from 560,000–325,000 BP (Edgell, 1990a; Rauert *et al.*, 1988). This conclusion is base on U/Th dated sinter deposits from the As Summan karst and it is probably safe to extrapolate to the areas of present-day dune accumulation. Clear chronologies do not really emerge until the Late Pleistocene when, both in the Rub' al Khali and in the Nafud lake deposits formed in two pluvial periods, the older dating from about 32,000–24,000 BP and the younger from about 9,000–6,000 BP. Episodes of dune formation and aeolian activity alternate with these periods of lacustrine deposition. Aeolian sand is found both above and below the older lake deposits but the extent of sand deposition prior to 32,000 BP is not clear. The main dune systems of Arabia seem to have been formed from 24,000 to 8,500 BP and in the short wet phase that followed there was some pedogenesis and stabilisation. The modern dune systems, which, as we shall see, often rest on the disturbed tops of older dunes, date from about 6,000 BP when the present hyperarid phase commenced.

6.9.1 Sand drift potentials and large scale mechanisms

In theory it is possible to estimate the amount of sand moved by the drag force of wind blowing over a loose surface once some threshold wind speed has been exceeded. The theory is a bit more complicated since surface roughness, grain size and moisture content of the sand are never actually known and assumptions have to be incorporated into the estimating procedure. Furthermore, only the potential amount of sand moving can be estimated since there is usually no information on sand availability. Fryberger (1980) calls this potential amount of sand movement Potential Wind Drift (PWD) and gives details of its calculation based on (i) standard tabulations of wind data (ii) theoretical vertical wind profiles due to Bagnold (1941) and (iii) a wind speed – sand transport relationship using the Lettau equation (Lettau and Lettau, 1969). The PWD is calculated for each segment of a wind rose and fed into a calculation of the resultant vector orientation (drift direction) and magnitude. One PWD vector unit is approximately 0.07 m^3 m-w^{-1} (metre width) A contour map of drift potentials in Saudi Arabia is shown in Figure 6.15.

The map reveals quite clearly the very high PWD in the north of the Kingdom and the relatively calm conditions in the Rub' al Khali. The PWD data in Table 6.1 provides a global perspective.

Fryberger and Ahlbrandt (1979) examined the large scale mechanisms for the formation of sand seas such as those in the Kingdom. They used Landsat imagery and available surface wind data to assess whether or not geologically significant

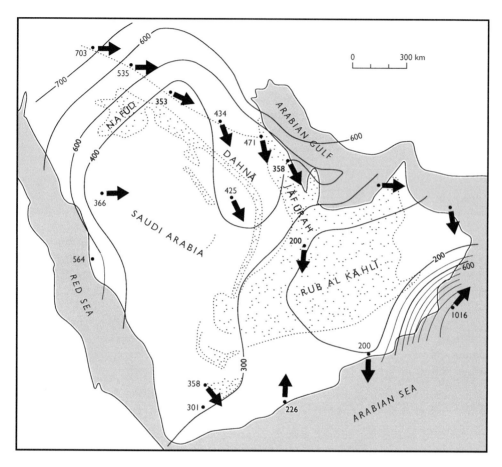

Figure 6.15 Contour map of drift potentials in Saudi Arabia (Modified from Fryberger, 1980).

quantities of sand are blown across modern deserts by wind action and concluded that present-day wind systems, and by inference past wind regimes, have sufficient energy to move geologically significant amounts of sand for long distances. According to these authors sand piles up to form sand seas in areas of low total or resultant wind energy in preference to areas of relatively high total or resultant wind energy. They go on to indicate that zones of low total or resultant wind energy can result from (a) the effects of positive or negative topography upon wind regime and (b) changes in the energy or directional properties of wind regimes along a direction of sand drift owing to changes of climates in the same direction. This is the case in Saudi Arabia where wind energies decrease steadily along the north-south arc defining the major sand seas. Thus while sandstorms are common in the northern deserts, they are almost totally absent in the Empty Quarter according to Thesiger. The presence of water can also play an important role by intercepting long range systems of sand drift but this is not important in Saudi Arabia even in *sabkhah* environments. However, the application of PWD data and

Table 6. 1 Average annual drift potentials for 12 desert regions

Desert region	Number of stations	Average annual PWD (in vector units)
High-energy wind environments		
Northern Deserts, Saudi Arabia, Kuwait	10	489
North-western Libya	7	431
Intermediate-energy wind environments		
Simpson Desert, Australia	1	391
Western Mauritania	10	384
Peski Karakumy and Peski Kysylkum, Uzbekistan	15	366
Erg Oriental and Erg Occidental, Algeria		
Namib Desert, Namibia	5	237
Rub' al Khali, Saudi Arabia	1	201
Low-energy wind environments		
Kalahari desert, South Africa	7	191
Gobi Desert, China	5	127
Thar Desert, India	7	82
Takla Makan Desert, China		81

Source: Fryberger and Ahlbrandt (1979).

the interpretation of Landsat imagery must be used with caution in any assessment of dune and sand movement in Saudi Arabia, and by implication in other major deserts. Whitney *et al.* (1983) make this point quite strongly and show that the large dunes in the Nafud are in fact quite stable, in spite of the windy conditions, and have been so for some time (see below).

6.9.2   Dune types

All dunes are bedforms in which granular sand interacts with surface winds capable of generating enough shear stress to overcome the sediment's shear strength – almost totally derived from the friction between grains. But whilst drift potentials are very useful more information is needed in order to understand the generation of individual dune forms. In particular, we need to know something about the seasonal direction of the wind and its variability. Thus a wind system blowing in a constant direction throughout the year will produce different types of dune from a system which has a bimodal or multimodal regime. But all this supposes that there is some equilibrium between the dune type and the present wind regime and this is not necessarily the case. Here, we have to use the concept of dune memory to appreciate the often complex nature of dune development through time. Consider the large linear dunes in the Rub' al Khali. They may have taken thousands of years to become aligned and may not be related precisely to present-day wind directions. In other words the large dunes may have a long memory. On the other hand, small blowout dunes on the coast may have formed as a result of just one particular storm and thus have short memories. The present-day orientation might tell us nothing about wind directions over, say, the last hundred years. With these thoughts in mind we can move on to formation of main types of dune found in the Kingdom.

Figure 6.16 Barchan dunes sweeping across the gravel plains west of Dhahran (Photo: author).

6.9.2.1   Barchan dunes

These crescentic dune are typically found sweeping up sand on gravel plains and pediments where they develop in unimodal wind regimes (Figure 6.16). They are sometimes thought to form from transverse, wavy, linguoid-barchanoid ridges of sand. Vortices develop in the lower linguoid nose which is then swept forward becoming detached from the barchanoid portions on either side. The vortices sweeping sand onto the margins of the barchan are then joined by a reverse vortex in the court area which lies immediately down wind of the dune's avalanche face. The side vortices elongate the arms, or horns, of the barchan because there less drag on the interdune surface and wind speeds higher. Barchan shapes are equilibrium bedforms and for the shape to be preserved all parts of the dune must migrate at the same rate. If too much sand saltates up to the crest steepening the form, flow divergence occurs lowering it. Barchan dunes are well developed in the Jafurah, the Nafud and Ad Dahna sand deserts.

6.9.2.2   Linear dunes

Linear dunes are found in both unimodal and bimodal wind regimes particularly on the edge of large anticyclonic cells. Winds need not be strong but must be consistent.

Spectacular linear dunes are found in the Rub' al Khali where they are often several hundred kilometres long (Figure 6.17). They can be as much as 200 m high, up to 1,500 m wide and the interdune corridors are up to 3,000 m wide.

Figure 6.17 Linear dunes in the western Empty Quarter south of Wadi ad Dawasir. The barchanoid slipfaces indicate sand drift diagonally across the linear dune axis. Ribs of small *zibar* dunes can be seen fringing the whole length of the linear dune (Photo: author).

Large linear dunes present many geomorphological problems most of which remain to be solved. For example, how are such large dunes initiated and the geometry repeated? And how do they grow?

One theory that goes back to the classic work of Bagnold (1941), suggests that as a barchan moves from gentle winds into zone of stronger winds from a slightly different direction one of its horns elongates and a small, linear, seif dune develops. Lancaster (1980) has found evidence in the Namib desert to support this model. However, barchans and linear dunes do not often co-exist (now!) and evidence for transitional forms are mostly lacking. Furthermore, where barchans do have elongated horns, they are not consistently on a particular side but they should be if their growth is the effect of one stronger wind direction?

In 1951 Bagnold produced a second theory which suggested that in a bimodal wind system, the sand drift would simply be aligned to the resultant wind direction. We can think of this geometrically and the resultant being viewed as the diagonal in a parallelogfram of forces. This theory has been confirmed in several desert areas but not in Saudi Arabia. In the Western Desert of Egypt the discrepancy between linear dune orientation and the resultant sand drift can be wrong by as much as 20°. Of course, this might be due to poor wind data or simply due to a large dune memory – the dune system having been oriented under another wind regime. Incidentally, although Bagnold wrote about Southern Arabia he never actually visited the Kingdom.

A development of Bagnold's theory was put forward by Tsoar (1978). In this flow diversion model, Tsoar accepts the resultant wind model but then goes on to hypothesise that the linearity was maintained by small vortices sweeping sand along the lee slope. The vortices develop as winds cross the dune's summit at some oblique angle and create an avalanche slope. Tsoar has experimented with smoke flares on dune crests and shows convincingly that these linear vortices do exist.

Secondary flow vortices on a much large scale are frequently hypothesised for the formation of linear dunes. These are rotary vortices which are said to develop in the boundary layer and maybe due to heating or perhaps a Coriolis effect. The base of these vortices touch down in the interdune corridors and in the Wahiba sands of Oman are thought to produce regular interdune spacing.

6.9.2.3   Star dunes

Star dunes are often extremely large and in parts of the Rub' al Khali are up to 300 m high. They typically have a pyramidal morphology with sinuous radiating arms rather like an amoeba. The arms are marked by multiple avalanche faces indicating very variable wind directions. Star dunes do not appear to move very much. One theory suggests that star dunes start from small linear dunes which undergo a seasonal reversal of wind, which together with secondary air flow over the dune develops multiple slip faces.

6.9.2.4   Dome dunes

Dome dunes are just that – domes of sand with no avalanche faces. Mostly they are just a few metres high but they can sometimes be extremely large and hundreds of metres high. Not much has been written about large dome dunes – but without slip faces sand transport cannot be large. It has been suggested that their initial development may be related to sites of local convection with surface winds drawing in sand which then piles up. The relationship between dome dunes and star dunes is puzzling. With climatic changes and windier regimes do dome dunes become star dunes? As with star dunes, dome dunes to not appear to migrate across the desert floor. Holm (1960) photographed dome dunes in the Empty Quarter during reconnaissance flights for Saudi Aramco.

6.9.2.5   Parabolic dunes

Parabolic dunes are small coastal blow-outs. They usually comprise a nose, pointing down wind, and two trailing arms. The dunes area rarely more than a few metres high and are usually stabilised by vegetation in areas of relatively high water tables. Anton and Vincent (1986) suggested that parabolic dunes become mobile when water tables fall and the vegetation dies off. As the nose migrates it sometimes detaches leaving behind two more or less parallel arms.

6.9.3 Ar Rub' al Khali

The Rub' al Khali is the world's largest contiguous sand desert (erg) with an area of about 640,000 km². The present-day erg is dominated by high crested longitudinal dunes, and seems to have originated about 6,000 years ago. It occupies a large embayment sag in the bedrock which developed in the mid-Oligocene when the Arabian Shield was uplifted (Edgell, 1990a). To the west the basin attains altitudes of about 1,200 m and descends progressively to the Gulf coast. Although some of the quartz sand of the Rub' al Khali is derived from the Dhana and Jafurah deserts, the greater part of it owes its origin to the development of the Shield's Tertiary drainage system when, in the pluvial and humid climate of the Mio-Pliocene, *widyan* broke through the Mesozoic escarpments of the Interior Homocline and deposited vast quantities of clastic sediments which now form the Miocene Hadrukh and Mio-Pliocene Hufuf Formations (Powers *et al.*, 1966). These Neogene formations are important because they later become one of the main sources of the huge volumes of quartz in the sand dunes of the Rub' al Khali. These texturally mature Tertiary sandstones have, in part, been eroded by extensive *wadi* systems in the more pluvial intervals of the Early Quaternary. Sand-filled *wadi* systems continue to be important sand sources. Edgell cites the Wadi Dawasir- Wadi Tathlith-Wadi Bishah systems as being particularly important in this regard. The intersection of these sand sources with the northerly *shamal* winds helps explain why so much sand has been delivered to the Rub' al Khali basin: once there it cannot easily escape over the mountains of the Shield to the west, and the Hadramaut to the south. It is also evident in the Najran region that the erosion of the Lower Palaeozoic Wajid sandstone outcrop must also have contributed to the Empty Quarter sands and there is some heavy mineral evidence to back this notion.

For an additional source of sand for the dune systems of the eastern Empty Quarter we can look to the Arabian Gulf. Prior to the formation of the relatively recent linear dune systems in the Empty Quarter it is thought that much of the region was covered by low dunes generated by *shamal* winds blowing across a dry Arabian Gulf. A eustatic fall in sea level of about 120 metres about 25,000 years ago meant that the coastline had retreated to the Gulf of Oman. The Gulf had become a riverine plain drained by rivers from Iran, Iraq and Arabia. Some evidence that sand was winnowed from these plains comes from the work of Al-Hinai *et al.* (1987) who observed submerged sand dunes in the Gulf of Bahrain formed during lower stands of sea level. In addition, the presence of wind-transported coastal ooliths and shallow water foraminiferids in the sands of southern Abu Dhabi and in the central Rub' al Khali is further evidence of the contribution of Gulf sediments to the sands of the Empty Quarter (Edgell, 1990a). Recently, however, it has become evident that a larger contribution of sand has come from deflation of the Hufuf Formation. These sands and gravels, with Precambrian clasts from the Shield, have isolated outcrops in Qatar and Bahrain indicating a considerable amount of erosion. Indeed, even the sands investigated by Al-Hinai (*op. cit.*) could have been derived from this source. With so many possible sources of sand, and low wind energies associated with the location of a semi-permanent high pressure centre it is perhaps not quite so surprising that this vast erg has accumulated.

Until the availability of satellite imagery our knowledge of the distribution of dune types in the Rub' al Khali was rather sketchy and based mainly on the travelogues of Thesiger. His descriptions were mapped and published by Bagnold (1951). Another early description which is widely cited in that of Holm (1960) who worked for Aramco and had access to oilfield information. In 1980, Breed *et al.* published a definitive map based on the interpretation of Landsat imagery and Skylab photographs. These images show that dunes in the Rub' al Khali are distributed by type into three principle areas:

i. North-east, the 'Uruq al Mu'taridah area (N23°, E54°), characterized by crescentic dunes (barchans)
ii. The eastern and southern margins of the erg characterized by star dunes
iii. The western half of the erg comprising mainly linear dunes

The giant crescentic dunes of the 'Uruq al Mu'taridah are up to 230 m high and developed on a widespread inland *sabkhah* foundation (Figure 6.18). They are called megabarchans by Edgell (1990a). Their slipfaces have very large horn to horn widths – in excess

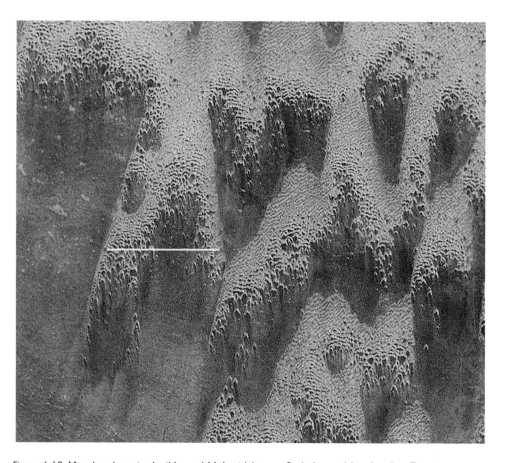

Figure 6.18 Megabarchans in the 'Uruq al Mu'taridah area. Scale bar = 6 km. Landsat 7.

of 6 km in many cases. The gentler stoss slopes are covered with smaller barchan dunes. Precisely what conditions separate these two sets of bedforms is an intriguing questions. These megabarchans are so large that they are thought to have originated in the Pleistocene by Glennie (1970). Although we do not know anything about the rates of movement of these dunes it is worth noting that the larger the dune the slower it moves.

The greater part of the Rub' al Khali erg is covered by various types of linear dunes or *'uruq* (singular, *urq* – also spelled *irq*) including: simple short; compound feathered; complex with star dune; complex with superimposed crescentic dunes. Some of the most spectacular linear dunes occur between in the western Rub' al Khali between As Sulayyil (southern end of Jabal Tuwaiq) and Najran. These dunes are spaced from 2 to 6 km apart and are up to 250 km long and commonly 150 km long. Edgell (1990a) refers to these gigantic dunes as draas but this term is very ambiguous. Smaller linear dunes, known as *seif* dunes, comprise a large part of the south western and central Rub' al Khali and sometimes form obliquely on the shoulders of much large linear dunes. Because the wind energies are not high in the Rub' al Khali, the interdune corridors (Arabic: *shuqqan*) between the parallel linear dunes are often not swept clear of sand and low ridges of coarse sand, having no slip faces accumulate – these are known as *zibar* dunes.

McClure (1978) notes that present high-crested linear dunes actually sit on to top of an older sand surface and the two sand bodies are separated by palaeosol marked by root and stem incrustations (Figure 6.19). The age of the older sands is not known for sure but is it supposed that dune formation was very active during the hyperarid period (17,000–9,000 BP) immediately before the onset of the Neolithic wet phase

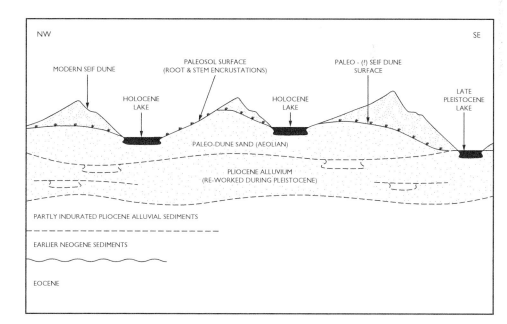

Figure 6.19 Schematic cross-section through the Quaternary sediments in the south-west Rub' al Khali (Modified from McClure, 1978).

(9,000–6,000 BP). Whitney *et al.* (1983) describe a similar story of an older and younger dune system in the Nafud.

Bagnold (1951) was particularly concerned about the relationship between the wind system and the orientation of the linear dunes. Were they aligned in the direction of the resultant (the vector sum) of the dominant sand drift or in the direction of the dominant wind? Even today few data exist to answer Bagnold's question. To confuse matters a little, on the map published by Breed *et al.* (1980), the major linear dune systems on the western part of the Rub' al Khali are not aligned with either of the suggested orientations. Bagnold's question does, however, presuppose that these vast dunes have a short memory and this cannot be the case. It would surely take several thousand years for such vast dunes to respond to changes in wind direction. Hence, wind data kept for a few decades might not be at all useful in interpreting sand dune systems as large as the linear dunes of the Rub' al Khali.

6.9.4   An Nafud

The Nafud erg overlies a nearly flat plain which dips gently to the north-east (Figure 6.20). The accumulation of sand is due to an interaction of the regional wind regimes and an abundant sediment supply from the Palaeozoic sandstones

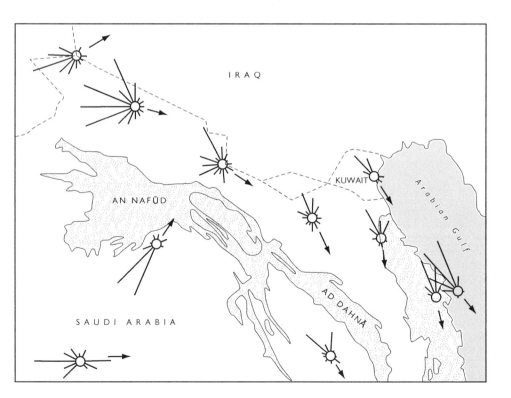

Figure 6.20 An Nafud and Ad Dhana sand seas in relation to sand drift potentials (Modified from Fryberger and Ahlbrandt, 1979).

exposed to the west. These calcareous sandstones are severely leached and they disintegrate easily.

So much sand is available, in fact, that Whitney et al. (1983) describe the erg as being 'choked'. In other words, the dunes are either closely space or actually overlap each other and it is rare to see the underlying surface.

According to Whitney et al. (ibid.), An Nafud is mainly composed of longitudinal, transverse and star dunes. The predominant dune type in the western and southern An Nafud is barchanoid and they are often linked together to form long transverse dunes. When viewed from the air Whitney and his co-workers describe these long ridges as looking like the cusped edge of a kitchen knife because the cresentic slip faces are separated by the pointed horns of the barchan dunes. In the northern, central and north-eastern Nafud Whitney et al. regard the dunes as being of an unusual type of linear dune. Instead of having a roughly triangular shape in cross section, as do the dunes in the Empty Quarter, the dunes are broad ridges having a chain of separated crescentic slip faces on the northern side of each ridge. It is thought that these rather unique dunes are formed by bimodal wind system from the south-west and north-west – the orientation of the slip faces implying that the south-westerly winds are dominant. This reasoning seems very similar to that developed by Bagnold for the formation of seif dunes from barchans (Bagnold, 1941).

In the eastern Nafud the dune trend turns south, and the longitudinal dunes become smaller and more widely spaced and more typically shaped. In the south eastern Nafud, star dunes are well-developed, some as much as 200 m high. These dunes have from 2 to 6 arms of active sand that project from a stable base and they rotate in response to local shifts in wind direction indicating that this part of the erg is subject to winds from several direction.

An Nafud, is located on the edge of the high pressure where a relatively strong, trade wind-type wind regime is dominant. Fryberger and Ahlbrandt (1979) comment on the apparent good correspondence between the trend of this sand sea and PWD directions, indicating that sand sea is extending down wind. This is, however, rather misleading because the vast majority of dune systems in the An Nafud and also in the Empty Quarter, are completely stablised and not active in the present arid climate. Theoretical PWDs do not, therefore, provide any insight into the location and formation of these ergs. There are active dunes, but these are small (less than 10 meters high) and these overly the large stable dunes which are commonly more than 100 meters high and according to Whitney et al. (1983) these smaller dunes cannot be detected using Landsat imagery which was state-of the-art when they made these comments. Smaller dunes can, however, been seen on aerial photographs having a scale of 1:60,000 or larger.

Lady Anne Blunt (1881) was the first to note the remarkable difference in the sand colour of the active and stabilized dunes in An Nafud. She described the colours as 'magnesia' and 'rhubarb' – today we would use Munsell Colour Charts. In the chart notation, the stable dunes have a Munsell colour of yellowish red (5YR 6/8 to 5YR 5/8) whereas the active dune sands are reddish yellow to yellow (7.5YR 8/6 to 10YR 8/6). The red hue of the stable dunes results from the oxidation of iron rich minerals in the sand and the addition of iron-rich clay minerals that infiltrate the sands by precipitation and dustfall. The active sands have partly lost this red hue due to abrasion during saltation.

The large dunes in An Nafud are partially stabilised by vegetation and there is sometimes a lag of pebbles from 5 to 35 cm thick which protects the shape of dune. The presence of this lag also implies a significant length of time during which there was the winnowing of fines. The lag deposits are also rich in calcium carbonate implying and least one phase of pedogenesis whereas in the active dunes no calcium carbonate is present and there is no surface lag of coarser sediments.

The clear differences between active and stable dunes in the Nafud suggests that they have been developed under very different wind regimes. The older stable dunes were evidently deposited under a more vigorous wind regime than exists today and was capable of moving vast amounts of sand. Although An Nafud is described as being relatively windy, most present-day winds cannot actually transport any sand which moves only during major storms. It is also noteworthy that the orientation of the stablised dunes does not correspond with the present-day prevailing winds which are from the north-east in the western side of the erg. The prevailing wind direction at the time of the formation of the stabilised dunes was from the west-south-west (Whitney *et al.*, 1983).

6.9.5   Ad Dahna

The Dahna is a narrow belt of complex linear dunes which emerge from eastern edge of An Nafud near the 43rd meridian and swing in a gentle 1,200 kilometer arc towards the northern Rub' al Khali. The topography comprises a cuesta landscape of cover rocks gently dipping eastwards towards the Gulf. Smaller nafud, such as the Nafud As Sirr, Nafud Qunayfidah and Nafud Ash Thuwayraz broadly mirror the arcuate trend of An Nafud and they too have accumulated in strike valleys.

Some sediments are supplied to Ad Dhana by An Nafud, but probably the major source of sand is alluvium from the Shield transported north-west in the Wadi Ar Rimah system. This *wadi* system becomes buried by sand in the Buraydah area and emerges from the sands of Ad Dahna as Wadi al Batin.

The parallel longitudinal dunes of Ad Dahna are quite widely spaced and lack crescentic slip faces because the westerly winds which dominate An Nafud to the north-west have much less influence on the dunes of Ad Dahna. The deflection of the longitudinal dunes to the south is in response to the *shamal* winds which can cause sand drift problems on the main highway between Riyadh and the Gulf coast.

In the southern Ad Dahna active barchan dunes have been observed moving across stable longitudinal dunes. In the Khurays area, Anton (1984) describes the development of a 0.4 metre thick palaeosol between reddish (Munsell dry 7.5YR 6/8), stabilised dunes and the sands of active dunes which lie above. Thus the dunes of Ad Dahna follow the general sequence observed in the Nafud and the Rub' al Khali with older, stabilised dunes overlain by younger active dunes.

6.9.6   Al Jafurah

The Jafurah desert is described by Fryberger, *et al.*, 1984 as extending along the Arabian Gulf from Kuwait in the north to the Rub' al Khali in the south, a distance of some 800 km though the sand desert proper does not reach this far north. It varies in width from 30 km at a choke point near Abqaiq to about 250 km near Al Ubaylah in the south. Most of the sand sea is relatively young compared to its westerly neighbour, Ad Dhana. This young age is reflected in the Jafurah's pale brown sands (Munsell, dry 10YR 7/3), though Anton and Vincent (1986) also note the existence of some large, elongate, stabilized, dome-like dunes which are probably much older. Anton (1983) suggests that most of the dunes are less than 4,000 years old. He bases his reasoning on two lines of evidence. First the present dune migration rates and their present distance from probable source areas and second, by the fact that at Ras Al Qurayah dunes are presently traversing raised beaches a few meters above sea level which, by analogy with ^{14}C dated sites mentioned by Al-Sayari and Zotl (1978, p. 56) are about 4,000 years old.

The summer months are intensely hot and characterized by the *shamal* which often blows with wind speeds of 40–50 km/h for several days on end. Gusts of up to 100 km/h generate sand and dust storms. Even during the *shamal* season, there is a sharp diurnal pattern to the wind regime, with quiet spells at night and the strongest winds in the day time. This diurnal pattern is probably due to the development of a low level inversion above the desert at night due to radiation under clear skies, and its subsequent breakdown due to convective eddies during the daytime.

Anton and Vincent (1986) indicate that some of the sand in the Jafurah sand sea has been winnowed from the major Quaternary fans developed by *widyan* transporting sands and gravels from the Shield towards the Gulf. Of particular importance in this regard is the vast fan, the Ad Dibdibah plain, which was developed by Wadi Al Batin – Wadi ar Rumah system in north-eastern Saudi Arabia (Powers *et al.*, 1966). There is also the possibility that some sand was derived from the Arabian Gulf during periods of low sea level during the Late Pleistocene.

According to Fryberger *et al.* (1984) the Jafurah desert comprises three broad zones: a northern zone of deflation where wind velocities are high; a central zone of dune transport where wind velocities less high; a southern zone stretching into the Rub' al Khali where wind energies are low and there are continuous dune fields. In the first two zones there are extensive sabkha plains and the dunes are more or less isolated. Throughout Al Jafurah, the dominant dune type is barchanoid and many magnificient barchans are to be seen traversing the sibakh. There are also many dome dunes and Fryberger *et al.* (1984) make the point that are many sites where dome dunes can be seen evolving into barchans and *vice versa*. Quite what changes in wind regime and sediment supply trigger these transformations is not known.

6.10 WIND EROSION

Little has been written about wind erosion and its possible role in the development of Saudi Arabian landforms. Indeed, in the relatively sparse geomorphological literature about on Saudi Arabia, wind is hardly mentioned. Brown *et al.* (1989) briefly mention wind erosion at the foot of the Nabatean tombs at Made'in Salih and the presence of some possible yardang troughs on the northern Shield but neither Al-Sayari and Zotl (1978) nor Jado and Zötl (1984), in their authoritative studies of the Quaternary Period in Saudi Arabia, make any reference to wind eroded features. Clearly, wind erosion has been effective in the dune areas where the sand grains are frosted due to

Figure 6.21 Yardang fleet north-east of Al Ula (Photo: author).

high impact collisions. It is unlikely, however, that erosion and fracture are important process in the production of dune sands themselves.

Today the Kingdom is not a particularly windy environment but there is abundant evidence of wind erosion features which probably relate to some former palaeoclimates. In particular, yardangs (Goudie, 1999) are very well developed north-east of Al Ula and form a fleet which progressively declines in size down wind in an easterly direction towards the western arm of An Nafud (Figure 6.21) Yardangs resemble inverted boat keels and much of the literature discusses their formation possibly as a result of deflation in soft sediments such as lake deposits. Those found north east of Al Ula are, however developed in hard Palaeozoic Saq sandstones and can only have been shaped by abrasion although they may have been initiated by fluvial incision (Vincent and Kattan, 2006).

There is progressive west to east denudation in the Al Ula region, from sandstone canyons to yardangs and finally to abraded rock pavements followed by dunes. One of the most remarkable features as one approaches the Nafud is the presence of hundreds of square kilometres of abraded rock surfaces (Figure 6.22). From the air, the sandstone jointing in the pavements is clearly defined by almost linear strips of vegetation, and more or less rectangular compartments of slightly raised, undulating grayish sandstone residuals – remnants of former yardangs. Two questions naturally

Figure 6.22 Wind abraded surfaces downwind of yardang zone north-east of Al Ula (Photo: author).

arise from these observations. First, is there a base level controlling the positioning of the pavement? Second, why are joints no longer exploited so as to maintain the yardang relief and some sort of equilibrium?

As far as the first question is concern two possibilities exist. It has already been mentioned the Saq sandstones contain conglomerate horizons which can be very extensive. These horizons can clearly been seen to armour the surface. It may well be that some pavements are perched on such protective horizons. Brookes (2001) notes similar surface armouring in his study of yardangs on the Libyan Plateau near Dakhla and Kharga, Egypt.

As to the second question, it appears today that the jointing in the Saq sandstone is not being exploited by wind erosion. Thus, there is a complete imbalance between yardang erosion in general and yardang corridor lowering in particular. This leads to the hypothesis that the rock pavements are extensions of yardang corridor floors whose base levels coincide with the base level of the original fluvial incision. No traces of this fluvial incision phase are now visible in the field. For example, old stream beds and ribbons of fluvial sediment are nowhere obvious. Clearly some surface lowering has been accomplished by the wind since the abraded horizontal surfaces marginal to the yardangs are whitish and polished, and yardang side walls less so.

The downwind lowering of the yardang fleet can be explained by two mechanisms which are here termed metachronous and synchronous development. Perhaps the most obvious explanation of the downwind decline in the size of the yardangs, the metachronous case, is to suppose that the stripping of the cover rocks from the Shield in the mid-Tertiary progressively exposed the Saq sandstone and hence large yardangs have developed on sandstone that has been more recently exposed than small yardangs. Whilst this is possible it is most unlikely because the cover rocks dip gently off the Shield to the east and stripping was from west to east whereas yardangs decline in size west to east, not east to west.

A more likely, and more interesting, case is that of synchronous yardang development. If yardangs decline in size downwind, and if the fleet was initiated more or less synchronously, then size must be related to abrasion rates. The rates themselves are not likely to vary with changes in downwind wind velocity because it is difficult to see why the wind climate should change so systematically. A more likely explanation is that the available sand load increases down wind away from the lava field. In short, yardang size in the study area is related directly to the changes in the abrasion load of the wind.

Although we have no information has to how long this denudation has taken observations by Doughty (1898) on the Nabatean tombs at Madain Salih about 40 km north of Al Ula is of some interest:

"We can see on the cliff inscriptions at Madain, that the thickness of your nail is not wasted from the face of the sandstone, under this climate in nearly 2000 years" (Doughty, 1989, p. 419).

Doughty's observations would suggest that erosion rates during the present arid climate are probably very low and that the yardangs correspondingly old. It is noteworthy that all speleothems in northern Saudi Arabia are older than 400,000 (Fleitmann *et al.*, 2004). This suggests that about this time arid conditions set in and conditions for yardang formation initiated. This, of course, is speculation and a programme of cosmogenic dating might produce some definitive answers.

6.11   KARST

There are only a handful of detailed investigations of the karst of Saudi Arabia but given that so much of the sedimentary cover comprises carbonates and sulphates it is not surprising that these surveys have found karst landforms to be surprisingly well developed. (Edgell, 1990b; Hötzl *et al.*, 1993; Peters *et al.*, 1990; Pint, 2000; Pint, 2003). For mining companies, karstified rocks are more than inconvenience and their records show that karstification to be very widespread, though often not noticeable on the surface. For example, the Al-Jalamid area, with its very important phosphate deposits, is a featureless, wind-swept, sand-covered plain about 40 km south of the Iraqi border, but extensive drilling and microgravity profiling in these dolomitic limestones has shown that this deceptively flat surface masks a karst terrain, honeycombed with cavities.

Much of the karst is fossil and related to former pluvial periods (Hötzl and Zötl, 1976) but some, associated with groundwater drawdown and collapse, is known to be recent. The most widespread karstified outcrops belong to the Umm er Radhuma Formation of Lower Eocene age and comprises thickly-bedded calcarenitic, non-porous, slightly dolomitic limestones with some marly interbedding. A good deal of the outcrop is covered by dunes of the Ad Dahna.

Peters *et al.* (1990) indicate that the principal karst landforms are: dry valleys and gorges, karst lakes, springs and doline fields. In addition there are many well-developed cave systems with presumably many more waiting to be discovered. An extensive tabulation of known caves in the As Summan plateau is given by Hötzl *et al.* (1993). Peters *et al.* (1990) suggest that the karst gorges exist where large wadyan have breached the Tuwaiq Escarpment. This, however, is probably not true. Karst, by definition, is associated with gravitational collapse, and not all gorges in limestone are the result of such as process. For example, many gorges in limestone are the result of antecedence, superimposition or tectonic movements. There is certainly no evidence of ceiling collapse or wall buttressing in the gorges breaking though the escarpment and these are best not thought of as karst features in a strict sense.

The As Summan plateau of north-eastern Saudi Arabia is of particular interest to geomorphologists. This plateau developed in the Umm er Radhuma limestone is not only known for its karst but also for its gravel ridges and inverted relief as mentioned previously. Edgell (1990) has made a particular study of the karst of the As Summan (*Arabic*: hard rock plain). He suggests that the area is mostly a holokarst (bare karst) but there are also patches of karst covered by clay-filled depressions and dunes. Edgell is attracted to a Mediterranean model of karst development and has identified large flat-floored poljes up to 2 km long. The floors of these sub-circular to irregular depressions are drained by sinkholes. Small residual hills or hums are also to be seen. He has also observed the present of uvalas with diameters of 100 m or more which have been formed by the coalescence of several dolines but Edgell also notes that the general rarity of uvalas which suggest that the karst cycle has not progressed passed an early stage. It is worth noting that the cyclic paradigm now has few adherents. Dolines (Arabic: *dahl*, pl. *duhul*) are said to be well-developed, especially around the village of Ash Shumlul (Ma'aqala).

Although the doline morphology is quite variable, Hötzl *et al.* (1993) and Edgell (1990) indicate that there at least two types, namely: solution dolines (German: *Lösungsdolinen*) and collapse dolines (German: *Einsturzdolinen*). Solution dolines, which have been formed by the dissolution of intersecting joins so as to form sub-vertical shafts, are the most common type on the plateau. Collapse dolines, such as Dhal Al Hashemi, have vertical, bottle shaped entrances and are floored with rock-fall and gravel. Edgell adds a third type, alluvial dolines, such as Rawdat Ma'aqala which he suggests are less easy to identify and probably exist under the alluvial-filled depressions (*rawdat*) which do not collect water even after rains. The terminology is, however, very confusing and this issue needs to be addressed within the context of the Saudi Arabian karst. For example, Bögli (1980) distinguishes between subsidence and collapse dolines, the former having been created by slow downward movements of a mass while the latter is formed by one rapid, usually single, occurrence, caused by a cavity which lies near the surface. This distinction is important from a hydrogeological point of view since collapse dolines indicate the presence of caves. Bögli notes that the alluvial doline is genetically an intermediate type between the pure solution doline and the subsidence doline and is often characterized by solution funnels in the alluvial fill.

Edgel has also described small-scale exokarst features such as widened surface joints (English: grykes or grikes; German: kluftkarren) and several types of solution groove (German: karren) such as rillenkarren and meanderkarren on outcrops of the Umm er Radhuma limestones but no systematic descriptions or measurements are available. The extremely detailed studies of the As Summan karst by a large team of scientists from the University of Karlsruhe in Germany, the Austrian Academy of Sciences in Vienna, and King Fahd University of Petroleum and Minerals (Dhahran) has been published in German (Hötzl et al., 1993). The central focus of the research was to assess the recharge or otherwise through the karst of the Umm er Radhuma aquifer but the team also carried out extensive geomorphological and speleological investigations.

Some of the most fascinating karst features are the lakes (Arabic: 'ayn – spring, lake or well) of central Saudi Arabia. For example, 'Ayn Al Burj, in the Al Aflaj oasis, is a large cylindrical lake basin about 100 m in diameter and 100 m deep. Peters et al. (1990) suggest that the lake basin was formed in the Plio-Pleistocene due to the collapse of anhydrite outcrops of the Arab Formation. 'Ayn Ad Dil, also in the Al Kharj oasis, is of similar dimensions and it, too, has probably been formed by the collapse of an anhydrite bed – the Hith Anydrite, since water samples from the water remaining in the lake are saturated with gypsum. Evidently, in historical times this lake was full, as witnessed by a series of steps leading to a water drawing platform cut several metres down the side of one wall of the 'ayn. Nowadays, the lake is almost dry, having been more or less emptied by pumps to irrigate the surrounding farms.

Many dolines and shafts in the Umm er Radhuma limestone of the As Summan, lead into extensive cave systems and a complete bibliography can be found at the Desert Caves Project site at http://www.saudicaves.com/index.html .

The caves systems on As Summan have been described by Pint and Peters (1985) and Benischke et al. (1987). Caves are up to 10 km long and are generally some 10 to 30 metres below the surface; cave passages are mostly vadose. Most caves show successively abandoned passages which appear to be due to the selective dissolution of chalk marl layers as groundwater levels fell in the Early and Middle Quaternary. This is well illustrated in the so-called UPM cave (N26°24', E47°15').

Most caves on As Summan have speleothems but they are not well developed. Good speleothem development is usually associated with high soil carbon dioxide levels associated with a thick vegetation cover, particularly forests, and even in pluvial periods this would not have been the case in north-eastern Saudi Arabia. Edgell (1990) describes the evolution of the karst of As Summan as having taken place in several stages. During the Paleocene and Early Eocene, the shallow water, marine Umm er Radhuma Formation accumulated and by the Late Eocene there was probably a widespread regression. Throughout the Late Eocene and Oligocene the Umm er Radhuma carbonates were exposed and underwent initial karstification. The upper surfaces of the Formation became weathered and a brown ferruginous cherty duricrust developed. In the Early Miocene, the Umm er Radhuma Formation was then buried by up to 250 metres of transgressive marls and sandy limestones of Late Tertiary age. By Early Pleistocene times much of this cover had been eroded away and during the warm pluvial which followed, from about 1.6 million to 700,000 years ago, the Umm er Radhuma karst developed the main landforms that can be identified to day, such as poljes and dolines. Where the protective cover of duricrust was also eroded caves and dolines are particularly frequent, as in the vicinity of Ash Shumlul (Ma'aqala) and Shawyah. In the Middle Pleistocene, arid conditions began to develop and since then there have been a number of short pluvials but probably little further karstification has taken place.

Rauert *et al.* (1988) have undertaken [14]Carbon and Uranium/Thorium (U/Th) dating on speleothems from the Summan caves and most are older than 270,000 to 350,000 BP indicating that the major karstification of north-eastern Arabia is old. Edgell (1990) argues for an Early and Middle Pleistocene age on the grounds that if the karst was Early to Mid-Tertiary the caves and poljes would contain Neogene sediments. On the other hand, Hotzl *et al.* (1993) suggest that the karst is essentially multi-staged, dating from more humid phases in the Mid- and Late Tertiary. Both models probably have some validity.

Of the younger ^{14}C dates recorded by Rauert *et al.* (1988), a number are interesting. First, bat guano at 7,090 ± 75 BP is significant in that bats no longer inhabit the dry caves of the As Summan plateau as the area is too arid. This date falls within the 'Neolithic Wet Phase' which lasted from about 9,000 to 6,000 BP. Second, several stalactites have dates ranging from 29,000 ± 2,200 to 36,600 ± 1,500 BP. This was a major wet phase in Saudi Arabia and sufficient water and vegetation must have been present to form speleothems. There is also one definite U/Th date of 72,000 BP on a cave sinter drape indicating a moister environment during an early phase of the Last Glacial Maximum.

Water from the Umm er Radhuma aquifer flows by gravity from the As Sulb region to important karst springs in Gulf coast region at Al Qatif and Al Hasa (*Arabic*: gravel). Peters *et al.* (1990) indicate that over one hundred and forty perennial springs water the 75 km^2 oasis at Al Qatif. The springs, which may be either non-flowing wells or artesian springs, are vertical pits several metres in diameter and up to 35 meters deep. The largest *'ayn*, 'Ayn Al Labaniyah, has an average discharge of 40 L/s. At the Al Hasa oasis, 16 important springs in the Dam Formation water some 200 km^2 of the oasis. The largest of these springs, the 'Ayn Al Khudud, has a discharge of some 1700 L/s. This spring has a vertical shaft some 7 m in diameter and is 11 m deep. It is fed by a system of phreatic tubes.

Immediately to the west of the Al Hasa oasis, east-facing escarpments in the Hufuf Formation have well-developed vertical fissure caves (Arabic: *ghar*). There is some speculation that the fissure caves are fossil marine caves formed during a period of higher sea level in the Arabian Gulf (Hötzl *et al.*, 1978a).

Other karsts of interest are the little-studied coralline karsts of the Farazan Islands in the southern Red Sea. These islands are underlain by an active salt diapir and the author has observed phreatic tubes and small caves in exposures of fossil raised reef. There is also black phytokarst development at shore level (Folk *et al.*, 1973). Guilcher (1955) who was on board one of Jacques Cousteau's famous Red Sea cruises of the Calypso, provides a sketch showing an abandoned dry valley floored with sinkholes. Presumably, in this example, a falling water table resulted from the diapiric uplift of a salt dome?

6.12 VOLCANIC LANDFORMS

In has already been noted that two separate periods of Cenozoic volcanicity have taken place in Saudi Arabia. An older period dating from 30–20 Ma is now seen as a series of very weathered and eroded *harraat* associated with the opening of the Red Sea. A younger group is oriented more or less along a north-south axis and is not associated with the opening of the Red Sea but with the development of a new continental rift, and axis of Shield uplift some 600 km long running from Makkah northwards to Madinah and on into An Nafud – the so-called MMN line. Distinct volcanic landforms are only associated with these newer *harraat*.

In addition to the *harraat* themselves there is a wide variety of volcanic landforms but few have been described or studied in any detail from a geomorphological perspective. The *harraat* are of particular interest for a number of reasons. First, their surfaces clearly reflect their age. Young *harra* are almost impossible to cross by vehicle and their surfaces have been little weathered and eroded, whereas older *harra*, such as Harrat as Sarat, the southernmost *harra* in the Kingdom, has been severely eroded and less than half of its former extent remains. The loss of this protective capping has exposed Tertiary saprolites and laterites over a wide area and in places the etched Precambrian surface can also be seen in may beautiful exposures.

Most *harraat* have well developed *wadi* systems on their surfaces. These are usually fed by ephemeral springs. Around the margins of *harraat* lava has sometimes flowed into and filled a pre-existing *wadi* which was part of a widespread Tertiary drainage system. The younger *harraat* sometimes have lava tube caves developed in their upper horizons. These form as a chilled ceiling develops over a more or less linear lava flow. The lava continues to drain down slope leaving behind an empty tube which opens out to the surface. John and Suzi Pint together with John Roobol (www.saudicaves.com) have made an extensive study of lava tube caves in the Kingdom and

Figure 6.23 Jabal Bayda' – a commendite maar on Harrat Kaybar (Photo: author).

Figure 6.24 Jabal Qidr – a stratovolcano on Harrat Kaybar (Photo: author).

have dated a number of animal bones from them. The tubes also form natural traps for wind borne loess.

Harrat Kaybar, north of Madinah has some of the most spectacular volcanic landforms in the Kingdom including stratovolcanoes, shield volcanoes, and maars. Maars have broad craters and have been formed by the explosive force of steam pressure as rising magma interacts with the water table. Of striking appearance is Jabal Bayda' (the White Mountain) which is a large maar crater formed of commendite (rhyolitic) ash (Figure 6.23).

Nearby is Jabal Qidr, a fine example of a stratovolcano with a central vent 200 m deep (Figure 6.24). Occasional fumerole activity and the presence of hot springs are signs that the field is not totally quiescent.

Less spectacular, but equally interesting, is the volcanic field of northern Harrat Rahat south of Madinah. This area has been mapped in detail by John Roobol at the Saudi Geological Survey and because of its proximity to Madinah has been designated as a possible geopark. There are excellent examples of pahoihoi and aa lava types accessible from good roads as well as lava caves and young lava flows including that which nearly reached Madinah in 1256 (Figure 6.25).

6.13 ESCARPMENTS AND CUESTAS

The word escarpment is a morphological term for a major, more or less linear, break of slope separating two topographical levels, and is usually associated with

Figure 6.25 Flows of pahoehoe lava through blocky aa lava – 20 km south-east of Madinah (Photo: author).

major, near vertical, cliff faces. Their foot slopes are frequently buried by talus. In simple terms escarpments can be formed by either faulting or erosion. Cuestas are found in sedimentary rocks and comprise an escarpment backed by a gentle dip slope.

The mid-Tertiary uplift of the Shield initiated some remarkable erosion along the line of normal faults running more or less parallel with the Red Sea. From Taif southward, erosion in steep, energetic *wadi* systems draining to the coast has produced a dramatic, cliffed, escarpment abutting against the Tertiary etchplain. The escarpment is now some 50–70 km east of the zone of normal faults which gives some indication of its retreat in the last 15 million years or so. Some recent apatite fission tracking research suggests an even younger uplift date for the escarpment which would make the rate of recession even more dramatic. Quite why the escarpment runs out north of Taif is not known. Almost certainly fluvial erosion is more effective as one goes south because of the influence of the Monsoon. However there may be an as yet unknown tectonic explanation.

The other spectacular escarpment in the Kingdom is the Tuwaiq Escarpment which runs in a gentle north-south arc to the west of Riyadh. This escarpment is not associated with fault systems but with drainage development on the gently dipping cover rocks. In fact, there are a series of escarpments and gentle dip slopes in the cover rocks which have been formed as a result of the incision of a trellised drainage system. Large, superimposed, consequent *widyan* developed when the Arabian plate was uplifted. As these *widyan* interacted more and more with the geology, tributaries flowing

Figure 6.26 A less often seen view of the southern section of Tuwaiq Escarpment shortly before it disappears under the sands of the Rub' al Khali (Photo: author).

more or less at right angle to the main *wadi* developed along the strike boundaries of different lithologies.

The escarpment is topped by a vertical free face some 20–30 metres high. In the Wadi ad Dawasir region spring sapping is an active process undermining the scarp face developed in the Jurassic limestone The escarpment becomes progressively lower to the south and eventually disappears under the sands of the Empty Quarter (Figure 6.26).

On the Gulf coast an interesting escarpment is located along the eastern edge of the Ghawar anticline. The escarpment runs for about 150 km more or less north from Harad towards Abqaiq and forms the western boundary of the Al Hassa (Hufuf) oasis. This escarpment lies about 160 m asl. and, as was noted earlier, is thought to have been formed by marine action during a period of high sea level during the Plio-Pleistocene (Sannah *et al.*, 2005).

6.14 QUATERNARY SEA LEVEL CHANGE

6.14.1 *Shurum*

The *sharm* (Arabic: pl. *shurum*) is a very characteristic feature of the Red Sea coast. It is narrow, deep water channel penetrating nearly perpendicularly into the coast for some distance and then throwing out deep lateral extensions, often at right angles,

Figure 6.27 Sharm north of Al Wadj cutting through raised coral reef (Photo: author).

so that the whole inlet resembles a T-shape or bottle shape. Sharm Jubbah north of Dhuba is a classic example (Figure 6.27). These inlets comprise the most important harbours along the coast and some, such as the Sharm Abhur just north of the centre of Jiddah, are favoured locations for urban expansion and provide prestigious waterfront sites for villas (Monnier and Guilcher, 1993).

Sometimes these features are also known as a *marsa* (pl. *marasi*) or *khawr* but these terms only occur occasionally. Guilcher (1954) uses yet another term *'churoun'* but this name seems not to have caught on in the literature. The designation of a place name component does not always have any genetic implication. For example, the channels running between the various islands of the Farazan archaepelago are also known as *khawr* but it is difficult to think of these channels as being similar to sharms in the accepted sense (Naval Intelligence Division, 1946) not least because the Farazan archaepelago is underlain by an active salt diapir, influencing local tectonics, and also because the island group lies some 20 km off the Jizan coast which is sharm free.

Sharms are usually connected to the *wadi* system draining from the escarpment but not always. The narrow throat at the entrance of the sharm is typically about 30 m deep where it cuts through the coralline ridge and can be some 10 m deep up to 5 km inland. Today there is no active erosion in the sharms and flood waters from the escarpment only rarely reach them. Beyond the coral reef sharm floors, on average, descend to about c. 60 m below sea level.

Early ideas on the formation of sharms suggested that they were contemporaneous with the coral growth now exposed in the flanks of the sharm throats, the idea being that silty water from *wadi* runoff inhibited the growth of corals in the central channel. However, other evidence suggests a later breaching of the fossil reef. For example, it has been argued that the bifurcation of the sharms behind the beach ridge represents lagoons that existed before the reefs were breached. Isotopic dating of the raised reefs breached by the sharms indicates that they are older that 40,000 years and maybe as old as 146,000 years BP.

Brown *et al.* (1989) recognize that the origin of the sharms requires a coincidence of low sea level and pluvial conditions such that *widyan* flowed to the Red Sea. Wadi flow to a sea level lower that at present would allow erosional breaching of the elevated reef and at the same time allow scour of some of the soft sediment built up in the lateral lagoons. The authors hypothesise that the sharms formed about 12,000 to 8,000 years ago during the last pluvial episode when sea level rose from –60 m to –20 m. At the maximun low sea level of –120 m, some 18,000 years ago, climate was too arid and after about 8,000 years BP sea level was too high for sharms to form. Sharms might well have formed during other interglacial periods but they have been filled in by coral growth.

An alternative explanation, at least for some sharms, has been put forward by Fricke and Landmann working on the Sinai. These authors examined the submerged section of sharms draining into the Gulf of Aqaba. Their findings indicated that sharm formation is probably an ongoing process sustained by recent gravity flows triggered by massive sediment shifts during occasional winter flash floods. The sub-aerial erosion hypothesis can, they argue, be tested by close inspection of the deepest Pleistocene base level. If the sharm cuts through the submerged terrace it must post-date it. This is the same argument as that put forward by Brown *et al.* (1989). Fricke and Landmann (1983) found that the sharms do indeed cut through submerged Pleistocene terraces and using their submersible and echo sounder they found scoured channels at –170 m floored with coarse sand and fragments of shallow water corals. They traced the erosional floor of these channels down to –320 m. Fricke and Landmann's observations do not disprove the ideas of Brown *et al.* (1989). Valleys submerged during the Holocene sea level rise could still have been progressively deepened by gravity flows triggered by flash floods. But subaerial erosion of the *sharm* by floods below the level of the deepest Pleistocene terrace is very unlikely. This would mean a water level depression far below the sill at Bab al Mandab (138 m below present sea level).

Fricke and Landmann's ideas are of interest from another point of view. By far the largest number of *shurum* on the Kingdom's Red Sea coast occur on the coast north of Yanbu. In this region the mountains descend almost to the sea and the coastal plain is extremely narrow if present at all. Flash floods in this region are unlikely to be soaked up by the coastal plain sediments whereas in the south this is much more likely. The convenience of flash floods feeding submarine gravity flows does however, break down for those sharms not connected to the *wadi* system. It is most likely that these sharms are controlled by faulting but no research has been done to supports this hypothesis.

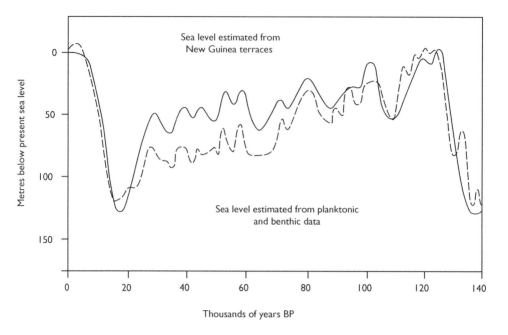

Figure 6.28 General pattern of sea level change (Modified from Bailey *et al.*, 2007).

6.14.2   Quaternary marine terraces

6.14.2.1   *Red Sea*

Marine benches cut across Quaternary reefs, in places linked to inland to *wadi* terraces, are common along much of the Red Sea coast. Their origin is far from straightforward and the complex interplay between eustatic sea level changes and local tectonics has led researchers to diametrical opposed views (Behairy, 1983; Jado and Zötl, 1984; Dullo, 1990; Sheppard *et al.*, 1992). To put these features in context we need first to examine the general pattern in relation to global eusatic sea level change (Figure 6.28).

At Bab el Mandeb, at the southern entrance to the Red Sea, the water is about 130 m deep but this is not the lowest possible sea level in the Red Sea because evaporation is thought to have reduced the level even further on some occasions. In Figure 6.28 it can be seen that sea level was last near present datum at about 110–140 Ka BP. This was followed by a long period from 110 to 30 Ka BP when sea level fluctuated but remained roughly between –30 and –60 m. During this period there were at least eight still-stands lasting between 1–2 Ka. According to Sheppard *et al.* (1992) not only is 1–2 Ka long enough for substantial coral reef growth but the sea levels of –30 m to –60 m corresponds to the depths of major sharm channels. From 30 Ka BP sea levels fell rapidly to about 17 Ka which is taken to mark the boundary between the Pleistocene and the Holocene. Low sea levels at about 17 Ka BP probably reduced the Red Sea to a hypersaline lake. In the Arabian Gulf, which is very shallow, the sea withdrew completely, apart from a small finger extending inward from the Gulf of Oman along the Iranian coast. At about 15 Ka BP global temperatures rose rapidly and was followed by a rapid rise in eustatic sea level to present levels about 7 Ka BP after which they may have been only minor fluctuations.

The exact number and age of raised reef terraces recorded varies from place to place and from author to author. Often three, and sometimes four, benches are observed cutting across raised reefs. North of Al Wadj more than twelve benches can be seen and one has been raised to more than 300 m asl. and has Precambrian clasts strewn on its surface. It is abundantly obvious, however, that none of these raised reefs are a record of former eustatic sea levels. Behairy (1983) records terraces on the southern Hijaz coast at 31 m asl. which he dates at 31 Ka BP. He then indicates periods of falling sea levels and still stands to account for lower marine terraces. But eustatic sea levels at this time were low not high! Dullo (1984) remarks that one reef terrace in the tectonically active Gulf of Aqaba coast is 98 m above sea level. At Haql the present author has observed reefs more than 100 m asl. and the main *wadi* terrace to the north of the town terminating abruptly at the coast is also at about the same altitude. In short, it seems very reasonable to suppose that the raised reefs and terraces are the result of the land going up not the sea falling to its present level. And if the sea level curve in Figure 6.28 is to be believed then all marine terraces on the Arabian Red Sea coast dating from about 7 Ka to 120 Ka BP must have been raised into their present position tectonically.

The whole of this northern Hijaz coastline is subject to tectonic uplift along the Gulf of Aqaba–Levant fault zone (Rowaihy, 1986; Williams *et al.*, 2000) and an added complication is that in some places, particularly in the tectonically active Haql region on the head of the Gulf of Aqaba, individual terraces, have been later subjected to normal faulting and uplift, giving rise to the paradox that two adjacent marine surfaces at different heights are exactly the same age. The flight of terraces at Haql are particularly well developed and is cut into by steep *wadi* fans composed of coarse, unsorted, pebble and boulder gravels.

6.14.2.2 Arabian Gulf coast

Reference has already been made to high sea levels associated with the Al Hassa escarpment. The retreat of the sea was not continuous and regressive phases created series of small cliffs a metre or so high leading down to the present day coast which is dominated by young *sibakh* a few metres above sea level.

6.14.3 Sibakh and sea level changes

The usual location for *sibakh* is a metre or so above sea level in low energy environments, although some can occur inland where the drainage cannot escape to the sea. A third type, occurring on the Gulf region is related to the high Plio-Pleistocene sea levels mentioned previously. Holm (1960) suggested that as this sea level fell it left behind numerous *sibakh* some of which are now more than 100 m above present sea level and well inland. The idea is a good one and *sibakh* of this type have been described in the United Arab Emirates. One inland *sabkhah* south-east of Harrad is possibly of this type.

Figure 6.29 Loessic alluvial silts in the Muhail basin derived from the stripping of deeply weathered saprolites. The sharp junction between the coarse gravels and silts marks an abrupt desiccation in the drainage basin (Photo: author).

6.15   LOESSIC ALLUVIAL SILTS

Silty, loess-like, sediments are common in the *widyan* of the Shield where they are sometimes 10 m or so thick. They form some of the most productive agricultural soils and are particularly extensive in the Muhail region, in Wadi Baysh and in the catchment of Wadi Tathlith. The silts are kaolinitic and their heavy mineral components indicate that the silts are related to the stripping of the deeply weathered mantle of saprolites and laterites which once covered the Shield. Optically Stimulated Luminescence dating indicates that the silts have been moved in a sequence of fluvial events, over at least the last 50,000 years, the most recent even being an historical flood a few hundred years old in the Sayl al Kabir (Arabic: the great flood) draining from the escarpment near Taif. In essence, the present-day silts can be compared to the 'river end' deposits in Namibia where the loessic alluvial silts are stranded in the river system because the river has simply dried up (Figure 6.29). Related to the loessic alluvial silts are true desert loesses which have been trapped in volcanic craters as at Tabah, near Ha'il and in lava tube caves such as those on Harrat Kishb.

Chapter 7

Biogeography

Contents

7.1 INTRODUCTION

Although there is now a good deal of information available for the Kingdom, it would be totally unrealistic to expect the existence of detailed distributional and ecological data that are available in Western Europe and North America. The biogeography of many taxa is still unknown and can only be guessed at. Furthermore, it is likely that some species have gone undiscovered. This is partly due to the exceptional difficulties of exploration, not so much in the desert areas, with their few ecological niches, but more particularly in the high recesses of the Asir where many mountain peaks and *widyan* have simply not been investigated.

In this overview I make no attempt to cover every aspect of the known biogeography of the Kingdom. Instead, I will try to highlight the general pattern of species' distributions and illustrate the main ideas by reference to well known taxa. The biogeography of mammals within the Kingdom is now reasonably well known, thanks partly to the devastation of many species through hunting and shooting. Much less is known about other taxa apart from birds and butterflies. Similarly, while there is now good general picture of the pattern of vegetation, our knowledge of the detailed distribution of flowering plants species is still poor and that of the non-flowering plants almost non-existent until recently (Kurschner and Ghazanfar, 1998).

From a conservation point of view benchmark distribution data are essential, if only to assess the impacts of climatic change, the consequences of habitat destruction and to obtain some measure of rarity. Lack of skilled personnel, a good mapping base and the sheer size of the Kingdom make the task of collecting such data quite daunting. This situation is very slowly being put right and in recent years a number of important books and research papers have been published which have made valuable contributions to the biogeography and ecology of the region. Of the specialist botanical books, the following are particularly important: *Flora of Eastern Arabia* (Mandaville, 1990), *Flora of the Arabian Peninsula and Socotra, Volume 1* (Miller and Cope, 1996) and *Vegetation of the Arabian Peninsula* (Ghazanfar and Fisher, 1998). Collenette's, *Wildflowers of Saudi Arabia* (1999) is also very useful and impressively illustrated. Sheila Collonette has also published a checklist of the 2,243 species of flowering plants for Saudi Arabia (Collonette, 1998). The Ministry of Agriculture and Water (1995c) published *A Catalogue of Vegetation Communities in Saudi* Arabia as Technical Annex D in *The Land Resources Atlas* and a botanical bibliography for the Arabian Peninsula as whole has been published by Miller *et al.* (1982).

The zoological record is now partly covered in the irregular series *The Fauna of Saudi Arabia* published by Pro Entomologica (Basle). The mammals of Arabia are described in detail in Harrison and Bates (1991), and Jonathan Kingdon's *Arabian Mammals, a Natural History* (1998) is copiously illustrated with sketches and basic distribution maps. As far as I am aware, the only detailed biogeographical mapping is that of the *Atlas of Breeding Birds of Arabia* (ABBA) project which is coordinated by Michael Jennings. The first product from the project is, *An Interim Atlas of the Breeding Birds of Arabia* (Jennings, 1995) whose publication was supported by the National Centre for Wildlife and Conservation Development (Riyadh). The definitive atlas will be published in the *The Fauna of Saudi Arabia* series. An excellent bibliography is also maintained at the ABBA web page: http://dspace.dial.pipex.com/arabian.birds/ Notes and commentaries on bird reports for the ABBA project are published regularly in *The Phoenix* details of which can be found on the ABBA web site. An excellent web site with short topical articles on the biogeography of the Arabian Peninsula as a whole is hosted at: http://www.arabianwildlife.com/home.htm .

One major task for biogeographers of the Arabian Peninsula is to piece together the pattern of distributional changes that have come about since the time of the tropical landscapes in the Early Tertiary. The general picture can be stated quite simply. The formation of the Red Sea and the uplift of the Zagros produced formidable barriers to species movements into the Peninsula. Many Afrotropical species which were widespread over the Nubian-Arabian shield became stranded on the newly formed peninsula (Leviton, 1986). In addition, the movement of Gondwana northwards and the climatic cooling of the late Tertiary and the Quaternary Period meant further problems of survival as vegetation belts moved south. The climatic cooling allowed an invasion plants and animals from Eurasia far into the south of the peninsula, and recent aridification has removed many suitable habitats for these species many of which have become relict and survive now only in moist, high altitude, locations in the south of the Peninsula. As the Kingdom gears up for a growth in ecotourism knowledge of the fauna and flora will become vital if this section of the tourist industry is to flourish.

7.2 THE FAUNA

In this brief survey I shall describe the distributions of mammals, birds and butterflies. Their biogeography is reasonably well covered in the literature and the factors governing their distributions of fairly general applicability. A biogeographical pattern which is probably reasonable for these and similar taxa, suggests that species richness increases towards the south-west of the peninsula. The Asir and Yemen are relatively well-watered, have a wide range of habitats and have had recent geological links with Africa.

7.2.1 Zoogeographical regions

According to the great English naturalist A. R. Wallace, the Arabian Peninsula lies at the junction of three faunal regions. To the west lies the continent of Africa and the Ethiopian faunal region, and to the east and north is Eurasia with its Palaearctic and Oriental faunas (Wallace, 1876). More recent knowledge has led to a slightly modified version of Wallace's broad regions and nowadays zoogeographers of the Arabian Peninsula use a four-fold faunal division of regional affinities (Delany, 1989):

a) *Afro-tropical Region*: the south, and especially south-west of the Peninsula, whose species have strong affinities to those in East Africa south of the transcontinental arid zone. Example: *Terpsiphone viridis* (African paradise flycatcher).

b) *Saharo-Sindian Region*: essentially the central desert zone of Arabia – part of a vast region linking the western Sahara to the deserts of Afghanistan and central Asia. Example: *Jaculus jaculus* (lesser jerboa) – found in desert environments throughout the Sahara from Morocco to SW Iran.

c) *Palaearctic Region*: broadly corresponds to Wallace's definition and includes northern Arabia, Mediterranean Africa, Europe and Asia north of the Himalayas. Example: *Canus lupis* (wolf) – formerly the whole of the Palaearctic region except for North Africa.

d) *Oriental Region*: follows Wallace's demarcation and includes the regions east of Pakistan and the Orient south of the Himalayas. Example: *Suncus murinus* (house shrew – probably introduced?) type locality – islands of the Orient.

Fluctuations in sea level from time to time and the amelioration of climate have provided invasion routes for animals to enter the Peninsula. According to Delany (1989) there is fairly strong evidence for at least one land connection between southern Arabia and Africa during the most recent decline in sea level during the last glacial maximum (*c*.18,000 BP). To expose a land link across the southern Red

Sea would have required a eustatic fall of sea level of about 140 m in the region west of Al Hanish al Kabir island. Such a large fall in eustatic sea level may have been unlikely, however a sea level fall of somewhere between 80 to 120 m would still have provided suitable invasion routes with short island hops between narrow marine channels perhaps just a few kilometres wide. In the Arabian Gulf region there was no such problem for invading species as much of the Gulf is less than 80 m deep and would have become a riverine plain during periods of low eustatic sea level. The dispersal of animals into the Peninsula from the Palaearctic zoogeographical region to the north has essentially been controlled by the Peninsula's arid setting and it seems likely that the mountains fringing the Red Sea have acted as the main corridor route to the south.

7.2.2   The mammal fauna

Nader (1990) has provided a checklist of the 98 mammals recorded from Arabia. Of these, 76 are found Saudi Arabia. Although there are no available distribution maps it is clear that the fauna of the south of the Peninsula, particularly areas watered by the northern edge of the Monsoon, is relatively species rich as compared with the deserts of the interior.

The origin of this southern fauna is of interest, particularly in relation to its links with Africa and the possible invasion routes taken. According to Delany (1989) of those mammal species now restricted to south-west Saudi Arabia 3 have Saharo-Sindian affinities and 11 have Afrotropical affinities. No species with Palaearctic or Oriental ranges are present. Delany suggests that the Afrotropical element could have arrived in Arabia either through Sinai in the north or from Eritrea-Somalia.

The northern invasion route for Afrotropical species has been researched by Tchernov (1981) who suggested that this element migrated into the Middle East via Sinai in three main waves. By the end of the first wave (c.500,000 BP) fossil records indicate that *Mastomys* spp., (multimammate rats) *Giraffa*, *Hippopotamus* and *Cryptomys (*mole-rats) had entered the Middle East. Of these, only *Hippopotamus*, survived into the Holocene as witnessed by McClure's finds of *Hippopotamus* teeth in the lake deposits of the western Rub' al Khali. This last conjecture needs to be qualified however. McClure's lakes were shallow and probably not long-lived. Furthermore, since no cave speleothems have been found in Saudi Arabia younger than 400,000 years old, it seems that aridity may have set in at least by the mid-Quaternary – hence the presence of *Hippopotamus* is problematic.

A second wave of mammals migrated into the Middle East during the Riss - Wurm cold stages (c.250,000–30,000 BP) including *Proclavia* (hyrax - now widely spread), *Phacocheorus* (warthog), *Alcelaphus* (hartebeest*), Felis leo, Arvicanthis ectos (*anteater*) and Mastomys batei* (mouse). More recent invaders include *Rousettus aethiopicus* (Egyptian fruit bat), *Acomys cahirinus* (spiny mouse), *Herpestes ichneumon* (banded mongoose) and *Eidolon helvum* (straw-coloured fruit bat).

A southern route across the Red Sea seems the most likely route for much of the Afrotropical mammal fauna in the south-west of the Peninsula as there are no fossil finds of these species in the Sinai region. Delany (1989) cites the following as examples: *Rhinolophus clivosus* (horseshoe bat), *Mellivora capensis* (honey badger) and *Acinonyx jubatus* (cheetah). For several species of bat southern Arabia, including the

Asir, appears to have acted as a corridor of suitable niches linking the Afrotropical and the southern Palaearctic regions – especially Iran. Among these species are: *Triaenops persicus* (Persian leaf-nosed bat), *Tadarida aegyptiaca* (Egyptian free-tailed bat), *Rhinopoma muscatellum* (Muscat mouse-tailed bat) and *Taphozous perforatus* (tomb bat). Presumably, these bats might once have occupied more northerly sites in Arabia but their distribution was pushed south as a result of the development of the arid interior.

Recent findings by Thomas *et al.* (1998) have thrown more light on the origins of the fauna with the discovery of vertebrate remains in interdune lacustrine deposits of the western Nafud desert. Although the fossil vertebrate fauna is restricted in diversity, its composition and state of evolution suggests an Early Pleistocene age, in contrast to the recent Pleistocene and Holocene lake beds in the central and southern Nafud. The Nafud fauna has clear African affinities whereas faunas from the Levant during the Middle Pleistocene onwards, contain typically European elements. Thomas *et al.* suggest that the Nafud fauna probably crossed into Arabia by way of Sinai.

Among the interesting species with African affinities discovered by Thomas *et al.* are: a tortoise – probably related to the large African tortoise *Geochelone sulcata*; bones reminiscent of the Pleistocene African elephant *Elephas recki*; fossil zebra bones resembling the *Equus* of Upper Bed II at Olduvai or of the 'Acheulian' of Garba III; buffalos akin to the giant fossil buffalos of Africa, *Pelovoris*. Other species present included a Bactrian-like camel, *Oryx* and a large freshwater osteoglossiform fish which would have originally been more than a metre long.

Thomas *et al.* (*ibid.*) used a recently developed approach of palaeoenvironmental reconstruction by using the ^{13}C isotopic signature of the *Pelovoris* tooth enamel. During the life of a mammal the mineralized enamel records the carbon isotopic signature of the diet which depends mainly on the photosynthetic pathways of the plants consumed by the herbivores. In tropical environments grasses are C4-plants whereas trees are C3-plants, with different isotopic signatures. Such signatures are evidently well preserved in tooth enamel even millions of years old. Thomas and his co-workers found that the enamel analysis indicated a C4-plant diet when compared with the modern African fauna and suggest, therefore, that the vegetation in the Nafud was an open savannah dotted here and there by large standing bodies of water.

7.2.3 Birds

Lees-Smith (1986) provides an interesting overview of the biogeographical affinities of the south-west Arabian avifauna and the impacts of environmental changes in the region. In assessing the faunal links Lees-Smith uses some very useful general environmental models to guide his thinking. In particular, he relates the present distribution of some bird species in the Kingdom to habitat changes due to aridification in the Holocene and species isolation in pockets of favourable habitat. For example, *Pica pica* (magpie) is a widespread Palaearctic species but in the Arabian Peninsula is found only in the highlands of the southern Asir where it nests in acacia and juniper trees 3–6 m above ground. Lees-Smith suggests that *Pica pica asirensis* is a Pleistocene relict now isolated in these high mountain environments. In isolation this magpie has lost almost all the gloss from body and wings and has acquired bill and feet which are proportionately larger than in northern populations. These are the kind of characters which can evolve in small populations with no gene flow from other populations.

Another Palaearctic species present in south-west of Arabia, and at the southern-most limit of its range, is *Gyps fulvus* (Griffon vulture). Lees-Smith argues that its presence in the south-west is probably due to descent from an ancestral population which was in the region at the end of the Pleistocene when most of Arabia was prob-ably moister than at present and carried savannah grasslands populations of large grazing and browsing mammals off which *Gyps* scavenged. As *Gyps* is not found in Afrotropical Africa it is probably a relict species in the Asir and Yemen. It is also worth noting that *Gyps* is found in the northern Nadj but Lees-Smith makes no com-ments about possible interaction between the two disjunct populations.

The biogeography of the endemic Arabian woodpecker *Dendrocopos dorae* (*Picoides dorae*) is fascinating but the details need to be worked out (Martins and Hirschfeld, 1998). It was discovered by St. John Philby and the specific epithet *dorae* is a reference to Philby's wife who often joined him on his travels. In the Arabian Peninsula this 'desert-coloured' woodpecker nests in acacia and juniper trees, and is distributed from the southern Hijaz down into the Yemen. It is not found in Oman and Lees-Smith suggests that its closest relatives are found today in the forests of the Himalayas and the Orient, implying that the areas between southern Arabia and the forests in Nuristan, Baluchistan and north-west India, now treeless except along water courses, were once covered by some sort of woodland during a favourable Pleistocene period which allowed the woodpecker to extend its range into southern Arabia. It is thought that the Arabian woodpecker is probably not related to the nearest juniper-dwelling woodpecker in Africa – the golden-backed *Dendropicos abyssinicus* of the Ethiopian highlands, but this is a matter for the taxonomists. Given the present-day distribution of *Dendrocopos dorae* such a relationship would certainly provide a simpler history. Other ornithologists regard the Arabian woodpecker as a Palaearctic species and per-haps Lees-Smith concept of the 'Arabian Pier' is relevant in this regard.

South-western Arabia has a number of endemics and relict species with northern affinities which could have reached the Asir-Yemen region along the chain of moun-tains bordering the Red Sea when a more favourable vegetation cover was present. At the height of the Pleistocene glaciations it is suggested that patches of warm temperate shrub and tree woodlands were strung out like beads from the Lebanon in the north to the Yemen in the south. Species with northern affinities which might have dispersed south down this Arabian Pier include: *Alectoris melanocephala* (Arabian Red-legged Partridge), *Strix butleri* (Hume's Tawny Owl), *Sylvia buryi* (Arabian Tit Warbler) and *Acanthis yemenensis* (Yemen Linnet). At the present time the Afrotropical influence is dominant in two disjunct ecological *islands* sustained by the Monsoon rainfall. One *island* comprises the montane south-west Arabia and the other Dhofar (Oman and eastern Yemen). The stability and extent of these two avifaunal zones in the Pleistocene must have been locked in to the north-south movements of the monsoon rain belt.

7.2.4   Butterflies

Torben Larsen (1984) provides a very thorough account of the zoogeography of the one hundred and forty-eight species of butterfly found in the Arabian Peninsula south of a line from Aqaba in Jordan to Basra in Iraq, and the interested reader is referred to this paper for a detailed account. Larsen reports that ninety-two (62%) of all the Arabian butterflies are of unequivocal Afrotropical origin. In contrast, only six of the Arabian butterflies are Oriental elements. None are typical of Oriental species and some seem to be related to the activities of man. For example, in Arabia *Papilio demoleus* is totally dependent on cultivated *Citrus* which was evidently only introduced to Arabia some seven hundred years ago.

Twenty of the Arabian butterfly species are of unequivocal Palaearctic origin and appear to have relict distributions. In Saudi Arabia eleven species are found in the mountains of the Asir and southwards into the Yemen. Several of these species are also found on isolated high mountains in East Africa. This is very enigmatic and Larsen finds no real solution to the origin of this puzzling, disjunct distibution. We might note here that a similar distribution is found in several plant species (see below). The gist of the problem is that there does not appear to have been a cool enough period in the Quaternary to allow a general north to south migration. On the contrary, the Sahara during the pluvial periods was encroached upon not by northern grasslands but by Afrotropical savannah. Nevertheless, Larson suggests that the presence of *Colias electo, C. erate, Lycaena phlaeas, Lasiommata felix* and *Argyreus hyperbius* (Indian Fritillary) in Africa and Arabia dates back to the pre-Pleistocene times when the Red Sea did not present a barrier. Only 8 species of butterfly in Arabia can be regarded as fully endemic and not a subspecies of African or Palaearctic taxa. Most of these species occur in the mountains of the south-west of Arabia.

The overall zoogeography of Arabia according to the distribution of butterflies is shown in Figure 7.1. The sandy desert (eremic) zone broadly coincides with the botanical Saharo-Sindian region of the Peninsula.

In an attempt to understand the broad pattern of butterfly distributions in Arabia Larson has ranked each of his ecological provinces by quality according to six selected indicators (Table 7.1).

He then plotted the number of butterfly species against this index of ecological quality and shows that there is an almost linear relationship. Although the findings are not at all surprising it does indicate the point that habitat and plant diversity are important features in butterfly conservation (Figure 7.2).

7.2.5   Fishes

I have included this very brief section on the biogeography of fishes in Saudi Arabia as it may come as a surprise to some readers that there are any species of fish at all in such an arid county with no permanent water courses.

Krupp (1983) has published an extensive account of the fishes of Saudi Arabia in which he not only gives details of the species found but also their biogeography. He makes the point that very few fish species inhabit the Arabian Peninsula and only ten taxa are recognized as valid. All are members of the Cyprinidae family (minnows) and inhabit *wadi* pools and springs particularly in the mountains of the Asir, the Yemen and Oman.

7.2.6   Mapping schemes – *Atlas of Breeding Birds of Arabia*

As far as I am aware the only attempt at biogeographical mapping is that undertaken for the *Atlas of Breeding Birds of Arabia* (ABBA) project coordinated by Michael Jennings and sponsored by the National Commission for Wildlife Conservation

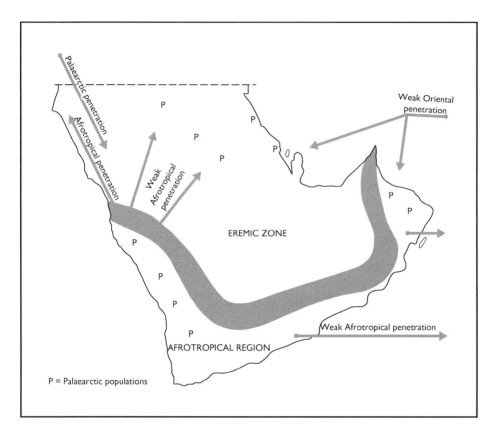

Figure 7.1 Zoogeographical regions according to butterfly distributions (Modified from Larsen, 1984).

Table 7.1 Ranking of the ecological provinces of Arabia by quality on six selected indicators

Ecological province	Rainfall	Predictability of rain	Presence of mesic sites	Vertical zonation	Habitat variety	Number of plant species	Total score
Yemen	10	10	10	10	10	10	60
Asir	9	9	9	9	8	8	52
Aden	8	7	8	8	9	9	49
Dhofar	6	6	7	5	7	7	38
Hadhraumaut	7	5	6	6	6	6	36
Northern Oman	5	8	5	7	5	5	35
Hijaz	4	4	2	4	4	4	22
Central Arabia	3	3	3	3	3	3	18
Eastern Arabia	1	1	4	1	2	1	10
UAE	2	2	1	2	1	2	10
Rub' al Khali	0	0	0	0	0	0	0

Source: Larsen (1984).

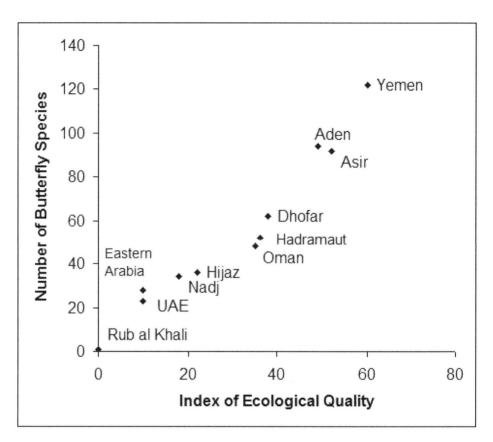

Figure 7.2 Index of Ecological Quality versus number of butterfly species (Modified from Larsen, 1984).

and Development (Riyadh). Jennings has divided the Arabian Peninsula into half degree grid squares which form the unit of survey. The base map for the grid is the 1:2,000,000 Geographical Map of the Arabian Peninsula (1984) published by the Deputy Ministry for Mineral Resources, Jiddah and uses a Lambert conformal conic projection. The use of this projection for such a large geographical area means that the 1,110 grid squares of ABBA are not equal area but the differences are not great and range from 55×56 km in the southern Yemen to about 48×55 km in Kuwait. Each square of the mapping scheme has a unique reference made up from a two letter longitudinal code and a numerical latitudinal code, both of which can be cross-referenced to latitude and longitude.

Contributions to the atlas are made by completing standard report forms, noting species, evidence of breeding and grid square. The complete dataset for the Atlas is stored in a relational database and provides the information for the maps. The project, which started in 1984, is enormously important from both biogeographical and conservation points of view and it is hoped that other taxa may soon receive the same treatment and eventually the whole database incorporated into a Geographical Information System.

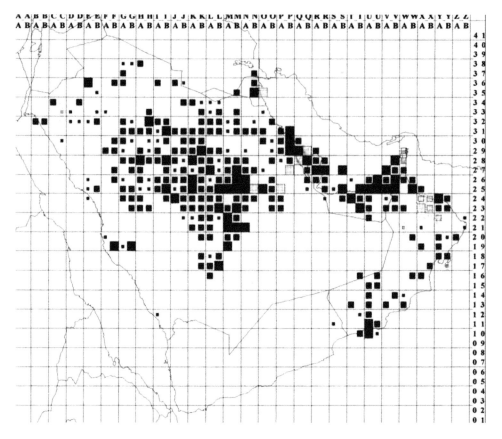

Figure 7.3 Distribution of the Eurasian Collared Dove (*Streptopelia decaocto*) (Courtesy: Michael Jennings).

An entry from the *Atlas of Breeding Birds of Arabia* is presented in Figure 7.3. This shows the distribution of the Eurasian Collared Dove (*Streptopelia decaocto*). This species is invasive and has expanded its range in north and central Saudi Arabia since it was first recorded in Kuwait in 1963. It is often commensal and associated with gardens and farms.

7.3   THE FLORA

Saudi Arabia's position between the continents of Africa and Asia is important in any attempt to understand the recent history of the fauna and flora and attention has been particularly focused on links with between the Asir and East Africa. The highlands of the south-west of the Arabian Peninsula are relatively species rich compared with the arid interior and low sea levels, and the relatively narrow barrier of the Red Sea, have allowed for a spill over of African species onto the Arabian mainland. Furthermore, some taxa are almost certainly relict from early Tertiary times prior to the rifting of the Red Sea.

7.3.1   Phytogeographical regions

The zonation of the globe into phytogeographical regions attempts to reveal the geological linkages or possible dispersal routes of plants over recent geological time. The regions themselves are centres of endemism having plants, or some percentage of plants, unique to that region. There are always problems regarding regional boundaries, and placing a line on a map should not imply some sort of rigid barrier since the number of endemic species will always decline away from regional centres. Furthermore, there are vast swathes of terrain about which little floristic information is available – but a line has to be drawn somewhere.

Michael Zohary's researches on the phytogeographical regions of the Middle East and the Arabian Peninsula have, for several decades, been a central source of information (Zohary, 1957, 1973). In Zohary's earlier paper he suggested that the boundary between two phytogeographical regions crossed the Peninsula. To the north lay the Saharo-Sindian region of the central, arid interior; to the south, and essentially the region affected by the Monsoon to some degree, lay the Sudano-Deccanian region (Figure 7.4). In his later work he extended the Sudano- Deccanian region into the Arabian Gulf as a narrow coastal strip as far north as Kuwait.

A more recent assessment of the phytogeography has been provided by White and Léonard (1991). These authors point out that Zohary's scheme needs to be modified to take into account modern conceptions of African phytogeography. In particular, White (1983) has subdivided continental Africa into regional centres of endemism separated by regional transition zones and regional mosaics. White defines a regional centre of endemism as a phytochorion (a plant geographical area) which has more than fifty percent of its species confined to it and more than one thousand

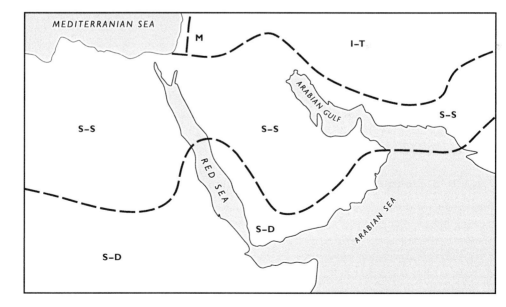

Figure 7.4 Main phytogeographical regions of the Middle East (Modified from Zohary, 1973).
S-S = Saharo-Sindian; S-D = Sudano-Deccanian; M = Mediterranean; I-T = Irano-Turanian.

species in total. Of course, it goes without saying that the whole exercise is subject to taxonomic niceties: new species can just as easily be created by taxonomists as by natural selection. White and Léonard's reassessment, based on the larger woody plants suggests that:

i) Most of the Arabian Peninsula should be regarded as the *Arabian regional* sub-zone of the Saharo-Sindian phytochorion.
ii) The Arabian regional subzone is bordered by a poorly defined narrow zone, along the Red Sea and Gulf coastline in particular, which is part of the Nubo-Sindian local centre of endemism.
iii) The concept of a vast, heterogeneous Sudanian Region should be replaced with a number of less ambiguous regions of endemism. The south-west and southern regions of the Peninsula below about 1,500–1,800 m should now be designated as part of the *Somalia-Masai regional centre of endemism* and represents an impoverished part of an African flora (Figure 7.5).
iv) The high mountains of the south-west of the Peninsula represent an impoverished outlier of the *Afromontane regional archipelago*. This archaepelago of mountain tops in East Africa include: Mt. Kenya, much of the Ethiopian Highlands and the mountains of Somalia (Figure 7.5).

7.3.1.1 Saharo-Arabian Sub-region

This is the largest phytochorion of the Arabian Peninsula. The flora is species poor and is probably derived from an early Tertiary flora distributed around the northern and southern coasts of the Tethys sea. Floral elements include *Anastatica hierochuntica* (Rose of Jericho/Resurrection plant), *Asteriscus pygmaeus, Diplotaxis harra, Helianthemum lippii, Paronychia arabica Savignya parviflora* (Figure 7.6) and *Stipagrostis spp.* The characteristic vegetation is a dwarf shrubland.

7.3.1.2 Nubo-Sindian local centre of endemism

The flora of these very dry formations of scattered trees include species such as: *Aerva javanica, Acacia raddiana* (Arabic: *talh*), *Acacia tortilis* (Arabic: *samur*), the salt-loving shrub *Halopeplis perfoliata* (Figure 7.6) and the grass *Panicum turgidum* (Arabic: *thumam*)

7.3.1.3 Somalia-Masai regional centre of endemism

According to White and Léonard (1991) the overwhelming majority of species occurring below about 1,500 m in the south-west of the Arabian Peninsula also occur on the other sided of the Red Sea and the Gulf of Aden in the original Somalia-Masai region as defined by White (1983). The flora include: *Acacia* species such as *A. asak, A. etbaica, A. hamulosa*, and *Commiphora* species including *C. foliacea, C. gileadensis* and *C. myrrha*. Figure 7.6 shows the distribution of *Euclea racemosa* subp. *schimperi* a leafy tree up to 5 m high which occurs sporadically in the southern Hijaz and Asir mountains. The characteristic vegetation is Acacia-Commiphora bushland becoming more evergreen at higher altitudes.

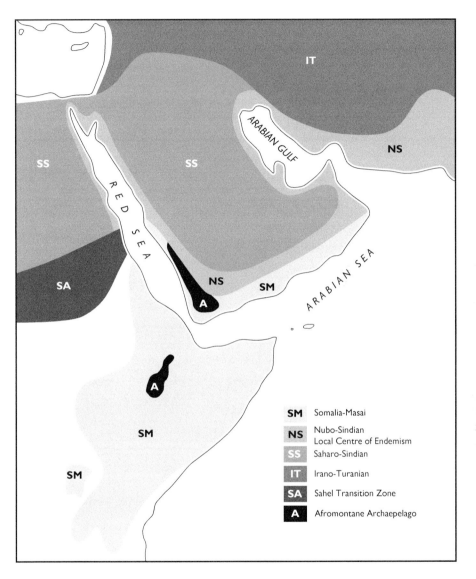

Figure 7.5 Regional phytochoria and local centres of endemism (Modified from White and Léonard, 1991).

7.3.1.4 *Afromontane regional archipelago*

The southern Afromontane flora of the Peninsula is extremely impoverished as compared with its African counterparts. The most abundant Afromontane tree is *Juniperus procera* which grows in abundance on the foggy slopes of Jabal Sauda the highest point in the Kingdom. Other woody species include *Buddleja polystachya*, *Rosa abyssinica* (widespread in the Asir and southern Hijaz), *Erica arborea, Dombeya*

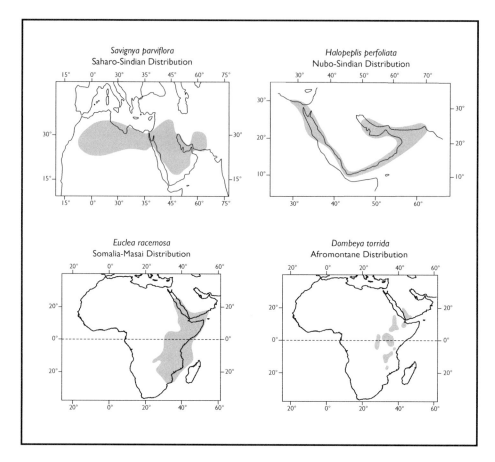

Figure 7.6 Main types of floral distributions involving Saudi Arabia.

torrida (Figure 7.6) and *Myrsine africana* found in scattered localities from Abha southwards.

7.3.2   Tertiary and Quaternary invasions

From the mid-Eocene to the Oligocene, and before the formation of the Red Sea, a Palaeotropical African vegetation extended onto the Peninsula as far as the shores of the Tethys sea on the eastern margin of the Shield. With the progressive withdrawal of the Tethys this vegetation extended eastward. On the coast, mangroves flourished and inland open savannah grasslands developed. By the Late Miocene, about 6 million years ago, the whole region had become arid. This was the so-called Messinian salinity crisis which dried up the Mediterranean basin and Red Sea. During this period of desiccation the tropical African vegetation must have all but disappeared apart from strongholds in the high mountains in the south-west of the Peninsula. From about 3.5 Ma to 1.2 Ma humid wet climatic conditions returned to the Peninsula and the major eastward flowing *widyan* such as Wadi ad Dawasir,

Wadi Birk, Wadi al Batin were reactivated and carried vast amounts of weathered material from the Shield onto the cover rocks. These vast silty valleys acted as invasion routes for a reintroduction of some Palaeotropical elements from their refuges in the Asir and Yemen. Relicts such as *Acacia gerradii* and *Suaeda monoica* would have migrated eastward in this manner.

7.3.3 Patterns of floral endemism

The Peninsula as a whole has about 3,400 species of flowering plants of which about 2,100 occur in Saudi Arabia. Of these only thirty-four or about 2 percent are endemic (Miller and Nyberg, 1991). Endemic species are particularly located in the south-west of the Peninsula in the well-watered highlands of the southern Hijaz, Azir and Yemen (Figure 7.7). Highlands provide a wide variety of ecological niches as well as some degree of ecological stability during periods of aridity and climatic change.

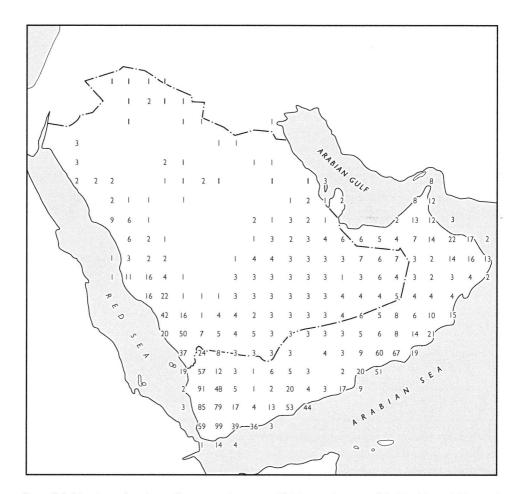

Figure 7.7 Number of endemic flowering plants per 100 km grid square (Modified from Miller and Nyberg, 1991).

Among the well-known endemics confined to the high Asir and not extending eastward into the interior is *Lavandula citriodora*. *Campylanthus pungens* has a similar distribution, though occupying dryer sites, and has a range that extends eastwards through the high Yemen into western Oman. Kürschner (1998) notes that this species has its closest relatives in Africa and presumably isolation as a result of plate movements has given rise to the endemism through speciation. Various species within the genera *Lavandula* and *Campylanthus* exhibit Arabian-Macaronesian range disjunctions. Such taxa have present-day distributions in south-west Arabia and archipelagos off West Africa such as the Azores and Canaries. It is presumed that these genera are part of a Tertiary African stock which became disjunct at the onset of aridity and the development of the Saharan desert.

Endemic taxa in the interior sandy deserts of the Peninsula are not common. Miller and Nyberg cite *Limeum arabicum* and *Calligonum crinitum* ssp. *arabicum* as examples. In the northern Hijaz, Jabal Dibbagh (N27°52', E35°43') is of particular interest as a local centre of extra-tropical endemism in Arabia. Five endemic taxa occur at high altitude on this granite mountain including *Delphinium sheilae* and *Nepeta sheilae*.

7.4   VEGETATION

As Miller and Cope (1996) suggest, the greatest problem in describing the natural vegetation of Saudi Arabia, and indeed the Arabian Peninsula as a whole, is the untold effect of several thousand years of animal husbandry. In more recent times there has also been wide scale destruction of woody species for fuel and the impacts of agricultural development and irrigation.

A number of generalised maps of the Peninsula's vegetation have been published and some are often more of a cartographic puzzle than a source of information. For example, Novikova's map has 14 choropleth classes and forty-one individually numbered vegetation units, none of which are based on any ground survey (Novikova, 1970). A less ambitious map was produced by Frey and Kürschner (1989) and in 1994 the Land Management Department of the Ministry of Agriculture and Water published useful poster-size map a simplified version of which is shown in Figure 7.8.

In this brief survey of the Kingdom's vegetation I shall broadly follow the material provided by Kürschner (1998), Miller and Cope (1996), and Watts and Al-Nafie (2003).

7.4.1   Coastal and Sabkhah vegetation

7.4.1.1   Mangal

Mangrove woodlands, *mangal*, are distributed more or less along the whole Red Sea coast of Saudi Arabia where their areal extent has been estimated to be 200 km². They are much more poorly developed in the Arabian Gulf region where they now occupy less than 4 km² of the Saudi Arabian coast. Ormond *et al.* (1988) suggest that mangroves in the Gulf may have been reasonably common 2000 years ago. Interestingly, it was from the Gulf that mangroves were first reported in the

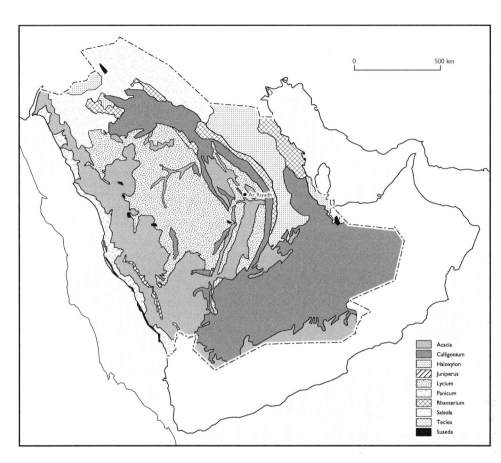

Figure 7.8 Generalized vegetation community map of the Arabian Peninsula (Modified from Ministry of Agricultural and Water vegetation map, 1994).

world literature by Nearchus and Theophrastus over 2000 years ago (Baker and Dicks, 1982). Both on the Red Sea and Gulf coasts the northerly limit seems to be about N 27°.

Of the three mangrove species known in the region *Avicennia marina* is by far the most common and can withstand high salinities (40–50‰) and low temperatures (the occasional frost). The other two species are *Rhizophora mucronata* and *Ceriops tagal*.

7.4.1.2   Sabkhah (Arabic: pl. sibakh)

Saline flats or *sibakh*, occur in three type of situations in Saudi Arabia: coastal, inland and interdune (Barth, 2002). Coastal *sibakh* are widespread along the Red Sea coast, particularly south of Jiddah. They also occur on the Gulf coast, south of Dammam. These coastal saline flats are usually near high tide level and are seasonally flooded. Inland sabkhat usually reflect some fossil hydrological condition. In the Uruq al

Mutaridah in the south-eastern Empty Quarter, inter-dune sabkhat occur between NW–SE trending linear dunes.

Sibakh are almost always devoid of higher plant life because of the very saline waters below the salt crust and the poor drainage. They may, however, have a zone of halophytes around their margins. Even thin patches of blown sand on the sibakh surface can provide improved drainage and suitable conditions for plant growth. Typical halophytic species are: *Zygophyllum qatarense*, *Zygophyllum mandavillei*, *Seidlitzia rosmarinus*, *Halocnemum strobilaceum* and *Salicornia europea*.

7.4.1.3 *Drought-deciduous thorn woodlands and shrublands*

Drought-deciduous thorn woodlands and shrublands often intermixed with xeromorphic grasslands are characteristic of much of the Tihamah coastal plain (Vesey-Fitzgerald, 1955, 1957; Demerdash and Zilay, 1994). In the northern Tihamah, the sand and gravel fans are dominated by *Acacia raddiana*, *A. tortilis* and *Retama raetam*. Further south, the *Acacia* woodlands have a Sudanian affinity and are accompanied by species such as *Belanites aegyptiaca* and *Maerua crassifolia*. On the Gulf coast open xeromorphic grasslands dominated by the perennial tussock grass *Panicum turgidum* are widely developed on well-drained sands and are an important grazing resource. Woody associates include *Calligonum comosum*, *Leptadenia pyrotechnica* and *Lycium shawii*.

7.4.2 Deserts and scarcely vegetated areas

7.4.2.1 *Rock and gravel deserts*

Rock and gravel deserts (*harraat* and *hamadas*) cover a large proportion of Saudi Arabia and are particularly extensive in the western Nadj. The deflation of fine material has left behind a complex mosaic of rock and gravel-strewn surfaces, here and there covered by thin patches of sand. The vegetation is difficult to classify but typically forms open, xeromorphic dwarf shrublands and after rains an annual 'meadow' often develops. Larger trees and shrubs, such as *Acacia* spp., *Lychium shawii* and *Tamarix* spp. are mostly restricted to *wadi* floors. Three broad communities are recognized:

A. *Rhanterium epapposum dwarf shrubland*

This community covers vast areas of northern central Arabia on shallow sand where the soil is well-drained. It is dominated by *Rhanterium epapposum* (*arfaj* shrublands) with *Astragalus* spp. *Fagonia* spp. and *Plantago* spp. These shrublands are of great importance as a grazing resource and the community is almost certainly sub-climax.

B. *Haloxylon dwarf shrubland*

Dwarf shrubland dominated by *Haloxylon salicornicum* (*rimth* saltbush shrubland) is found mainly on shallow sand and gravel plains and occasionally on rocky surfaces where the ground water is fairly shallow and rainfall averages *c*.100–140 mm. The community is found from Kuwait south to the edge of the Rub' al Khali, and on the Arabian Shield along the edges of the *harraat* where it dominates moderately saline, sandy and clayey depressions and plains. Associated species are *Acacia* spp. *Anabasis lacantha* and *Agathophora alopecuroides*.

C.   Anastatica, Anvillea, Belpharis community

The rocky pavements of the cuesta regions of central Nadj are characterized by a sparce vegetation of *Anastatica hierochuntica*, *Anvillea garcinii*, *Blephasis ciliaris* and small grasses such as *Oropetium* spp. and *Tripogon multiforus*.

7.4.2.2   Sand deserts

1.   The Jafurah

The coastal Jafurah desert stretches south from the Dammam region to merge with the Rub' al Khali (Anton and Vincent, 1986). The coastal plain is low lying and in many places the water table is high enough to give rise to extensive *sibakh*. Much of the desert comprises thin sandsheets and sporadic barchan dunes. Active barchan dunes are nearly barren of vegetation but the sandsheets are dotted with wild or naturalised date palms which are dense in the spring-fed oases such as Al Hasa. Tamarix bushes are common, too. The margins of the *sibakh* are often zoned by halophytes. Much of the natural vegetation of the Jafurah has been destroyed by infrastructure developments associated with the oil and gas industry and the phenomenal population growth of the Eastern Region.

2.   An Nafud

The word *nafud* simply means a large sandy area and at 72,000 km², An Nufud (the Great Nafud) is exactly that. Chaudhary (1983) describes the vegetation of An Nafud as comprising a series of overlapping communities, the nature of the community depending mainly on the physical composition and depth of sand. The perennial vegetation comprises open, dwarf shrubland and after winter and spring rainfall desert ephemerals cover much of the area. The main vegetation types are:

A.   ANNUAL VEGETATION COVER

Except for wind eroded areas near the summit of dunes the Nafud bears dense flushes of annual herbs following winter and spring rainfall. Of the 100 or so annual species listed by Chaudhary (1983) the majority exhibit typical desert ephemeral life cycles with rapid germination and quick development and flowering, sometimes after only a few centimetres growth. Others have adopted different adaptations to this uncertain environment. *Neurada procumbens*, for example, carries several seeds, but reportedly only one of these germinates with each rain shower, if the first seedling fails to become established, another one comes up with the next rain shower. The most important annual species in terms of biomass production are listed in Table 7.2.

Table 7.2 Important annual plant species of An Nafud

Anthemis deserti	Lotononis platycarpa
Arnebia decumbens	Medicago laciniata
Astragalus hauarensis	Neurada procumbens
Astragalus schimperi	Paronychia arabica
Cutandia memphitca	Plantago boissieri*
Emex spinosa	Rumex pictus
Ifloga spicata	Schimpera arabica
Eremobium aegyptiacum	Schismus barbatus
Hippocrepis bicontorta	Silene villosa
Launaea capitata	

* Often the most important contributor to the biomass.
Source: Ghazanfar and Fisher (1998).

B. HALOXYLON PERSICUM – ARTEMISIA MONOSPERMA – STIPAGROSTIS DRARII COMMUNITY

This is one of the most characteristic communities of An Nafud and characteristically occupies deep sand on the shoulders of sand dunes and shallow hollows (Schulz and Whitney, 1986a). It nearly corresponds to the *ghada* shrubland described by Mandaville but for the presence of *Artemisia* which is mostly absence further south (Mandaville, 1990). Removal of *Haloxylon persicum* for use as fuel is common practice especially around *badu* encampments, and in such areas only the two other components are seen. Erosion of *Stipagrostis drarii* by sand-laden winds or by heavy gazing may often leave *Artemisia* as the only visible component. Two other perennials associated with this community are *Moltkiopsis ciliata* and *Monsonia heliotropoides*.

C. CALLIGONIUM COMOSUM – ARTEMISIA MONOSPERMA – SCROPHULARIA HYPERICIFOLIA COMMUNITY

This open shrubland is the most widespread community in An Nafud and corresponds to Vesey-Fitzgerald's (1957) 'central Arabian red sand vegetation' and the *'abal-'adhir* sand shrubland described by Mandaville (1990). It is found near the mid-upper parts of dunes and also on undulating sheets of deep sand. Removal of *Calligonium* for fuel is widespread as it is long- and clean-burning. *Moltkiopsis ciliata, Monsonia heliotropoides Stipagrostis drarii, Centropodia fragilis and Cyperus conglomeratus* are commonly associated with this community according to Chaudhary (1983).

3. Ad Dahna sand belt

The vegetation of these red sands is generally similar to the *Calligonium comosum – Artemisia monosperma – Scrophularia hypericifolia* community of An Nafud. After rains, the northern part of the Dhana supports flushes of annuals concentrated at the bases of dunes and in hollows. Baierle and Frey (1986) list the common species in the annual herb layer as: *Eremobium aegyptiacum, Astragalus schimperi, Astragalus hauarensis, Plantago cylindrica* and *Neurada procumbens*.

4. The Rub' al Khali

The vegetation in the vastness of the Empty Quarter is difficult to characterize in terms of a single geographical unit and the MAW (1995c) suggest that it should be considered as two, and probably three, distinct ecological regions: the *Eastern Empty Quarter*, the *Central Empty Quarter* and the *Western Empty Quarter*. These divisions do not quite coincide with the descriptions and map provided by Mandaville (1986) (Figure 7.9).

A number of general characteristics are worth pointing out. First, vegetation is almost omnipresent and it would be wrong to think that the Empty Quarter is such a harsh environment that plant life is absent. Second, there is a close relationship between the vegetation and the sand terrain. Sand actually has good moisture-storing properties and is well drained. In this sense it is a better medium for plant growth than, for example, salt-encrusted *sibakh*. Mandaville (*op. cit.*) suggests that, apart from the barren north-central region the Rub' al Khali carries more plant cover than some of the rocky deserts and gravel plains of northern Arabia. Third, the vegetation is very limited floristically with hardly 10 species of importance, clearly the products of rigorous selection for hyperarid conditions (Table 7.3).

A. CORNULACA ARABICA COMMUNITY

This is a very widespread community marked by the presence of the endemic Chenopodiaceous bush *Cornulaca arabica*. This prickly shrub is an important grazing

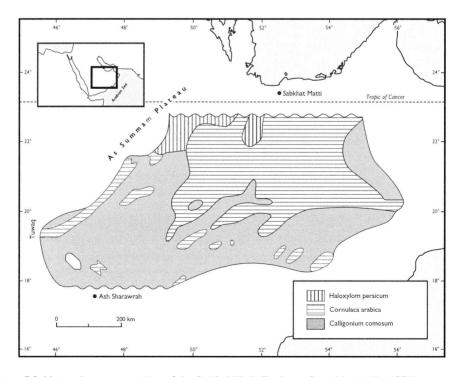

Figure 7.9 Major plant communities of the Rub' al Khali (Redrawn from Mandaville, 1986).

Table 7.3 Dominant plant species in the Rub' al Khali

Cornulaca arabica
Calligonum crinitum subsp. arabicum
Dipterygium glaucum
Cyperus conglomerates
Limeum arabicum
Tribulus arabicus
Haloxylon persicum
Zygophyllum mandavillei
Stipagrostis drarii

component. It is found mainly in the central and eastern region of the Empty Quarter immediately south of Sabkhah Matti. The vegetation in this region is often subject to morning and evening fogs, heavy dews and very erratic rainfall. The annuals as a rule are missing in this area. In the central region *Conulaca* is spaced at distances of 20–100 m or more and is accompanied only by occasional tufts of *Cyperus conglomeratus*.

B.   CALLIGONUM CRINITUM AND DIPTERYGIUM GLAUCUM COMMUNITY

This is a rather poorly defined and variable community found over wide areas where other dominants such as *Cornulaca* or *Haloxylon* are absent. In the Rub al Khali, *Calligonum crinitum* subsp. *arabicum,* is a woody species up to 2.5 m high. It is characteristic of well-drained sands and is often accompanied by *Dipterygium glaucum* and *Cyperus conglomeratus*.

C.   HALOXYLON PERSICUM COMMUNITY

This is a well-defined and conspicuous community that is restricted to the northern and north-west edges of the Rub alKhali. A typical *Haloxylon* stands comprises large woody shrubs up to 3 m high, spaced 10–30 m apart on prominent hummocks. Associates may include *Dipterygium, Limeum, Cyperus,* and *Stipagrostis* and a few sand-adapted, annuals such as *Eremobium* and *Plantago boissieri*.

7.4.3   Montane woodlands and xeromorphic shrublands

Undegraded woodland remnants are now restricted to the more inaccessible parts of the Hijaz and the Asir. In the Hijaz the plains and lower slopes are dominated by *Acacia* spp. shrubland which is often stunted due to the continuous nibbling by goats. Where *Acacia* has been cut down soil erosion is common and the site invaded by *Calotropis procera* (El-Ghani, 1996). South of Makkah an *Acacia-Commiphora* shrubland is dominant and at higher elevations semi-evergreen shrubland replaces drought-tolerant communities. Both in the Hijaz and the Asir there is a reasonable altitudinal zonation of vegetation.

Hijaz – altitudinal zonation (Vesey-Fitzgerald, 1957b). Transect running eastward from a point between Jiddah and Yanbu

1. Tihamah: very open drought-deciduous shrubland with *Acacia tortilis*, *A. asak* and *Maeura crassifolia* (*Acacia-Commiphora* associations are absent from the coastal plain this far north).
2. Western foothills of the Hijaz (up to 800 m): open drought-deciduous shrubland with *Acacia hamulosa*, *A. totilis*, *A. ehrenbergiana* and *Commiphora spp.*
3. Western slopes of the Hijaz (800–1,500 m): drought-deciduous shrubland with *Acacia asak* and *A. etbaica*.
4. Western slopes of the Hijaz mountains (above 2,500 m): open semi-evergreen shrubland with *Olea europaea* subsp. *africana* and *Juniperus phoenicea* (*Juniperus* usually occurring only on isolated peaks).
5. Mountain plateau (not exceeding 1,500 m): very open drought-deciduous woodland with *Acacia asak*.
6. Lower altitudes on the eastern slopes of the mountains: scattered trees of *Acacia tortilis*.
7. Dry eastern slopes of the Hijaz (down to *c*.1,000 m): scattered trees of *Acacia tortilis* and *Maerua crassifolia*.

Asir – altitudinal zonation (König, 1986)

1. Tihamah (0–250 m): Acacia-Commiphora drought-deciduous woodland with *A. ehrenbergiana*, *A. tortilis*, *A. mellifera*, *A. oerfota*, *A. hamulosa*, *Commiphora myrrha* and *C. gileadensis*.
2. (250–400 m): *Acacia tortilis*- *Commiphora* drought-deciduous woodland with *Commiphora myrrha*, *C. gileadensis*, *C. kataf*, *Euphorbia cuneata*, *E. triaculeata*, *Acacia hamulosa*, *A. mellifera*, *Maerua crassifolia*, *Cadaba longifolia* and *Dobera glabra*.
3. (400–1,100 m): *Acacia asak*– *Commiphora* drought-deciduous woodland with *Commiphora myrrha*, *C. gileadensis*, *Euphorbia cuneata*, *Dobera glabra*, *Moringa peregrina*, *grewia cillosa* and *Acacia tortilis*.
4. (110–1,350 m): *Acacia asac* drought decididous woodland.
5. (1,350–1,600 m): *Acacia etbaica* drought deciduous woodland.
6. (1,600–2,000 m): sclerophyllous scrub with *Acokanthera schimperi*, *Teclea nobilis*, *Pistacia falcata*, *Tarchonanthus camphoratus*, *Aloe sabaea*.
7. (2,000–2,400 m): *Juniperus-Olea* forest with *Juniperus procera*, *Olea europaea* ssp. *africana*,*teclea nobilis*, *Celtis africana*, *Rhus retinorrhoea*, *Maesa lanceolata*.
8. (above 2,400 m): *Juniperus procera* woodland with *Acacia origena*, *Dodonaea viscose* and *Euryops arabicus*.
9. (2,650–2,500 m): *Juniperus procera* or *Acacia origena* (around terraced cultivation) woodland.
10. (2,450–2,300 m): low shrubland with *Euphorbia schimperiana* and *Lavandula dentata*.
11. Below 2,300 m the vegetation thins and eventually merges into dwarf desert shrubland.

The Juniper woodlands of the high Asir are the only dense woodlands in the Kingdom and are of particular importance from a conservation point of view. Of particular concern is the widespread decline of *Juniperus phoenicea* and *Juniperus procera* on exposed lower slopes with some stands comprising mostly dead or dying trees. Fisher

(1997) has made a special study of the problem at the Raydah Reserve and puts forward four hypotheses for the decline juniper woodlands in Arabian Peninsula as a whole. All four hypotheses involve climatic changes though operative at different spatial and temporal scales.

Hypothesis 1

Overgrazing by domestic livestock has altered the local vegetation structure, causing a decline at lower altitudes due to changes in microclimate

Hypothesis 2

The global temperature rise in the twentieth century with elevated spring temperatures in the Middle East (Nasrallah and Balling, 1993) is causing woodland decline through temperature-induced dieback at the lower juniper ecotone.

Hypothesis 3

Dieback is caused by periodic droughts combined with long regeneration cycles, the effects of which are more marked at lower, hotter, elevation.

Hypothesis 4

The present arid phase, which began between 4,000 and 6,500 years ago (Sanlaville, 1992) is still developing, causing woodland dieback through long-term increased in aridity.

7.4.4   Wadi communities

Widyan are one of the most important plant habitats in the Kingdom. Their floors are often filled with well-drained silts, sands and gravels which may hold considerable amounts of moisture. Kürschner (*op. cit.*) describes *wadi* vegetation as azonal. This is because they make it possible for some species to penetrate into otherwise harsh environments and indeed, the major *widyan* of the Nadj in particular were important plant migration routes eastwards towards the Gulf as Tertiary and Quaternary sea levels fell.

Three types of *wadi* community are recognized:

The first type of community is found in the large *widyan* of the central part of the Kingdom (e.g. Wadi ar Rimah, Wadi al Batin, Wadi Hanifa, Wadi as Sabha, Wadi ad Dawasir and Wadi al Hamd). The dominant indicator species are scattered trees of Saharo-Sindian origin: *Acacia gerrardii, A. raddiana, A. tortilis,* with an under storey of *Astragalus spinosus, Chrysopogon plumulosus* and *Cymbopogon commutatus* as the main associates.

A second type is found in *widyan* of the northern and north-western parts of the Kingdom where the climate is not so arid. Here the indicator species are shrubs of Mediterranean and Irano-Turanian origin such as *Dyerophytum indicum, Nerium oleander* and *Pteropyrum scoparium*.

A final type is associated with *widyan* in the south-west of the Kingdom dominated by plant species of Sudanian and xero-tropical African origin such as *Acacia ehrenbergiana, Salvadora persica, Hyphaene thebica, Tamarix* spp. (König, 1986). El-Demerdash and Zilay (1994) also note the presence of *Pandanus odoratissimus* and *Phoenix dactylifera*.

Chapter 8

Soils and soil erosion

Contents

8.1   INTRODUCTION

Apart from exhumed saprolites and laterites, most soils in Saudi Arabia are young, and to the untrained eye there seems little evidence of pedogenic development. Indeed, who would think of dune sand as a soil? To the pedologist, however, soils are simply a medium in which plants can grow. From the very moment that some alteration of the parent material takes place, perhaps by the accumulation of organic material or the removal of soluble salts during a downpour, soil development can be said to have started – no matter how slight. In the case of Saudi Arabia, soil horizon development due to the downward translocation of soluble minerals, fine particulates, and organic material is limited by the dearth of rainfall and the sparsity of vegetation cover. Even some upward translocation of salts occur because of the intense evaporation and shallow water tables in coastal areas and *wadi* floors. In the scorching heat of the desert, soil development is forever being halted as wind erosion and wind deposition repeatedly disturb profile development.

Soil is a precious resource and particularly in an arid country such as Saudi Arabia and soil surveys have been part and parcel of the agricultural expansion that took place from the 1960s onwards. Several reconnaissance soils surveys involving the survey and profile descriptions were carried out in the 1960s as part of the hydro-agricultural assessment of the Kingdom by the Ministry of Agricultural and Water. Beginning in 1966 soil surveys were undertaken in areas showing the most potential for irrigated agriculture and by 1981 as *Schematic Soil Survey Map of the Kingdom of Saudi Arabia* was completed at the scale of 1:2,000,000 (Land Management Staff, 1981). In the 1980s the Land Management Department of the Ministry of Agriculture and Water (MAW) conducted high intensity surveys of *widyan* catchments such as Sahba, Hanidh, Urayi'rah, Jabrin, Aflaj, Saqiua Khatmna, Idima, Sharqa and Hanifa. There have also been surveys associated with small project areas investigated for different types of development but most of these reports are not in the public domain. However, to fulfil the needs of other government agencies the MAW conducted a low density soil survey for the whole of the Kingdom in the early 1980s and published the details in a splendid atlas entitled *General Soil Map of the Kingdom of Saudi Arabia* (1985).

The primary reason for the production of the *General Soil Map* was to distinguish agricultural from non-agricultural land and eighty-three percent of the Kingdom was surveyed for this purpose. Only inaccessible parts of the Empty Quarter were left out of the survey. The mapping methods of the survey were essentially those of the National Cooperative Survey of the United States of America. Mapping of the Kingdom was done on LANDSAT images at a scale of 1:250,000 with most of the photo interpretation being done in the field. Apart from mountainous terrain and major dune fields, mapping coverage was obtained by linear transects 10 to 40 km apart. Field teams, with pickup-mounted power augers sampled soils at 2 to 12 km intervals along each transect. At selected sites, pits were also dug and the soils profiles classified and described in detail.

The system of classification adopted for the *General Soil Map* is that described by the United States Department of Agriculture (1975) in *Soil Taxonomy*. The system relies to a large extent on the recognition of *diagnostic horizons*. In the context of

Saudi Arabia six diagnostic surface or subsurface horizons are regarded as important. These are:

> *Calic horizon*: a subsurface horizon that has accumulated calcium carbonate or calcium-magnesium carbonate. Calciorthid and some Gypsiorthid soils have a calcic horizon. (from the Latin *calxis*, lime).
>
> *Cambic horizon*: a subsurface horizon altered to one or more of a) redder or greyer colours, b) obliteration of rock structure, c) in some soils, the formation of soil structure, d) the removal of most of the carbonates. Camborthid, Eutrochrept and Haplaquept soils have a cambic horizon. (from the Latin *cambiare*, to change).
>
> *Gypsic horizon*: a non-cemented or weakly cemented horizon enriched with secondary sulphates. Gypsiorthids which do not have a pan have a gypsic horizon. (from the Latin *gypsum*).
>
> *Petrogypsic horizon*: a cemented subsurface horizon enriched and cemented with sulphates. Gypsiorthids which have a pan have a petrogypsic horizon. (from Greek *petros*, rock, and the Latin *gypsum*).
>
> *Salic horizon*: a subsurface horizon which has a secondary enrichment of salts more soluble in cold water than gypsum. Salorthids have a salic horizon. (from the Latin *sal*, salt).
>
> *Ochric horizon*: a light coloured surface horizon that is medium to low in organic matter. It is hard or massive when dry. The major soils in Saudi Arabia have this kind of surface layer. (from the Greek *ochros*, pale).

The *Soil Taxonomy* is a divisive classification, soils being progressively assigned to orders, suborders, great groups, subgroups, families and series. For the most part, the *General Soil Map of the Kingdom of Saudi Arabia* classifies soils as far down as great groups.

A unique feature of the *Soil Taxonomy* is its nomenclature which seems daunting at first but is quite logical and based on Latin or Greek roots, as noted above. Soil orders are assigned a descriptive syllable which provides a clue to the genesis of the soil. In Saudi Arabia three soil orders are distinguished: *Entisols* (*ent*, taken from the word 'recent'); *Inceptisols* (*ept*, from the Latin *inceptum* meaning beginning); *Aridisols* (*id*, from the Latin *aridus*, meaning dry). Suborder and great group names are built up in a similar manner.

8.2   ENTISOLS

Entisols show little or no evidence of horizon development. The upper surface is an ochric epipedon. In the Kingdom, entisols are found on actively eroding slopes, especially in the Asir highlands, the sand deserts, and alluvial terraces in *widyan*. Three suborders and five great groups have been mapped (Table 8.1).

8.2.1   Psamments (*psamm* – from the Greek *psammos* – sand)

Two great groups are recognised – Torrispsamments and Udipsamments.

Table 8.1 Types of entisol in Saudi Arabia

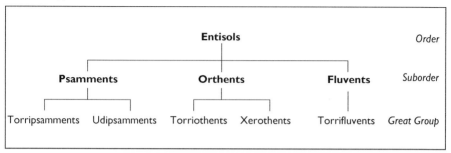

8.2.1.1   Torripsamments

Torripsamments (torri, from the Latin *torridus*- hot and dry) are entisols which have formed on well-sorted dune sands (Figure 8.1). They are mostly non-saline and deep. Torripsamments are widespread in Saudi Arabia and occur extensively in the Rub' al Khali and An Nafud but also occur where thin sand covers rock outcrops. The profile details of a typical Torripsamments are shown in Table 8.2.

8.2.1.2   Udipsamments

Udipsamments (udi, from the Latin *udus*, meaning humid) are best developed on low-lying coastal sands where the water table is high, often within 50 cm of the surface, more or less throughout the year. They are strongly saline, highly permeable, soils and have a very high concentration of salts in the upper few centimetres which form a dry crust. These crusts are, however, not thick enough to qualify as a salic horizon. In the Kingdom, Udipsamments occur along the Tihamah and Gulf coastal plains and are

Table 8.2 Profile characteristics – Torripsamment

Physiography: rolling sand plain 2 percent slope

Parent material: aeolian sand

Location: N19°40', E45°05'

Remarks: An area of sandy deep soils, 0–5 percent slopes Soil slightly moist from18 to 90 cm, desert pavement of coarse sand and fine gravel. Soil horizons arbitrarily separated for soil sampling

C1 – 0 to 18 cm; brownish yellow (10 YR 6/6) coarse sand, single grained; loose; common fine roots; slightly effervescent, very slightly saline, diffuse smooth boundary

C2 – 18 to 90 cm; brownish yellow (10 YR 6/6) coarse sand; single grain; loose; common fine roots; slightly effervescent, very slightly saline, diffuse smooth boundary

C3 – 90 to 100 cm; brownish yellow (10 YR 6/6) coarse sand; single grain; loose; common fine roots; slightly effervescent, very slightly saline, diffuse smooth boundary

Source: Modified from MAW (1985).

Figure 8.1 Typical Torripsamment profile (Photo: MAW).

Table 8.3 Profile characteristics – Udipsamment

Physiography: depression in coastal plain; slope <1 percent

Parent material: water reworked aeolian sand

Location: N26°45', E49°57'

Remarks: In an areas of Udipsamments and Torrifluvents; marine flats and Dunes. No mottles; water table at 80 cm. Electrical conductivity of Water in pit was 118 mmhos/cm.

Cz1 – 0 to 5 cm; very pale brown(10 YR 7/3 – moist) loamy sand; single Grained; loose, strongly saline (EC 250 mmhos/cm); abrupt smooth boundary.

Cz2 – 5 to 30 cm; brown(10 YR 5/3 moist) loamy sand; single grain; loose; Strongly saline (32 mmhos/cm); gradual smooth boundary.

Cz3 – 30 to 65 cm; very pale brown (10 YR 7/4 moist) sand; single grained; Loose; moderately saline (EC 8 mmhos/cm); gradual smooth boundary.

Cz4 – 65 to 100 cm; yellowish brown (10 YR 5/4 moist) loamy sand; Single grained; loose; strongly saline (EC 74 mmhos/cm).

Source: Modified from MAW (1985).

mapped as a minor Association along with Torripsamments. The profile details of a typical Udipsamment are shown in Table 8.3.

8.2.2   Orthents (*orth* – from the Greek *orthos* – true or common)

Two great groups have been mapped in the Kingdom, Torriorthents and Xeroorthents.

8.2.2.1   *Torriorthents*

Torriothent soils are mostly developed on colluvial and residual parent materials often on actively eroding slopes and in material resistant to weathering. Some also develop on coarse river terrace alluvium. Torriorthents are mostly shallow, loamy and skeletal. They range from non-saline to saline types. One type, the Lithic Torriorthent, is very stony and is developed on agricultural terraces on the steep hillsides of the Asir and to a less extent the Hijaz (Figure 8.2). Torriorthents are widely developed on steep slopes throughout the Kingdom but their greatest extent are the steep hillsides of the Shield.

8.2.2.2   *Xerorthents*

This soil group has a *xeric* moisture regime. The Xerorthents are commonly deep loamy soils on agricultural terraces constructed to collect water running down slope. In favourable situations enough organic matter accumulates to develop a mollic epipedon – a dark, friable and generally fertile horizon (Latin *mollis*, soft). Xerorthents are not extensive in Saudi Arabia.

Figure 8.2 Torriorthent soil profile (Photo: MAW).

8.2.3   Fluvents (*fluv* – from the Latin *fluvius* – a river)

8.2.3.1   Torrifluvents

These soils develop on alluvial deposits associated with intermittent streams. In profile the soils appear stratified due to deposition during times of flood (Figure 8.3). On broad flood plains, such as those in the Tihamah, the stratification is difficult to detect by eye. Torrifluvents are mostly deep soils and are usually non-saline. They have a *torric* moisture regime.

Figure 8.3 Profile of torrifluvent soil (Photo: MAW).

Table 8.4 Classification of Inceptisols

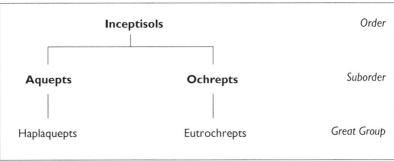

8.3 INCEPTISOLS

Inceptisols show only minor alteration of the parent material and usually have a cambic horizon. They exclude soils having a hot, dry *torric* moisture regime. In Saudi Arabia inceptisols are not extensive. Two great groups have been mapped, both of which typically occur in topographically low positions where the water table is high (Table 8.4).

8.3.1 Aquepts

8.3.1.1 Haplaquepts

Haplaquepts are poorly drained soils. They are developed in deep sandy loams and loams which have the water table at or near the surface unless drained. They are usually quite strongly saline.

8.3.2 Ochrepts

8.3.2.1 Eutrochrepts

Eutrochrepts are better drained that haplaquepts but the water table can still be as high as 75 cm from the surface. They are developed best in loams and sandy loams and are usually quite saline.

8.4 ARIDISOLS

As the name suggests these soils are dry and have little moisture available for mesophytic vegetation for long periods of time. All aridisols in Saudi Arabia have an aridic moisture regime except the Salorthids which are have intermittent high water table regimes and develop a salic horizon. Orthids are the dominant aridisol in the Kingdom although small areas of argids also occur. Four great groups have been mapped (Table 8.5).

Table 8.5 Major types of Aridisols in Saudi Arabia

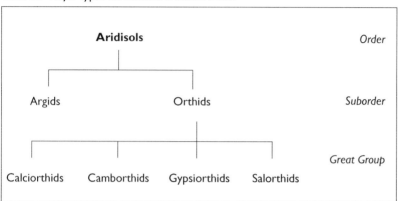

8.4.1   Calciorthids

These Aridisols have accumulated secondary carbonates in a calcic horizon that has its upper boundary within 1 m of the surface (Figure 8.4).

Rainfall is generally not sufficient to leach carbonates from the upper 18 cm of the soil. As a result, these soils are calcareous down to the lower limit of the calcic horizon and sometimes beyond. The calcium carbonate equivalent is usually about 15 to 40 percent. Calciorthids are generally loamy but where developed over lavas can be quite stony. They are the most widely developed Aridisol in the Kingdom and are found extensively on the calcareous cover rocks of the sedimentary Nadj and occasionally on the eastern Shield. Table 8.6 shows the profile characteristics of a Typic Calciorthid near Jabal Tuwaiq.

8.4.2   Camborthids

These Aridisols have a cambic horizon and in Saudi Arabia they are deep loamy or sandy loamy soils. Although they are not extensive in the country as a whole they are the most important soil type in the Tihamah.

8.4.3   Gypsiorthids

These soils have a gypsic horizon within 1 m of the soil surface. In Saudi Arabia, many Gypsiorthids also have a calcic horizon above the gypsic layer. They are usually loamy and saline. As a group, these soils are widely scattered throughout the Kingdom and are particularly extensive in the Eastern Region. (Table 8.7 and Figure 8.5)

8.4.4   Salorthids

Salorthids have a salic horizon. They are associated with wet basins where capillary rise and evaporation draw salts to the salic horizon. In times of rain the soils are often

Figure 8.4 Calciorthid soil profile (Photo: MAW).

flooded and they frequently develop a salty crust on drying. Salorthids are often deep soils and have clayey and clay loam textures. They are found throughout Saudi Arabia but are particularly common on the coastal flats along the Gulf coast. Some profile features are described in Table 8.8.

Table 8.6 Profile characteristics – Typic Calciorthid

Physiography: nearly level plain. 2 percent slope

Parent material: old alluvium

Location: N23°05', E46°58'

Remarks: In an area of Calciorthids, 0 to 5 percent slopes;
loamy-skeletal, deep, saline soils, 100 percent of surface covered with gravel

Akz – 0 to 3 cm; light yellowish brown (10 YR 6/4). Moderate to fine subangular. Blocky structure; few fine roots; violently effervescent; calcium carbonate equivalent to 30.7 percent; about 5 percent gravel; neutral; abrupt wavy boundary

Bkz1 – 3 to 28 cm

Bkz2 – 28 to 52 cm

Bkz3 – 52 to 90 cm; red (2.5 YR 4/8) very gravelly sandy loam; weak coarse subangular blocky structure; very hard; violently effervescent; calcium carbonate equivalent to 15.2 percent; strongly saline; about 45 percent gravel; neurtral; diffuse smooth boundary

Bkz3 – 90 to 160 cm; red (2.5 YR 4/6) very gravely loam; weak coarse subangular structure; very hard; strongly effervescent; strongly saline; about 45 percent gravel; neutral

Source: Modified from MAW (1985).

Table 8.7 Profile characteristics – coarse-loamy Gypsiorthid

Physiography: nearly level plain. 1 percent slope

Parent material: old alluvium

Location: N24°08', E49°10'

Remarks: In an area of Gypsiorthids, loamy soils; gypsym pan; 0 to 3 percent slope

Akz – 0 to 18 cm; reddish yellow (7.5 YR 6/6); fine sandy loam; moderate fine subangular blocky structure; soft; few fine roots; violently effervescent; calcium carbonate equivalent to 20 percent; strongly saline; mildly alkaline; abrupt wavy boundary

Bkyz 1 – 18 to 30 cm; strong brown (7.5 YR 5/6); sandy loam; moderate fine subangular blocky structure; hard; few fine roots; violently effervescent; calcium carbonate equivalent to 14 percent; strongly saline; mildly alkaline; clear smooth boundary

Bkz 2 – 30 to 80 cm; reddish yellow (7.5 YR 6/6); sandy loam; moderate fine subangular blocky structure; hard; few fine roots; violently effervescent; calcium carbonate equivalent to 20 percent; strongly saline; mildly alkaline; clear smooth boundary

Bkyzm – 80 to 160 cm; white (10 YR 8/2) loamy sand; massive; weakly cemented; violently effervescent; calcium carbonate equivalent to 28 percent; strongly saline; moderately alkaline.

Source: MAW (1985).

Figure 8.5 Loamy Gypsiorthid (Photo: MAW).

Table 8.8 Profile characteristics – fine Salorthid

Parent material: alluvium

Location: N25°17', E45°37'

Remarks: in an area of sabkhah Salorthids; loamy and clayey, deep saline
soils; 0 to 1 percent slope; no mottling observed; soil moist below
8 cm and saturated below 120 cm. Horizon breaks for sampling arbitrary from 8 to 120 cm

Az – 0 to 8 cm; very pale brown (10 YR 7/4) silty clay; massive; slightly hard; slightly effervescent;
calcium carbonate equivalent to 13 percent; strongly saline, visible salt crystals present throughout;
mildly alkaline; abrupt smooth boundary

Bz1 – 8 to 20 cm; dark yellowish brown (10 YR 4/4) silty clay; medium subangular blocky structure;
friable; violently effervescent; calcium carbonate equivalent to 27 percent; strongly saline; mildly
alkaline; diffuse smooth boundary

Bz2 – 20 to 50 cm; dark yellowish brown (10 YR 4/4) silty clay; medium subangular blocky structure; friable; violently effervescent; calcium carbonate equivalent to 31 percent; strongly saline;
mildly alkaline; diffuse smooth boundary

Bz3 – 50 to 85 cm; dark yellowish brown (10 YR 4/4) silty clay; medium subangular blocky structure; friable; violently effervescent; calcium carbonate equivalent to 36 percent; strongly saline;
mildly alkaline; diffuse smooth boundary

Bz4 – 80 to 120 cm; dark yellowish brown (10 YR 4/4) silty clay; medium subangular blocky structure; friable; violently effervescent; calcium carbonate equivalent to 38 percent; strongly saline;
mildly alkaline; diffuse smooth boundary

Source: MAW (1985).

8.5   SOIL MAP ASSOCIATIONS

For mapping purposes the *General Soil Map of the Kingdom of Saudi Arabia* has
employed the concept of the map unit which represents an area of landscape and
comprises one or more soils at the great group taxonomic level. This method avoids
the confusion that can arise in areas of complex topography where the extent of individual soil types may be limited and impossible to map. In homogenous terrain the
map units comprise one soil group – these map units are known as *consociations*. In
more complex terrain the map unit will comprise several soil types. This type of map
unit is called an *association*. The *General Soil Map* identifies forty-nine map units
but for the purposes of this chapter, and following the practice of the Ministry of
Agriculture and Water, these have been grouped into 8 major soil associations and
1 consociation (Figure 8.6).

The general characteristic of the Kingdom's soil associations are shown in
Table 8.9.

8.5.1   Calciorthid-Camborthid association

The Calciorthid-Camborthid association is found mainly on gently undulating plains
on the western edge of the sedimentary Nadj (Figure 8.6). The Association has a
complex pattern where crossed by *wadi* channels and comprises about 50 percent
Calciorthids, 30 percent Camborthids and 20 percent other soil types. These deep

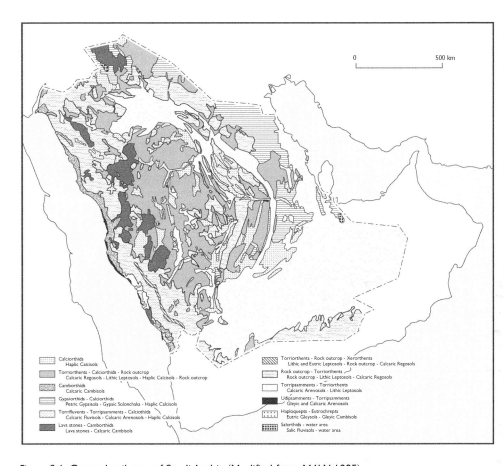

Figure 8.6 General soil map of Saudi Arabia (Modified from MAW, 1985).

loamy soils are usually associated with rangeland but where groundwater is available they are suited to large scale irrigation. Camborthids are also important on the Tihamah.

8.5.2 Calciorthid-Torripsamment association

The Calciorthid-Torripsamment association has been mapped mainly on sedimentary rocks of southern central Saudi Arabia north of the Rub' al Khali. The deep loamy soils are best developed on alluvial plains, broad *wadi* terraces and thin sand sheets over alluvium. In remote areas these soils are under rangeland but elsewhere are extensively irrigated.

8.5.3 Calciorthid – rock outcrop – Torriorthent association

This is the most extensive soil association in Saudi Arabia and has been mapped over vast areas of the eastern Shield and crystaline Nadj (Figure 8.6). The Calciorthids

Table 8.9 Properties of major soil associations

Soil association	Topography	Hydraulic conductivity*	Available water capacity**	Salinity***	Land use suitability
Calciorthids-Camborthids (deep, loamy arable soils)	nearly level plains and terraces	Calcio: moderate Cambor: moderate	moderate moderate	slightly to moderately saline slightly saline	especially large scale irrigated agriculture; rangeland
Calciorthids-Torripsamments (deep loamy, sandy, arable soils)	alluvial plains; terraces; low dunes	Calcio: moderate Torrip: rapid	moderate low	slightly to moderately saline slightly saline	high potential irrigated agricultural; rangeland
Calciorthids-Rock outcrop-Torriorthents (deep, loamy, arable soils and rock outcrops)	alluvial plains; fans and interfluves	Calcio: moderate Torrio: moderate	moderate moderate	slightly to strongly saline do	rangeland; limited irrigated agriculture
Gypsiorthids-Calciorthids-Torripsamments (non-arable plains and sibakh; soil deep arable soils)	plains, footslopes; sand sheets; sibakh	Gypsio: moderate Calcio: moderate Torrip: rapid	moderate to low moderate low	slightly saline do do	rangeland; small areas suitable for irrigation
Torripsamments (non-arable dunes)	sand sheets; low to steep active dunes	Torrip: rapid	low	slightly saline	rangeland; unsuited for irrigated agriculture

Salorthids-Torripsam-ments (*non-arable sibakh and saline soils*)	nearly level *sibakh*; *wadi* floodplains	Sal: slow to very slow Torrip: rapid	low low	strongly saline slightly to moderately saline	rangeland; severe limitations for ir-rigated agriculture
Calciorthids-Rock Out-crops (*non-arable saline soils and rock outcrop*)	nearly level to strongly slop-ing, gravelly and cobbly plains	Calcio: moderate	moderate	strongly saline	rangeland
Rock Outcrop-Calciorthids-Torriorthents (*rock outcrop and loamy non-arable soils*)	footslopes and *widyan*	Calcio &Torrio: moderate	moderate	slightly to strongly saline	rangeland; potential for small scale irrigation

Source: Based on Ministry of Agricultural and Water (1984).

*Hydraulic Conductivity Class	Downward movement cm/hr
Rapid	>4
Moderate	0.04 to 4.0
Slow	<0.04

**Class	Available Water (cm per cm of soil)
Low	<15.0
Moderate	15.0 to 22.5
High	22.5 to 30

***Class	Electrical Conductivity mmhos/cm
Slightly	<8
Moderate	8–16
Strongly	>16

are located in the interfluve areas and footslopes and are sometimes intricately mixed with exposed bedrock. The Torriothents are often found on large gravel fans where *widyan* debouch from mountainous confines. This association mainly supports rangeland. There is some irrigated agriculture where rock outcrops are not too numerous.

8.5.4   Gypsiorthid – Caliorthid – Torripsamment association

A vast swathe of the Eastern Region from the Kuwait/Iraq border south to the sands of the Empty Quarter is mapped as an association of Gypsiorthid, Calciorthid and Torripsamment soils (Figure 8.6). The topography is one of nearly level and gently sloping plains, low hills, *sibakh* and sand sheets. Although some elements of the association are suitable for irrigated agriculture they often have a complex pattern in the landscape and are non suitable for large scale irrigated agriculture. The Gypsiorthid component is moderately deep and loamy but is highly saline.

8.5.5   Torripsamment consociation

The Torripsamment consociation consists of sand dunes, undulating sandy plains and eroded rock outcrops. As can been seen from Figure 8.6 this soil unit covers practically the whole Rub' al Khali basin and the Nafud desert. On the sand sheets and sand dunes drainage is excessive and there is little possibility of irrigated agriculture apart from in some interdune corridors. In readily accessible areas the soils are used for rangeland.

8.5.6   Torripsamment – rock outcrop association

This Association differs from the Torripsamment consociation in that it has some 25 percent rock outcrop. Agricultural potential is very low.

8.5.7   Salorthid – Torripsamments association

In the south-east of the Kingdom inland *sibakh* and rolling sand sheets are mapped as a Salorthid – Torripsamment association. The Salorthid unit limits the possibility of any form of agriculture unless the salt is first leached out. The vast Sabkhah Matti, south of Qatar and the inland Sabkhat Badur south-east of Yabrin are mapped as this association.

8.5.8   Calciorthid – rock outcrop association

This association comprises nearly level to strongly sloping, gravely plains with bedrock interfluves. It is best developed on the Precambrian rocks of central Shield area where the drainage pattern is composed of endless small *widyan*. Nearly all this association is used for rangeland. The rock outcrops and saline nature of the soil prohibit agricultural development of any kind.

8.5.9   Rock Outcrop – Calciorthid – Torriorthent association

The Rock Outcrop-Calciorthid-Torriorthent association is dominated by rock outcrops which comprise about 65 percent of the cover. It is found in two large belts of country. One running along the Red Sea Escarpment, the other on the escarpments of Jabal Tuwaiq (Figure 8.6). Most of the association is used as rangeland but some irrigated agriculture is carried out. The major limitations are steep slopes and rocky outcrops. Flooding is a problem on the *wadi* floors.

8.6   SOIL MINERALOGY

Relatively little is known about soil mineralogy over vast areas of Saudi Arabia. Pedogenic processes in such an arid environment are slow and the clay minerals are likely to be inherited from the parent rocks. At the continental scale, three papers are of interest and provide much basic information. The mineralogy of soils in the Asir region between Makkah and Abha has been described by Abu-Husayn *et al.* (1980). Viani *et al.* (1983) have dealt with the central alluvial basins – particularly the Wadi ad Dawasir and Wadi Najran areas and Lee *et al.* (1983) have examined the mineralogical properties of soils along a 500 km east-west transect of the eastern region from Riyadh to Damman.

In the Asir, mineral soils are rich in kaolinite regardless of parent material. This presumably implies some inheritance from former hotter, moister, conditions perhaps before the Shield uplifted and when it was further south. Laterites and saprolites are widely exposed where not protected under a cover of lava. In many *widyan*, the loessic alluvial silts are also derived from tropical weathering products and are rich in kaolinite.

Soils of the central alluvial basins described by Viani *et al.* (1983) are higher in smectites than elsewhere in Saudi Arabia. These authors suggest that the smectites have been transported into the basins from the Asir highlands rather than having formed *in situ*. The accumulation and concentration of smectites in *wadi* sediments in arid and semi arid zones is widely observed. However, deep weathering products are widespread in the Asir and the genesis of smectites as suggested by Viani *et al.* needs to be examined further.

The mineralogical variations among the eastern Saudi Arabian soils are due mainly to the differences in the sedimentary rocks. The clay mineral assemblage comprises palygorskite, kaolinite, mica and smectite. Lee *et al.* (*op. cit.*) suggest some interesting trends in the clay mineral composition with palygorskite dominating the assemblage in the central zone and declining towards Jabal Tuwaiq and the Arabian Gulf (Figure 8.7). The palygorskite is believed to have been derived from the Mio-Piocene sedimentary rocks. Lee *et al.* indicate that palygorskite is not present in Cretaceous or older sedimentary rocks and that palygorskite in the soils of the Jabal Tuwaiq area may have been inherited from younger rocks which have now been eroded away. Not all palygorskite is detrital. El Prince *et al.* (1979), for example, report the pedogenic development of palygorskite in oases soils at Al Hasa.

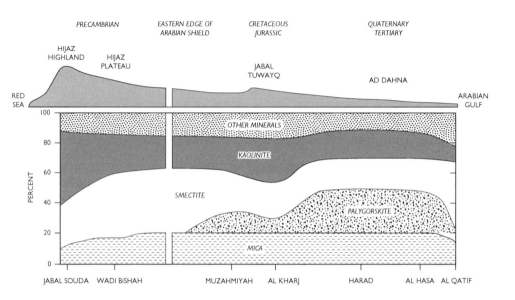

Figure 8.7 Distribution of clay minerals in a cross section of the Kingdom (Modified from Lee *et al.*, 1983).

8.7   PALAEOSOLS

Scattered through the literature are accounts of palaeosols which provide hints as to former pedogenic conditions. There has been little systematic research on these fossil soils apart from an interesting short monograph by Shogdar (1998) who studied the palaeosols on the Tihamah between Jiddah and Rabigh.

8.7.1   Late Cambrian saprolites

Deeply weathered saprolites, the base of lateritic profiles, are exposed on the Precambrian Shield where they have been exhumed at the outcrop edge of both the Saq sandstones of the northern Shield and their counterpart, the Wajid sandstones, in the south. The extent of the saprolitic cover is not known. In the field the saprolites are sometimes mottled and up to 3 metres thick. An upper lateritic zone was probably eroded away as fluvial systems forming the Lower Palaeozoic sandstones eroded the landscape. The saprolitic formation is related to an extensive period of etchplanation of the Nubian-Arabian Shield.

8.7.2   Tertiary lateritic horizons (Oxisols)

There is some terminological confusion regarding the use of the term laterite and much depends on whether or not one is a geologist or pedologist. Here, I assume that a lateritic profile at one time comprised a basal saprolite leading up through a mottled zone into a lateritic horizon. In general, laterization is characterized by silica being leached from the system as well as the common mobile ions. In addition, organic accumulation is inhibited. The effect of these processes is the accumulation of high concentrations of hydrated oxides of iron and aluminium. In some cases the silica combines with available aluminium to form kaolinitic clays but where silica is more extensively depleted by leaching the excess alumiunium is usually present as gibbsite $Al(OH)_3$ or more rarely boehmite $(AlO(OH))$. Strictly speaking, the term laterite should be reserved for a nodular or continuous layer or oxides or hydroxides or iron and/or aluminium. From a pedogenic point of view such horizons usually develop in the upper profile of Oxisols. Between the lateritic layer and the parent material is a clay rich zone. The upper part of this zone is often mottled with iron oxide; the lower part, known as the pallid layer is bleached and has low iron content. In the description below, the term laterite is used for the whole profile following the geological usage of Overstreet et al. (1977).

Red, yellow and white laterite underlies a thick section of basalt in the as Sarat Mountains between N17°45'–N18°20' and E43°00'–E43°30'. The area covered by the laterite is about 1,000 km². The mountains lie close to the Red Sea escarpment, between Abha and Najran, and have altitudes generally ranging between 2,200 and 2,400 m.

The laterite was discovered by Harry St. J.P. Philby in November 1936 and samples were sent to the British Natural History Museum. A misunderstanding of the Museum's analyses of the samples led Philby to think the as Sarat site was an enormous bauxite deposit which he calculated to be about 7,000 million tons. He published the erroneous details in his book *Arabian Highlands* (1952, pp. 382–383). Since Philby's discovery there have been a number of investigations of the site by the United States Geological Survey.

The laterite lies above Precambrian crystalline rocks and beneath Tertiary basalts. The presence of laterite-basalt outliers indicates that the *harra* was considerably more extensive than it is now and it is clear that erosion of the escarpment from the west has removed significant parts of the outcrop. Indeed, some geologists think that as much as 2/3rds of the *harra* has been eroded since the Late Tertiary. Overstreet et al. (*op. cit.*) conducted a mineralogical investigation of the laterites and they report that gibbsite and boehmite (the bauxite minerals) are absent from the profile and that kaolinite and quartz are present.

Brown (1970) reported that potassium-argon isotopic ages for the As Sarat basalts range from 29 My at the base to 25 My at the top and given the generally uneroded junction between the laterite and the basalt it is suggested that the laterite was still forming at the time the lower basalts were erupted, implying a Late Oligocene age. Overstreet et al. (1977) suggest that the laterites in the as Sarat indicate a relatively warm wet climate during this period. The Precambrian rock surface on which the laterite has developed appears to be essentially planar because the base of the laterite is everywhere without marked relief. Laterites of approximately the same age underlie Tertiary basalts in the Yemen, Ethiopia and the Sudan. It appears that during and before the Late Oligocene the Red Sea area experienced little tectonism, had little relief and had a warm tropical climate. Unless protected by basalt widespread erosion of the lateritic cover would have taken place as the Red Sea rift developed.

Of course, the laterite development could be older than Oligocene. A hint of this comes in the description by Collenette and Grainger (1994) who, in describing the 8.5 m thick Az Zabirah bauxite deposit suggest it is part of an early Cretaceous *in situ* paleolaterite profile developed on the profound unconformity between the mid-Cretaceous sandstones of the Wasia Formation and the Triassic Minjur Formation. The Az Zabirah deposit (N27°56', E43°43') lies between 580 and 650 m asl. and is part of the cuesta region of central Saudi Arabia. The former extent of the laterite is not known since we do not know how much Permo-Triassic cover has been removed from the eastern Arabian Shield nor do we know the former extent of the Cretaceous transgression. However its unconformity with the Permo-Triassic rocks suggests a good deal of erosion.

A recent investigation of the as Sarat laterites was undertaken by Roobol *et al.* (1999) as part of an investigation of the palaeosol's gold bearing potential. Roobol *et al.* concluded that the laterite is early Miocene and had little gold content (Figure 8.8).

Figure 8.8 Laterite developed on Precambrian rocks under Harrat as Sarat (Photo: author).

8.7.3 Late Pleistocene palaeosols

Anton (1984) notes the presence of thin palaeosols developed on older dune sands which are presently covered by mobile dunes in the Dahna sand field west of the town of Khurays. He describes a thin upper palaeosol of about 0.2–0.3 m in which there are calcified root channels. A lower palaeosol is thicker, about 0.3 0.4 m and is better developed. Anton suggests that the lower palaeosol developed in the Late Pleistocene humid phase, 36,000–17,000 BP when dunes were stabilized by vegetation growth. El Prince *et al.* (1979) in their study of the soils of the Al Hasa oasis also remark on the presence of palaeosols under sand dunes. These palaeosols may correspond with those in the Dhana to the west. Several palaeosols marked by organic horizons are visible in quarry sections in *wadi* sediments north of Ad Darb in the southern Tihamah. They are almost certainly Holocene but await dating.

Schmidt *et al.* (1981) in their study of gold placer deposits of the Jabal Mokhyat area, southern Najd, noted the presence of a weakly developed buff grey soil with

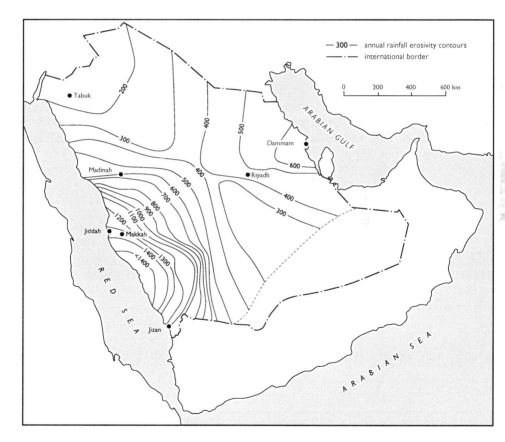

Figure 8.9 Annual rainfall erosivity (Modified from Mohammad and Abo-Ghobar, 1992).

humic layers developed in loessic silts. A ^{14}C date of 5,930 ± 300 BP for charcoal from near the top of the loessic silt suggests that some soil development was underway during the Holocene humid phase identified by McClure (1976).

8.8   SOIL EROSION

Soil erosion by both wind and water is widespread in Saudi Arabia to the extent that less than 10 percent of the area is actually erosion-free. However, little of this is apparently man-induced accelerated erosion (MAW, 1995). This conclusion does, however, seem a little strange given the vast amount of overgrazing in the Kingdom. The MAW also suggest that erosion is not much of a problem in cultivated areas which are either under a crop or are in some way protected from wind and runoff. In fact, there are few data on soil erosion and given the amounts of dust in the atmosphere it seems likely that erosion is generally serious where crop cover is not permanent. In the Asir, where villages are becoming deserted and young people are leaving the land for a life in the city, the system of agricultural terraces is slowly being abandoned. Walls have collapsed, soil has washed away into *widyan* and once fertile slopes are denuded of soils.

Soil erosion by water is a function of both rainfall erosivity and land erodibility (Hudson, 1981). Erosivity is an aggregate measure of rainfall intensity. Mohammad and Abo Ghobar (1992) have calculated erosivity indices (R) for 12 stations (Table 8.10). These data, supplemented by a further 8 stations enabled the authors to establish the following regression relationship between the annual erosivity index (R) and the average annual rainfall (P).

$$R = 286.15 + 7.08 \ P$$

This relationship was then used by the Ministry of Agriculture and Water to produce a rainfall erosivity map (Figure 8.9). This must be highly speculative, however, as there is no direct relationship between annual precipitation and erosivity since rainfall intensity was ignored. Further details regarding soil erosion and salinization can be found in the next chapter.

Table 8.10 Rainfall erosivity indices for selected stations

Station	Yearly R	Monthly R (MJ mm ha⁻¹ hr⁻¹)											
		Jan	Feb	Mar	Apr	May	Jun	Jul	Aug	Sep	Oct	Nov	Dec
Tabuk	160.1	3.2	7.0	11.5	18.5	–	–	–	–	–	3.0	21.0	95.9
Najran	569.9	6.0	13.9	75.9	238.2	11.5	–	52.9	21.8	32.4	130.6	13.7	–
Bishah 1	2647.1	270.9	229.2	471.3	659.2	367.5	6.5	65.1	132.5	0.5	46.1	132.0	206.5
Bishah 2	2187.4	0.8	9.2	125.5	1317.4	40.7	–	–	136.0	–	–	22.6	532.2
Buraydah	656.2	49.3	39.3	39.6	64.4	496.8	–	–	–	–	–	–	5.1
Riyadh	220.2	24.6	11.9	45.7	114.9	18.3	–	–	–	–	–	–	4.8
H.Sudair	396.1	34.7	12.1	92.5	25.7	72.4	–	134.4	–	–	27.2	86.6	30.5
Jiddah	1941.9	108.0	10.6	10.6	31.2	3.3	297.9	934.3	63.9	4.6	128.3	76.8	272.2
Malaki	4081.5	182.2	169.6	130.5	153.1	302.0	20.0	415.6	1186.9	392.1	603.1	331.9	195.6
Kwash	2384.1	65.1	56.0	114.9	40.6	267.5	235.6	190.4	470.5	561.6	250.5	36.8	94.7
Taif	1103.4	36.1	25.8	364.4	76.3	186.7	1202	–	28.8	35.6	165.0	20.4	44.3
Madinah	561.8	29.7	33.0	34.6	40.8	87.7	–	230.7	0.5	–	73.9	18.4	12.5

Source: Mohammad and Abo-Ghobar (1992).

Chapter 9

Environmental impacts and hazards

Contents

9.1   INTRODUCTION

In this chapter I shall try to provide a general overview of various environmental impacts and hazards in the Kingdom. Perceptions of the Saudi Arabian environment are not normally associated with geohazards, pollution or even general environmental degradation. But with a burgeoning population, and a sensitive environment prone to many rapid, and sometimes dangerous disturbances, more and more attention is being paid to hazard detection and mitigation. Much of the research in this area is undertaken by the Saudi Geological Survey and recently the Survey produced an important geohazards map for the Kingdom at a scale of 1:3,000,000 (Al-Rehaili *et al.*, 2002), and the accompanying explanatory booklet contains an excellent bibliography compiled by Roobal and Bankher (2002).

9.2   GEOHAZARDS

9.2.1   Blowing sand and dunes

Blowing sand and mobile dunes are a particular hazard in the Eastern Province of the Kingdom where both the Dahna and Jafurah dune belts are traversed by important road and rail links. The mobile barchans of the Jafurah sand sea are also a threat to the vast Al-Hasa oasis with its hundreds of spring-fed date farms, and also the petrochemical installations and airfields of the coastal zone from Jubail southward. The northerly winds in the Jafurah, especially in the *shamal* season create immense sand control problems and it has been reported than annual drift rates reach 30 m³/m width (Fryberger *et al.*, 1983, 1984). Not surprisingly, this conflict between man and nature has spawned a good deal of pioneering research on techniques of sand and dune management dating back to the classic work by Kerr and Nigra (1952).

Watson (1985, 1990) has written thorough reviews of the sort of preventative measures that can be undertaken to minimize the potential hazard. He suggests that there are four main approaches:

1.   Promotion of sand deposition upwind of the problem area

This is usually achieved by the construction of vegetation barriers and fences. In this regard, a good deal of research has been undertaken at the Research Institute of the King Fahd University of Petroleum and Minerals on the design and placement of fences (Figure 9.1). Notwithstanding the sophistication of plastic fences designed as a result of wind tunnel experiments it seems that fences made from palm branches are quite effective and also ecofriendly. Vegetation barriers have been successfully established in the Al-Hasa oasis using non-irrigated *Tamarix* trees.

2.   Enhancement of sand transport within the problem area

To prevent sand accumulation on embanked roads, the camber can be adjusted so that the surface slopes gently into the wind roughly at the same slope angle as the windward slope of a barchan. At important sites surfaces can also been treated with chemicals such as synthetic latex and asphalt. The creation of a smooth surface

Figure 9.1 Sand fence control experiments conducted by the Research Institute, King Fahd University of Petroleum and Minerals (Photo: author).

discourages the lodgement of sand particles in cracks which might absorb some of their kinetic energy as they saltate.

3. Reduction of the sand supply

Where the hazard is severe the surface can be covered with a lag of gravel which blankets the surface thus reducing deflation. It is worth bearing in mind, however, that there can be considerable environmental disturbance when heavy equipment is used to lay the gravel. A less environmental friendly technique is to spray on a chemical stabilizer or even crude oils or asphalt. The biological impacts of such treatments are not known but are probably adverse.

4. Deflection of the moving sand

Fences with low porosity can be used to deflect moving sand though, as Watson (*op. cit.*) notes, they are not advocated over large areas or along roads. Fences can either be aligned at a deflecting angle of about 45° to the moving sand or arranged as a V-shaped barrier pointing into the wind. On the whole, both types of fence ultimately encourage sand accumulation and need to be cleared regularly.

9.2.2 Dust storms

Dust in the atmosphere can have many environmental impacts apart from the obvious reduction in visibility. Middleton (1986) cites its effect on climatic change, air pollution, and crop growth but in addition, industrial complexes requiring clean air, and vehicle and jet engines can all be affected. Middleton's analysis of meteorological data, mainly from airport sites, shows that Saudi Arabia experiences dust storms throughout the year although there is some seasonality. Jiddah, for example, has about 5 dust storms per year mainly delivered by northerly winds in March and April, whilst Jizan's dust storms are mainly during the summer months when the Monsoon is active.

Behairy et al. (1985) have examined dust in the coastal area north of Jiddah and conclude that mineralogically they are dissimilar to fine sediment in the near shore environment. In particular, they show that the coastal dust contains high concentrations of heavy metals, particularly cadmium which, they suggest, is derived from cement factories north of Jiddah and by local oil combustion. Apparently much of the dust is deposited during periods of rainfall and the high levels of cadmium in the dust, up to 36 ppm, pose a potential health hazard.

9.2.3   Subsidence and collapsing loessic sediments

The formation of earth fissures and subsidence due to excessive pumping of ground water has been noted by several researchers (Roobol et al., 1985; Al-Harthi and Bankher, 1999; Bankher and Al-Harthi, 1990). Roobol et al. studied the village of Tabah on Harrat Hutaymah near Ha'il. The village had a population of about five hundred and was situated in an ancient volcanic crater whose vent is filled with sands and gravels eroded from tuffs. The central zone of the crater is filled with loessic silts and supports a number of farms, mainly growing dates. The vent fill is a good aquifer and local road building schemes outside the crater margin in the early 1980s drew heavily on this water supply lowering the water table from 50 to 120 metres below the surface. As a result of this excessive draw down ground shrinkage, subsidence and earth fissuring developed and the village was moved to a safer site just outside the crater rim (Figure 9.2).

A similar story is told by Al-Harti and Bankher (op. cit.) in their studies of earth fissuring and subsidence in Wadi Al-Yutamah some 75 km south of Madinah. Here, many of the farms have been abandoned due to the excessive lowering of the water table. The authors suggest that the earth fissures in the wadi were mainly developed as a result of land subsidence which was caused by the compaction of the loessic sediments and the formation of fissures over bedrock highs. Later the fissures were opened and enlarged at the surface by flooding and possible further hydrocompaction.

9.2.4   Karst collapse

Sub-surface and surface karstic solution features are widespread in the Eastern Province particularly in the younger cover rocks. Cavernous weathering and dolines, often hidden by fills of blown sand, form particular hazards in both the Arab and Hith Formations where the karst is developed in anhydrite and gypsum, and also in the chalky limestones of the Dam, Dammam, Rus and Umm er Radhuma Formations (Amin and Bankher, 1997).

The hazard is now more important than in the past for two reasons. First, the rapid urban development in the Eastern Region has meant that buildings are larger

Figure 9.2 Earth fissures in loessic silts leading to building collapse, Tabah crater (Photo: author).

and foundations have to be much stronger. Amin and Bankher (*op. cit.*) describe cave systems under a Saudi Aramco site north of Dhahran in which solution chimneys extend to within 1 m of the surface. Second, the off-road coastal zone in the Kuwait border, is frequently used for army training with heavy vehicles and tanks. The detection of the shallow karst cave systems and sand-filled dolines has become important in order to avoid serious accidents. John Roobol, William Shehata and Ian Stewart of the Saudi Geological Survey have recently conducted extensive research into the problem (Figure 9.3).

9.2.5 Anhydrite dissolution

Beds of anhydrite ($CaSO_4$) are very common in the Upper Jurassic sequences of the Interior Homocline and their solution poses serious geotechnical problems. Under field conditions anhydrite normally hydrates first to gypsum which dissolves by dissociation

$$CaSO_4 + 2H_2O \rightarrow CaSO_4 \cdot 2H_2O$$
anhydrite + water $\rightarrow$ gypsum

Figure 9.3 Karst collapse doline near the Kuwaiti border probably due to the lowering of the local water table by farmers (Photo: John Roobol).

When buried beneath 200–300 metres or more of overburden most gypsum is converted to anhydrite. The dehydration of gypsum to anhydrite occurs at pressure of $18–75 \times 10^{-5}$ Pa and results in a volume reduction of about thirty-eight percent (Ford and Williams, 1989). Overlying strata are often brecciated as a result of this sulphate volume reduction. Hydration of anhydrite probably occurs within 100 metres or so of the surface and occurs usually as a very irregular front. What happens next is a matter of debate and there are doubts regarding the efficacy of expansion. One suggestion is that the loading of the newly created gypsum causes its rapid deformation and flow giving rise to a tightly corrugated or small scale block-faulted topography.

Sharief *et al.* (1991) have made a study of these formations in central Saudi Arabia and have compared rock sequences in the Riyadh area, where anyhdrite beds have been dissolved, with complete sequences recorded further east at Khushum al Halal. They come to the remarkable conclusion that about 270 m of anhydrite has been removed by solution in the Riyadh area. This seems to have caused intensive brecciation, minor folds and fractures in Arab and Hith Formations as well as in the overlying Sulaiy Formation. The west-facing escarpment of the Salaiy Formation is extensively slumped and has migrated eastward as a result of the dissolution of the underlying anhydrite. To the south-east of Riyadh, solution of the Hith Formation anhydrite has given rise to the spectacular sink hole at Dahl Hit (N24°29', E47°00') which is about 120 m deep and 60 m wide at its mouth. The brecciation and weathering of limestones caused by the anhydrite collapse pose particular problems for the development of high rise buildings and deep basement structures in the Riyadh area. Adding further to the problem is the fact that the dissolution of the anhydrite releases large amounts of calcium sulphate into the groundwater, and sulphate attack on cement structures is well known. The most significant damage appears to be the reaction of the calcium aluminium hydrate in the cement with the sulphate ions in the shallow groundwater giving rise to a volume increase due the formation of ettringite (Cooke *et al.*, 1982).

Similar sinkhole development and collapse occur in the al Uyan area near Al Kharj some 40 km south-east of Dahl Hit. Other impressive examples of anhydrite solution occur at Al Aflaj (N22°10', E46°45') where solution of the Hith anhydrite lead to the development of a series of lakes , the largest of which is some 400 m long and 150–200 m wide. The lake water was clear but very bitter. The lakes are now drained.

9.2.6   Halite dissolution

Rock salt diapirs occur both on the Dammam coastal zone of the Gulf and also in the southern Tihamah and Farazan Islands in the southern Red Sea. Erol (1989) provides an interesting account of the effect of one such diapir on the buildings in the town of Jizan. Much of the old town is built on the diapir, which is about 50 m asl. and some 4 km² in area. Surrounding the diapir there are extensive coastal *sibakh* which have severely restricted the growth of the town.

The Jizan diapir pierced the overlying country rock of the Baid Formation in the Late Tertiary/Early Quaternary as evidenced by raised corals. The irregular top of the salt outcrop lies within 6–15 m of the surface and is capped by sand and silts (Figure 9.4). Halite or rock salt has a high solubility (360 kg/m³ at 10°C) and numerous buildings in the city have been damaged beyond repair as a result of dissolution of the diapir by circulating ground waters. The dissolution appears to be erratic and on the surface is marked by bowl shaped depressions. To confirm that building damage was the result of dissolution Erol conducted an experiment by flooding part of the foundations of a disused factory with 50 m³ of water. At this site he estimated the cover of sand and silt to be about 10 m thick. After four weeks the surfaces had been

Figure 9.4 Salt diapir just beneath the surface doming up sands and silts – Jizan Hill surmounted by Turkish fort (Photo: author).

lowered by about 8 mm. At the end of that period 12 mm of rain fell and in week five of the experiment the surface had lowered some 16 mm in total. Dissolution processes in Jizan are still active and the irregularity of the solution is possibly due to the pattern of domestic effluent discharge and variation in the thickness of the overlying windblown Quaternary sediments. Jizan, like many towns in the Kingdom has no piped domestic sewerage system and controlling waste water disposal is the only real solution to the problem.

9.2.7 Flash floods

Flash flooding is a serious problem in south-west Arabia and many hundreds of lives have been lost over the last few decades. The reasons for the flooding are a combination of both climatological and pedological conditions. First, in the steep catchments of the high escarpment in the Asir, there is very little soil to soak up water and so infiltration is very limited and runoff is almost immediate. Second, warm moist air rising over the escarpment, particularly in the Monsoon season, induces high intensity rain often associated with super cells which produce spatially concentrated precipitation.

One of the few detailed studies of a flash flood is that of the 1982 flood affecting Wadi Dellah which flows southwestward down the escarpment from Abha to Ad Darb (Ministry of Communications, 1992). The upper section of the *wadi* is more than 2,000 m asl. And a further 55 km down *wadi* has an altitude of about 200 m before it runs into the sands of the Tihamah. The storm which generated the runoff was recorded by a gauge at Abha and had a total duration of 15.3 hours and an accumulated precipitation of 166.6 mm. Of some importance is the fact that the day before the storm some 15 mm of rain fell over a period of 10.5 hours evidently flooding any cracks and joints in the rocks of the watershed.

The flash floods caused extensive damage and many bridges and sections of road were swept away (Figure 9.5). Bridge damage seems to have been due to three causes. Overflow of the deck, boulder and debris impact on columns and underscoring of the foundations. In some critical parts of the *wadi* sediment transport seems to have been akin to a mudflow due to the presence of very large boulders. Although the *wadi* was not gauged, estimates of discharges during the flood even were obtained by measuring slopes and the heights of flood debris above the *wadi* floor. The maximum discharge in the main section of the *wadi* was estimated at 3,715 m³/s – in excess of the average discharge of the River Nile!

9.2.8 Earthquakes

Although it is sometimes perceived that the Arabian microplate is virtually aseismic, there have been twenty-two earthquakes recorded in the Kingdom with intensities in the range IV–IX on the Modified Mercalli Intensity (MMI) scale during the period 627–1884 AD. The MMI scale is a non-instrumental scale and is assigned according to its perceived effects, such as damaged to buildings and geological effects. Most of the epicentres of large MMI events are concentrated in the Gulf of Aqaba region and along the central axis of the Red Sea (Ambraseys *et al.*, 1994; Poirier and Taher, 1980; Roobol *et al.*, 1999) (Figure 9.6).

Figure 9.5 One of many road bridges destroyed during the 1982 flash flood in Wadi Dellah (Photo: author).

To judge by the amount of recorded damage over a large area, at least three very large earthquakes are thought to have occurred in the last 1,200 years: September 873 AD; 11th March 1068 AD and 4th January 1588 AD. The earthquake of 873 AD had its probably epicentre near Tayma but damage was widespread and reported from both Makkah and Madinah. The earthquake which struck on the morning of March 18th 1068 reportedly killed some 20,000 people and had its epicentre between Tayma and Aqaba. Damage was reported in Madinah and on the coast at Yanbu. Three new springs issued in Tabuk and the port of Elat in Israel was completely flattened. The large 1588 earthquake affected the whole of the Sinai Peninsula and was strongly felt in Cairo. In the Kingdom it was felt as far south as Makkah, and in Tabuk, the castle collapsed.

Since the 1980s a national network of thirty-six seismic stations has been installed and run by King Saud University, six of which are located near the Gulf of Aqaba. In addition, two local seismic networks monitor seismic activity around Madinah and Makkah. At 04:15 on the 25th November, 1995, the national seismographic network was overwhelmed by a major earthquake which occurred in the Gulf of Aqaba along a major left-lateral strike-slip not far off shore from Ash Shurayh. Theoretical calculations indicate an average dislocation of 3 m over a length of some twenty five km at a depth of fourteen km. The surface-wave and body-wave Richter magnitudes were estimated at 7.2 and 6.2, respectively. A very strong series of aftershocks followed the main earthquake and persisted throughout the rest of the year. Groundwater radon

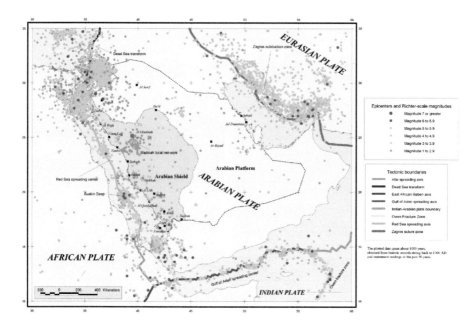

Figure 9.6 Seismic activity map constructed by the Saudi Geological Survey (Courtesy: Saudi Geological Survey).

emissions peaked at about 500 Becquerels per m³ about one hundred and ten hours after the main shock.

Generally speaking, the reinforced concrete in the modern buildings in the towns withstood the tremors very well but in the traditional adobe brick buildings in the rural areas there was more structural damage. Only one person is reported to have died as a result of the earthquake and that was because of the collapse of a poorly constructed un-reinforced outside wall of a traditional house in Al Bad near Haql.

From a disaster relief perspective the experience of tremors as far south as Madinah are of some importance given the fact that during the *hajj* hundreds of thousands of pilgrims visit the city, and there is minimal public awareness of the effects of earthquakes. Furthermore, it is quite likely that any local emergency plans would be completely overwhelmed if vast numbers of pilgrims were in the vicinity during an earthquake event.

Although the damage to buildings in Madinah was minimal, the tremors lasted for about 15 seconds and many people ran out into the streets when hanging lights started to sway and pictures fell off walls. The tremors were particularly felt in the Al Awali and Quba'a districts. The city is mostly located in a gravel-filled *wadi* between two lava tongues of the Harat Rahat where the shallow water table can be tapped. Roobol *et al.* (1999) have described very nicely the relationship between the local geology and the zonation of the effects of the tremors (Figure 9.7).

The field effects of the earthquake are summarized in Table 9.1 and an isoseismal map of MMIs for the 1995 earthquake is shown in Figure 9.8. The trends shown

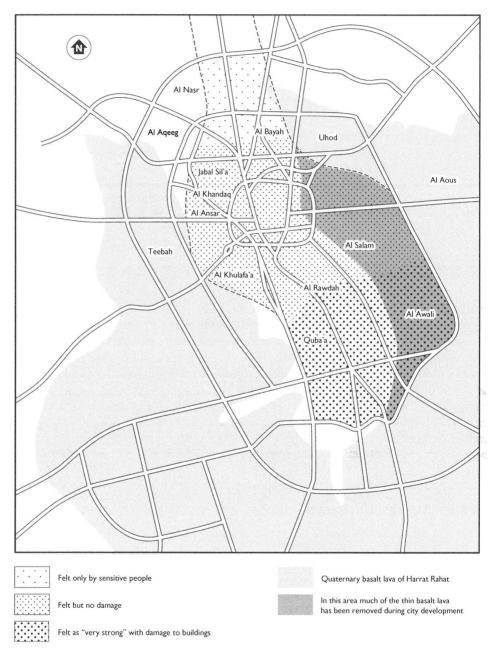

Legend:

Felt only by sensitive people

Felt but no damage

Felt as "very strong" with damage to buildings

Quaternary basalt lava of Harrat Rahat

In this area much of the thin basalt lava
has been removed during city development

Figure 9.7 Relationship between local geology and tremor intensity in Madinah. Gulf of Aqaba earthquake, 22nd November, 1995 (Modified from Roobol *et al.*, 1999).

Table 9.1 Field Effects of the Gulf of Aqaba Earthquake, 22nd November 1995

Modified Mercalli intensity	Comments
Less than III	Some residents of Umm Lajj and Al Wadj thought they had felt a shock in retrospect
III–IV	Minor tremors felt in Tayma, Madinah and Yanbu
V	Damage to buildings in Halat Ammar and Tabuk
VI	Observed at Al Bad. Many buildings damaged but none collapsed. Some shear cracks observed
VII	Rock falls on cliffs behind Coastguard Station at Tayyib al Ism. Development of earth fissures and some damage to reinforced-concrete buildings
VIII	11 km rupture zone developed 30 km south of Haql. Cumulative uplift estimated at between 75 to 90 cm. Presence of small sand volcanoes and development of a small tsunami causing sea level to rise 1m above high-tide mark

Source: Roobol *et al.* (1999).

probably reflect the location of damage assessment studies in, what is, a sparsely populated area rather than the strike-slip direction of the causative fault.

Earthquake risk maps have recently been produced by Al-Hadad *et al.* (1994) and Al Amri (1999). These maps place the Jizan-Abha, Jiddah and Gulf of Aqaba regions in the highest risk category. In the case of Jiddah, the Kingdom's second largest city, the risk probably needs to be mitigated by the fact that most buildings are new and built to a fairly high standard. On the other hand, the city's foundations are almost totally on Tertiary and Quaternary sediments and the water table is high. In such circumstances it would be prudent to enforce a seismic-resistant building code and develop proper emergency plans especially since Jiddah is the main entry point for most *hajjis*.

9.2.9   Rising saline water tables

Rising water tables are an inherent problem in many urban areas in Saudi Arabia. The reason is simple. Growing cities such as Riyadh and Jiddah import vast quantities of water for domestic and industrial use and much of this is then disposed of directly into the ground. Most of Riyadh's drinking water, for example, is piped hundreds of kilometres from desalination plants on the Gulf coast. Jiddah has its own vast desalination plants. Neither city has a fully integrated drainage or sewerage system and both liquid and solid waste is added to the groundwater on a large scale although massive investment will soon put this right. In the northern parts of Jiddah the water table is now only 30 cm from the ground surface and is drawn up to the surface along walls and Tarmac roads whose materials act as wicks.

The water table in the Jiddah region slopes westwards, from the low Precambrian foothills fringing the Tihamah, towards the coast. There is little fresh water recharge and the groundwater is very saline. In places, particularly near the coast, the water table is now within 60 cm of the ground surface and large areas of north

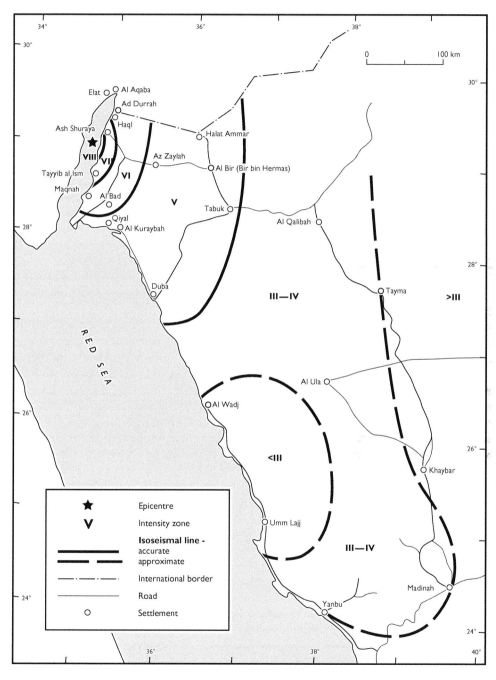

Figure 9.8 Isoseismal map of the Gulf of Aqaba earthquake 22nd November, 1995. (Modified from Roobol *et al.*, 1999).

Jiddah have damp salty efflorescences readily visible. These shallow, salty basins are actually urban *sibakh*. Evaporation concentrates the salts at the surface and, if not washed away by rain water or floodwater, they accumulate and develop a crust often several centimetres thick. This usually overlies damp muddy sediments which have relatively little load bearing capacity. This makes them completely unsuitable for construction purposes, as does the very high concentration of corrosive salts (Erol, 1989). There are few plants capable of growing in such saline conditions other than a few salt marsh species and *sibakh* are formidable barriers to urban growth. Rising water tables and the development of new *sibakh* in Wadi Fatimah immediately to the southeast of Jiddah has ruined much farmland.

In general, the high water tables in the Jiddah region have little to do with proximity to the sea or sea level rise. Neither have water tables risen due to increased recharge by precipitation or increased natural *wadi* flow both of which make negligible contributions to the groundwater. Basamed (2001) has researched this in some detail and has examined data for surface depths to water table based on sixty-one wells in the Jiddah region between 1996 and 1999. Notwithstanding ephemeral fluctuations in water table depth after infrequent rainfall, Basamed concludes that the average water level rise was about 0.37 m yr⁻¹ during this period. Dramatic as this estimate is, it is considerably less than the alarming figure of 0.5 m yr⁻¹ suggested by Abu-Rizaiza *et al.* (1989).

Regardless of which figure is correct, water tables are rising, and relatively quickly. It is generally suggested that the real culprits accounting for most of the rise are the absence of an effective sewerage system, leaky drinking water pipes/tanks and effluent lakes in the foothills to the east of the city. The rapid rise in the urban population and consequent household water use is clearly relevant, too. However, these reasons might not be the whole picture. Both immediately north and south of Jiddah air photograph and satellite surveys have revealed a remarkable growth in the extent of *sibakh* over the last fifty years (Qari and Shehata, 1994). Quite what has made these salty areas expand is open to question but if global warming and sea level rise have played some part then Jiddah's situation is extremely serious. Clearly, in these areas of new *sabkah* growth the influence of Jiddah's water and waste disposal in raising the water table can be ruled out as a cause since they are well beyond any urban influence. It is also unlikely that the relative sea level rise and *sabkhah* growth is due to subsidence since the coastal plain, as a whole, has been a zone of uplift in the Quaternary and several raised marine terraces can be traced along the coastal zone. Increased deflation is another possible process which might have removed unconsolidated sediments and exposed the salty water table. Whatever the causes the sea level changes in the Jiddah region need to be monitored carefully.

Given the rapid growth in the urban population it is not at all surprising that septic tanks and cesspits in Jiddah deliver considerable quantities of waste water to the water table in those situations where they still remain above the water table. Abu-Rizaiza (1999) provides some interesting comments in relation to this recharge. He calculates that the total urban water consumption is about 267,759 m³ per day. This is based on estimates of the amount of desalinated water produced. He then suggests that some twenty percent of this water leaks from the septic tanks, corresponding to a 55,550 m³ contribution to the groundwater (Bayumi *et al.*, 2000). This is almost certainly an underestimate because the desalinated water distribution pipes themselves are known to leak as much as thirty percent of the volume water carried. In addition there are unknown recharge contributions from the irrigation of public parks, roadside verges and leakage from domestic underground water storage tanks. More recently there has also been a contribution to the groundwater recharge from a very large effluent lake to the east of the city although precise amounts are unknown.

Sabkhah development is a serious problems facing urban development in north Jiddah. These damp surfaces are encrusted with salts and unless special precautions are taken severe building damage can ensue. House walls can act as wicks and draw groundwater vertically up several metres where the salts crystallize and have enough compressive energy to burst open bricks and plaster. Using reinforced concrete for the walls is not a solution since the iron rods themselves succumb to chemical attack and simply rot *in situ*. Adversity really is the mother of invention in Jiddah's building trade and local house builders now often lay a course of bitumen coated bricks near ground level to stop the capillary rise of salt laden water. Whether this turns out to be effective only time will tell.

The groundwater in Jiddah is generally high in sulphates (SO_4), averaging about 1,400 mg/L. The sulphates react with normal Portland cement bondings of concretes to form expansive ettringite. When the expansive stresses become greater than the concrete's tensile strength, it begins to crack thus allowing even more sulphate-rich water to enter.

One simple solution to sulphate attack seen on many Jiddah building sites is the use of a carpet of polythene sheeting under concrete foundations but its effectiveness is probably only temporary and the method is reliant on the sheeting remaining waterproof under load. On high cost buildings expensive geotextiles are sometimes used in place of plastic sheeting. These have better load bearing properties and are designed as to be permeable in a downward direction only.

More permanent, but costly, solutions against sulphate attack require the use of special Portland-Pozzoland cements (Portland cements blended with reactive silica). The concrete surfaces are then wrapped in stiff plastic sheeting and coated with bitumen. Iron reinforcing rods are also protected from corrosion by a coating of epoxy resin (Figure 9.9). All these safeguards must add considerably to the cost of building construction and although the materials themselves are not expensive the whole operation is quite time consuming.

Anyone driving through residential neighbourhoods such as Al Marwah or Al Nahfah in north Jiddah immediately south of King Abdulaziz International airport, cannot but help notice how rutted many of the road surfaces are. At its worst the road surfaces are buckled into linear troughs and ridges, particularly on the edges of building sites where the movement of heavily laden trucks is concentrated. At first glance it might be supposed that it is simply the fact that the load bearing capacity of the road surface (technically called the pavement) is not sufficient. That, however, would only be a partial answer. In fact, the problem arises because of the present of high water tables and the effect of this saline water on the asphalt-concrete mix used in road construction. This has the effect of stripping the asphalt bonding from the concrete aggregate and thus reduces the overall tensile strength of the road (Shihata and Abu-Rizaiza, 1988; Wahhab and Hasnain, 1998). Under compression, the weakened road surface buckles up and folds into long ridges as high as 10 cm or so.

Figure 9.9 Building foundations, north Jiddah, protected from salt damage by a cover of epoxy resin (Photo: author).

One fascinating finding in Wahhab and Hasnain's study of asphalt-concrete durability was the fact that soapy water was more effective in stripping the bonding than either sea water or fresh water. The implication of this finding is that the domestic wastewater leaking into the groundwater from septic tanks is inadvertently contributing to the weakening of the road surfaces.

In one or two neighbourhoods temporary solutions to the presence of high water tables include building up the pavement surface with further layers of asphalt aggregate topped by a new skim of Tarmac. The net effect of this procedure has been to invert the pavement relief whereby the central reservation has now become lower than the pavements on either side. This is clearly more than a design disfigurement and does nothing to help road safety (Vincent, 2004).

9.2.10   *Sabkhah*

As has been mentioned previously, a *sabkhah* (plural: *sibakh* – but *sabkaht* when used with a named *sabkhah*) is a salt-encrusted shallow basin with a puffy surface (Figure 9.10). In the Eastern Province of Saudi Arabia, such features are most common in low-lying coastal plains, but older ones can be found in places as far inland as the edge of the Summan Plateau, some 125 km from the coast. Non-saline silt flats composed entirely of

Figure 9.10 The north-western edge of the remote inland *Sabkhat al Budu* one hundred km south of Harad, Eastern Province (N23°15′, E49°21′). Finds of Palaeolithic tools have been found around the shores of this once freshwater lake (Photo: author).

silt, fine sand, and clay, and which lie well above the water table, also occur in depressions and *widyan* beds and are called *faydah* (pl. *fiyad*). If vegetation is present, they are called *rawdah* (pl. *riyad*).

Sibakh probably form where wind erosion removes surface materials down to the water table. Water is always associated with *sibakh* in the form of flooding, run-off accumulation, capillary rise, and tidal fluctuation. The sediments that fill *sibakh* comprise sand, silt, clay, and salts. Their flat surfaces mark the elevation to which soil moisture rises above the static water level: below this surface, the materials are damp, wet, or saturated; above, they dry out and blow away. Many *sibakh* are covered with sand sheets or with dunes. Areas that appear to be inter-dune flats are commonly *sibakh* concealed by windblown sand. Sand around *sibakh* margins is usually veg-etated and hummocky.

According to Holm (1960), there are two types of *sabkhah* on the Arabian coast: (1) arenaceous, or sand filled, and (2) argillaceous, or clay filled.

1. The arenaceous *sabkhah* is formed by the in-filling of embayments of the sea with wind-borne sand, and the subsequent reworking of this sand by waves and currents. This type of *sabkhah* forms in areas downwind of major sand sources, i.e., dune fields, and is especially typical of the Eastern Province of Saudi Arabia. Many of the coastal *sibakh* show growth lines in subparallel arcs that mark former shorelines. The surface of the arenaceous *sabkhah* is flat and smooth, with gradients of less

than 0.5 m per km. During dry periods, capillary water in the sand evaporates and concentrates to a saline brine, with eventual precipitation of salts near the surface. Water rising through pore spaces of the sand increases lubrication, and produces a soft quicksand of low load bearing strength. Evaporation of standing water forms a coating of salt crystals, which can thicken into a thick armour plate of salt. If buried under sand, this plate can be preserved until dissolved by a rise in water level.

2. Argillaceous *sibakh* form in sheltered coastal embayments by the manufacture of calcareous mud by algae or other organisms. This mud is wet, soft, sticky, has a low load bearing strength, and flows under pressure. Wind-borne sand sticks to the mud and forms layers mostly less than 1 m thick. These layers can support the movement of a light truck. Once wet, the sand softens so that a heavy vehicle, and sometimes a light one, will sink through until it rests on the frame. A thin sandy crust can appear safe, but should be tested before use. When not covered by sand, an argillaceous *sabkhah* dries in the summer heat, and where used as a roadway, hardens to an asphalt-like consistency. Argillaceous *sibakh* are typical of coastal Arabia south and east of Qatar.

Driving with heavy vehicles across a *sabkhah* surface is rarely safe even in summer, and if a vehicle breaks through the salty crust it will be bogged down in the quick-sands beneath. In the Dammam region sabkhat are being reclaimed on a grand scale for factories and new roads.

9.2.11   Radon

Radon (Rn-222) is a noble gas and one of three known isotopes produced by the radioactive decay of natural radium (Ra-226) which itself is a decay product of the natural uranium (U-238) series. Epidemiological studies have shown that long term exposure to Rn-222 and its immediate, short-lived, daughters (Po-218, Pb-214, Bi-214, Po-214) is associated with an increased risk of lung cancer. In the open atmosphere, Rn-222 diffuses naturally from the soil and is dispersed but in mines and houses with poor ventilation Rn-222 can build up and become a health risk. The risk from Rn-222 in water supplies is low as diffusion into atmosphere is generally very fast.

Uranium occurs naturally at low concentration in most rocks of the Shield and a number of ground and airborne radiometric surveys have been conducted. Collenette and Grainger (1994) note a number of conspicuous uranium anomalies including the Ar Rawdah area in the north-central Shield where rhyolite has been intruded by alkaline granite ring complexes and the whole affected by intense tectonic activity of the Najd fault system. Here the volcanic rocks of the Umm Misht Formation contain as much as 12,500 ppm U, the highest recorded values so far found anywhere in the Shield. Many unconsolidated sediments in *widyan* are also mentioned as having anomalously high levels of detrital U.

Although detailed epidemiological studies are still to be carried out there is no doubt that lung cancer cases are well above expected norms in western Saudi Arabia and these are probably linked to Rn-222. Data for Saudi Arabia are few and far between. Abu-Jarad and Al-Jarallah (1986) report one study of 400 houses which were surveyed using passive alpha-track detectors. Radon concentrations were found to be from 0.13–0.98 picocuries/L (pCi/L) with a mean value of 0.43 pCi/L. These levels are rather low and depressed by the fact that only seventy-two houses were located on the Shield (Taif – 15; Jiddah – 57). However, recent preliminary work by James Elliott and Hisham Hashem of the Saudi Geological Survey indicates that that Rn-222 is potentially a serious hazard. They have investigated the Jabal Said mineralized granite consisting of an aplite-pegmatite radioactive body approximately 35 km north-north-east of Mahd adh Dhahab (N23°50', E40°57'). They used a sophisticated soil gas probe to sample Radon-222 in the overburden (soil, sand and gravel). Values ranged from 55 to 32,182 pCi/L with the extremely high values associated with alluvium derived directly from the aplite-pegamatite body. The level of Rn-222 in the poorly ventilated village houses along the floors of the *widyan* draining this body clearly need to be investigated as a matter of some urgency. Similar mineralized granites occur at Jabal Tawlah, Ghurayyah west of Tabuk, Jabal Hamra, Umm al Birak and Ha'il (Drysdall *et al.*, 1984).

9.2.12   Impact craters

In 1932 Harry St. John Abdullah Philby was exploring for the famed city called "*Ubar*", that the *Qur'an* claimed had been destroyed by God because its people were sinful and leading unrepentant lives. Philby erroneously transliterated Ubar as Wabar. He did not find Ubar, but did find two meteorite impact craters or astroblemes – a small, third, has been recently discovered. The two reported by Philby measure 116 m and 64 m wide respectively. Prescott *et al.* (2004) have recently dated the impact using luminescence techniques and found the age to be 290 ± 38 years BP. They suggest that an impact as young as this has implications for the assessment of hazards from small meteorites.

9.3   BIOLOGICAL HAZARDS

9.3.1   Desert locusts

The desert locust (*Schistocerca gregaria*), the locust of the Bible, was first described and name scientifically by Pehr Forsskål (see – Chapter 1) in 1763, and the Middle East in general has had infestations since time immemorial. Some of the earliest research on the desert locust, which also produced a great deal of general ecological information, was conducted by British-led scientists in the 1940s when the British Government became concerned about the possibilities of wartime famine in the region. This research was carried out in Arabia under the leadership of Desmond Vesey-Fitzgerald. The Saudi Arabian government took over the anti-locust operations subsequently and has continued to play a major rôle in the prevention of serious regional outbreaks (Uvarov, 1951). The Red Sea basin is a major focus of desert locus outbreaks and is monitored by the FAO's Desert Locust Information Service.

   In its gregarious state, the locust forms flying swarms which can measure 300 km² or more and cover enormous distances. During the infestation of 1943–4 over 70,000 km² were found to be infested and some 1,200 tons of poisonous bait was laid down.

This was obviously effective in killing off the locusts but must also have done untold damage to other insect life and animals higher up the food chain.

The behaviour of desert locusts is triggered by rains which bring on a plentiful supply of vegetation. This allows solitary hopper phase locusts to increase rapidly in numbers. During the following dry period the locusts crowd into smaller areas and this population pressure causes the locusts to adopt a swarming, winged phase.

Because water is so important in the phase changes of the desert locust the natural breeding grounds in Saudi Arabia are along the Red Sea Coast. Infestation from these sites can spread rapidly westward across the Red Sea into North Africa and also east-ward in the interior of the Kingdom. The threat to the areas of centre-pivot irrigation and cereal production in the western Najd are thus quite real.

Cressman (1998) provides a detailed account of the infestations in Saudi Arabia during the 1996–7 outbreak which was a classic example of a desert locust upsurge (Figure 9.11). In the Spring of 1996 small scale gregarization occurred in the Tihamat Asir and especially in the Al Qunfudhuh area where about 500 ha were treated by ground teams using pesticides. This treatment greatly reduced the locust population. In June 1996 a cyclone from the Indian Ocean brought widespread rain to Oman and the Yemen which was followed by heavy rain over the northern Red Sea in November. These rains triggered a regional upsurge in locust numbers with Saudi Arabia being particularly affected.

Light to moderate rains on the coastal region from Jiddah to Al Wadj occurred during the winter of 1996 and spring of 1997 and temperatures were lower than usual allowing vegetation to remain greener than in hotter, dryer years. Solitary adults were noticed along the Tihamah in November 1996 and were thought to have come from the Yemen on southerly winds drawn northward by the persistent depression over the northern Red Sea. Egg laying was reported to be heavy at a number of infestation sites from Al Lith in the south to Al Wadj in the north.

By late May 1997 temperatures increased rapidly and the vegetation in coastal region was dying off and fourth and fifth instar gregarious hoppers were present in densities up to 50 per m². Some adults had also moved inland and were reported in the important agricultural areas around Ha'il.

There were also unconfirmed reports of locusts as far east a Riyadh on 31st May. Major migrations westward across the Red Sea into Sudan also took place at about this time. Locusts were also reported migrating eastward onto the Tihamah from the Sudan during the period October 1997–December 1998. Ground control operations began in mid-February 1997 against mature adults groups and a total of 1,100 ha were treated at Al Lith, Khulays and Rabigh. By mid-June, when the campaign had come to and end, some seventy ground teams and four aircraft had treated 140,000 ha with 340,000 litres of pesticide much of which must have entered the food chains and possibly been magnified at higher levels.

9.3.2   Urban groundwater contamination

For most urban areas in Saudi Arabia the lack of any sewerage system means that the only way to dispose of human effluent in septic tanks is to pump it into tankers and dump it in the nearby desert. Much liquid effluent also drains directly into the ground water. In the case of Jiddah, the waste is disposed of inland about 20 km

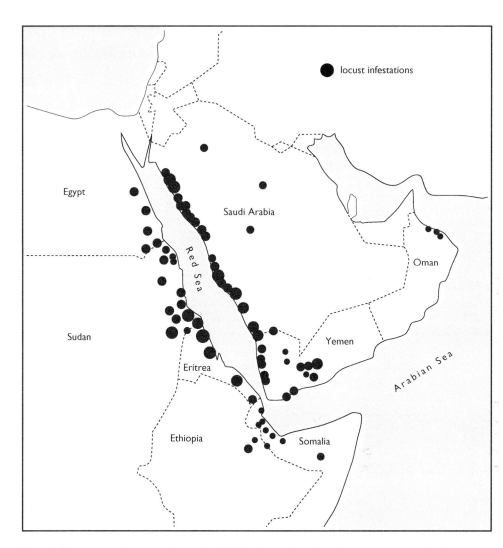

Figure 9.11 Locust infestations during the 1997 outbreak (Modified from Cressman, 1998).

east of the city into vast lagoons. In Riyadh domestic waste has transformed a *wadi* immediately to the south of the city into a semi-permanent river. These might seem practical solutions to a growing problem but the only environmentally sound solution is for proper effluent treatment. From a distance, the desert lagoons may look green from the rich algal bloom that develop, and the heat of the desert undoubtedly kills off many harmful bacteria.

The problem seems to be that the general flow of groundwater from the escarpment to the coastal plain is bringing some of the effluent, which seeps into the shallow groundwater, back towards the city thus causing a health hazard. Research is underway to see if the return flow can be traced using satellite imagery. Although the epidemiological link is unproven it is of note that more than eight hundred cases of amoebic dysentry are recorded each year in Jiddah – several orders of magnitude more than in any other urban centre in the Kingdom. Moves are now underway to lay a complete sewerage system for Jiddah. It is likely that that slums of south Jiddah, housing thousands of ethnic minorities and illegal immigrants, will be the last to be linked to the system although here the health needs are highest.

9.4 INDUSTRIAL HAZARDS

9.4.1 Cement dust and heavy metal pollution

There is a good deal of evidence to suggest that dust particles generated by the cement manufacture and oil combustion are enriched with heavy metals such as cadmium, selenium and arsenic (Förstner and Wittman, 1979). With this in mind, Behairy *et al.* (1985) analysed the composition of dust falling on the coastal zone 30 km north of Jiddah in the vicinity of Sharm Obhur – a low energy tidal creek. They estimate that the average amount of dust falling over the Sharm Obhur area to be 1.09 t/km²/month which they note to be significantly lower than the dust falling over Jiddah, estimated by MEPA to be 9.07 t/km²/month. Of particular interest were the remarkably high levels of cadmium in the dust samples, averaging twenty-three ppm. This compares with levels of about 1 ppm in the marginal shelf sediments from the Red Sea and 2.6 ppm from the creek sediments. Cadmium is very toxic and easily concentrated in food chains. Behairy *et al.* concluded that the cadmium pollution of the local coastal ecosystems is the result of the cement factories on the coast north of Jiddah, particularly at the Arabian Cement plant at Rabigh. Since the prevailing wind direction is from the north for most of the year there is the obvious implication regarding cadmium polluted dust fallout over Jiddah down wind of the cement plants.

9.4.2 Desalination thermal pollution

With the growth of the Kingdom's urban population over the last thirty years has come an almost insatiable demand for drinking water. For example, water use in Jiddah has risen 12.8 percent in five years to 986 million m³ in 1997. The Kingdom's approach is to supply this demand from desalinated water with ground water being used only as a backup. In 1970, at the start of the First Five Year Plan, the quantity of water supplied by the Kindom's desalination plants was 4.6 million US gallons per day (mgd) and by 1991, this had increased to 409.8 (mgd), representing an annual growth rate of 21.4 percent.

The production of potable water by desalination processes is, however, a potential source of environmental pollution. In coastal desalination plants, water pollution is the main problem, while in inland plants, attention has to be paid to the disposal of the rejected concentrated brine. In the case of multistage flash thermal distillation (MSF), air pollution problems also arise since the plants employing this process often burn high sulphur content fuels (Al Mutaz, 1991). As a guide, MSF desalination raises the water temperature by about 9°C and the salinity by about twelve percent.

Water pollution by desalination plants is caused by the disposal of hot brine which has both thermal and saline impacts. Al Mutaz (*op. cit.*) made a detailed study of the pollution at the Jiddah desalination plant and estimates that the total salts added to the sea amount to some 8.3×10^9 kg/year. The regulatory bodies appear to have done little to assess the serious impact on marine life in the area of desalination outfalls and environmental damage could be minimized by dispersing the rejected brine. With regard to air pollution, MEPA studied the effect of the Jiddah desalination plant on nearby housing from January 1st until March 31st, 1986. About seventy-six violations in sulphur dioxide concentration were recorded above the one hour average MEPA standard of 0.28 ppm. During the measurement period the maximum one hour concentration of sulphur dioxide was 0.757 ppm. Most crude oil in Saudi Arabia has a sulphur content in excess of 3.2% and desulphurization of heavy crude oil is difficult and expensive. Although there are electrostatic precipitators used to control the emission of particulates at the Jiddah plant there are no means of controlling air pollution from the 150 m high boiler stacks.

9.4.3   Oil spills

Oil spills are a major threat to both the Red Sea and the Arabian Gulf. In particular, the Gulf has experienced a number of moderate-to-large oil spills over the past 20 years. During the Iran-Iraq war from 1980 to 1988, oil tankers in the Gulf were attacked, resulting in thousands of barrels of oil spillage. This damage was dwarfed by the catastrophic effects of oil spilled during the Arabian Gulf War: on January 23, 1991, when Iraq intentionally began pumping crude oil into the Gulf from the Sea Island supertanker terminal 15 km off the Kuwaiti coast. Approximately 5.7 million barrels of oil were dumped.

9.4.4   Air pollution

Air pollution is a cause for concern in both major urban areas and industrial zones. A major source of pollution is the transportation sector. Motor vehicles emit nitrogen dioxide and volatile organic compounds which, in the presence of sunlight, convert low-level ozone into smog. The Saudi Government announced plans to produce lead-free petrol in 2001 which has improve matters. As noted above, desalination plants also account for a major portion of the sulfur dioxide and nitrogen oxide emissions from fixed sources. For example, the plume from the Jiddah desalination plant lies directly over residential areas and several important government facilities, including a hospital. This is an almost constant hazard given the almost continuous sea breeze.

Jiddah's Industrial City, in the south of the city and beyond the built up area, also contributes significant amounts of airborne pollution as do Jiddah's refineries and King Abdul Aziz International Airport, both in the northern suburbs. The Government has ordered major industrial projects to conform with international air standards in order to limit emissions, but there has been little in the way of monitoring and enforcement. With the ban on the production and use of halons (Saudi Arabia is a signatory to the Montreal Protocol), the Government encourages environmental agencies and the private sector to participate in protecting the environment by not releasing halons, which deplete the ozone layer. The USECO Halon Recycling Factory in Dammam was the first factory established in the Middle East to recover, condition and recycle halon gases.

Figure 9.12 Air pollution – Ghawar oilfield north of Harad (Photo: author).

Air quality in Saudi Arabia's Eastern Province has benefited greatly from several initiatives. Aramco's Master Gas System, which has significantly reduced the need for flaring, now recovers more than 3,500 tons of elemental sulfur per day from gas produced in association with crude oil. All is not perfect, however, as can be seen in Figure 9.12.

Saudi Arabia's total carbon emissions have risen in the past 20 years, but not at the same rate as the country's energy consumption. In terms of carbon emissions per capita, Saudi Arabia is not a regional leader. In 2003 carbon emissions per capita were 13 metric tons as compared with Qatar (63) and Kuwait (31). By comparison, emissions from the United States were 19.8 tons – though of course its population is vastly larger.

Saudi Arabia is a signatory to the London amendment to the Montreal Protocol (1985), which calls for phasing out chlorofluorocarbon (CFC) gases harmful to the ozone layer by 2010. Saudi Aramco, in keeping with the protocol on ozone-depleting substances, has been identifying its CFC-based cooling systems and converting them to safer alternatives. However, under the United Nations Framework Convention on Climate Change, Saudi Arabia, a non-Annex I country, is not required to reduce its emissions below 1990 levels. Saudi Arabia ratified the Convention and became a signatory to the Kyoto Protocol in 2004.

9.5   SOIL SALINIZATION AND IRRIGATION

Large areas of Saudi Arabia are now irrigated by centre-pivot irrigation. With an irrigation arm of c.250 m some 19.6 ha can be watered by a single unit. In the course of an irrigation season approximately 5,500 m³ of irrigation water would typically be used per hectare. About eighty percent of this water evaporates and the rest recharges shallow aquifers. The central problem is that the groundwater often used for irrigation has been in close contact with bedrock, often for many thousands of years, and may be charged with large quantities of sodium and bicarbonate salts. The effect is to create saline soils which have poor pedological properties. The soil pH is raised and the infiltration rates are lowered. Eventually the soil texture becomes hard and difficult to plough, and the soil loses friability. This in turn leads to poor root development and poor soil aeration. Irrigation with fresh water can control salt accumulation somewhat but this is hardly practicable in most situations in Saudi Arabia.

Al-Abdela *et al.* (1997) investigated the irrigation water used for wheat and alfalfa in the Al Qasim region of central Saudi Arabia. The water had an average salinity of 2,375 ppm but was as high as 8,200 ppm. They calculated that between 16.6 and 83 tons per hectare of salt was deposited in the soil per season. Indeed, the water was so corrosive that it damaged the centre pivot pipelines!

Notwithstanding the damage to soils by salt-laden irrigation waters drawn from depth, the shallow aquifers which contain some potable water are also locally contaminated by salts leaching down from the surface and by fertilizers used to maintain crop yields. It is interesting to note that in March 2007 the Agriculture Minister, Fahd Balghunaim, proposed the establishment of a National Commission for Organic Farming. Certainly, reduction in the fertilizer load would be welcome but it is difficult to see, given the very poor soils in the Kingdom, how commercial crops can be grown without reliance on fertilizers.

9.6 DESERTIFICATION

Desertification is a process of land degradation which reduces its productivity. It can be measured by: the reduced productivity of desirable plants; undesirable alterations in the biomass and the diversity of the micro and macro fauna and flora; accelerated soil deterioration; increased hazards for human occupancy. There are various factors involved including climatic variations and human activities. In Saudi Arabia, Barth (1999) suggested that Holocene land degradation probably began around 5,500 BP caused by a southward shift of the Monsoon winds and their associated precipitation (Sirocko, 1996). With less precipitation, particularly in the Eastern Region, grasslands were probably replaced by open shrub land and dune systems reactivated.

In the last hundred years or so land degradation has intensified due to rapid social and economic changes and particularly those associated with the development of the oil industry. Traditional husbandry has all but died out in most parts of the country and government subsidized foodstuffs have allowed an increase in flock size and the overexploitation of range resources. Today overgrazing is the key factor in the continued desertification of the rangelands of Saudi Arabia. Although no precise figures are available, Chaudhary (1989) estimated that about eighty percent of the total rangelands of the Arabian Peninsula is used for domestic camels, donkeys, goats, cattle and sheep most of which roam more or less free.

A number of studies have attempted to compare the actual number of animals to that which can be supported by the natural vegetation in terms of livestock unit demands. Shaltout *et al.* (1996) indicate that about sixty percent of the rangelands of Saudi Arabia are serious degraded due to overgrazing and overcutting. They indicated that the expected optimal carrying capacity is 1.04 million LSUs compared with an actual number of about 6.7 million. One LSU (Livestock Grazing Unit). requires about 14 kg of forage (dry biomass weight) in a Saudi Arabian-type environment and is equal to one camel or about 9–10 sheep/goats.

Hajara and Batanouny (1977) quote some stocking rate information for the Nafud Basin produced by the Basil Parsons group in 1966–67. In this region the safe carrying capacity of the rangelands was estimated at 191,500 LSUs while the average number of animals was 270,000 – an overstocking by some 40 percent. An interesting finding of the survey was that the local *badu* did not believe that few healthier animals was preferable to more undernourished ones. There is a common belief than social status is achieved simply though numbers not quality.

The only known actual livestock census was undertaken by the Government in 1965. This recorded 2.8 million sheep, 1.4 million goats, 0.6 million camels and 0.27 million cattle, representing 8.5 million LSUs – more than twice carrying capacity of 3.89 million LSUs estimated by the McLaren International Consulants' report (1979). As the census took place after a severe seven year drought the data are probably quite conservative.

An alternative way at getting at the scale of the problem is to use breeding stock numbers from slaughter rates at official abattoirs. Al-Saeed (1989) provides data for 1987 and these are shown in Table 9.2 together with the breeding stock required to support the slaughter numbers. The data excludes all cattle and also the large number of private slaughterings. The data presented in Table 9.2 are crude estimates but are the only data available. Even so, they indicate the severity of the overstocking on the Kingdom's rangelands – some 2.79 times the sustainable stocking rate estimated by McClaren International Consultants.

Rangeland animals are often in poor condition because of the overstocking but supplementary feeding can sometimes be devastating as it can result in the severest form of overgrazing. Sheep in particular have an inbuilt habit to keep their rumen full, thus better fed animals will still make the maximum use of any available vegetation. Because the animals are kept alive by supplementary rations, grazing intensity does not decline as the available plant food declines with the result that damage to the vegetation accelerates and can lead to its total destruction.

In a country literally swimming in oil resources it is ironic that so much wood is grubbed up for fuel. Hajara and Batanouny (*op. cit.*) describe one survey of the

Table 9.2 Breeding stock estimates to support 1987 slaughter data

	Slaughtered		Stock Estimate	
	Millions	*LSUs millions*	*Millions*	*LSUs millions*
Goats	3.3	0.66	9.9	1.98
Sheep	7.4	1.48	22.2	4.44
Camels	0.4	0.40	4.4	4.44
Total	11.1	2.54	36.5	10.86

Source: Al Saeed (1989).

volume of wood consumed in Buraydah and Unayzah. It was found that about 3,800 loads of wood were used in each city per year. An average load of wood might contain the roots from almost 400 plants. In other words, the number of plants grubbed up annually around these two cities was c.1,555,000. It is not surprising that in Saudi Arabia as a whole, the grubbing up of trees and shrubs leaves some 12,000 ha/year almost devoid of woody vegetation. Since much of the damage is done by the *badu* as they move their flocks from place to place one simple solution is simply to provide cheap kerosene stoves for cooking. To the rural poor a small truck load of wood sold at the roadside is a ready source of income and can fetch as much as SR500. It is, however, illegal sell such wood and it carries a fine of up to SR1,000 though this seems to be rarely imposed.

The ecological carrying capacity of Saudi Arabia's rangelands depends largely on the permanent, perennial vegetation. Continuous grazing and browsing stress affects both regeneration, growth and general vitality of the vegetation as the young shoots are selectively grazed. Annual plant species are only available after rains and under severe grazing pressure are eaten before they can set seed. Furthermore palatable species are particularly sought. Goats in particular will browse on almost anything and can even be seen eating cardboard boxes! By instinct they ignore plants which are bitter or contain poisonous alkaloids, such as *Rhazya stricta, Nerium oleander, Calotropis procera and Solanum incanum*. The abundance of these species in an area is a good indication of overgrazing (Ghazanfar and Fisher, 1998).

With no common management of range resources heavily grazed areas may not have time to recover (Figure 9.13). As there is no control on the sensitive equation between carrying capacity and flock size – a classic 'tragedy of the commons' scenario developes. It is no wonder that woodland regeneration in the Asir is so slow when so much of it is cropped without common management. Of particular relevance here is the fact that livestock, even in quite remote areas, can now be supplied with supplemental fodder and water from tankers and herding practices which let traditional grazing areas recover are slowly being assigned to history.

Hard scientific data regarding the effectiveness of stock control on vegetation recovery are quite few and much more research is necessary. The difficulty of maintaining fencing in trial areas is a real problem. One study to look at the comparative ecology of vegetation in a protected *hima* (see Chapter 10) as compared with a grazed, unprotected, range was conducted by Hajar (1993) at the Hima Sabihah, in the al-Bahah region. Hajar showed that the protected *hima* range maintained higher species diversity and contained more palatable plants such as the grass *Cymbopogon schoenanthus*. The grazed area had a lower plant cover composed mainly of non-palatable species such as *Asphodelus fistulosus* (onionweed) and *Psiadia punctulata*. Hajar suggested that soil erosion in the grazed area was also a serious problem.

Barth (1998) paints a worrying picture of the actual impacts of overgrazing in his study of an area north of the industrial city of Jubail. In this area traditional husbandry has been, in many cases, transformed into a commercial ranging system and the provision of concentrated foodstuffs has replaced dependency on the open range. One of his most dramatic findings was that the size of active dune-fields may double in 12 to 15 months as a result of continuous heavy grazing and soil disturbance, particularly by excessive off-road driving. Few *badu* now use camels for transport and Toyota half-trucks are common. As a result, the once isolated *badu* can now shop and

Figure 9.13 Calotropis procera – Wadi Tathlith. An indicator of overgrazing (Photo: author).

trade with ease in towns and villages, and the routes out to the flocks in the surrounding desert are a maze of anastomosing vehicle tracks.

Jubail suffers from blowing sand particularly during the *shamal* season and to investigate the impact of overgrazing Barth examined the vegetation cover on the open range and compared it with that inside the fenced Khursaniyah Cement Factory. Some of Barth's results are presented in Table 9.3. The percentage cover of palatable annual plants clearly varies between the fenced and open sites. *Erucaria crassifolia* has been completely grazed out on two of the open range test sites. In contrast, the unpalatable perennial *Calligonum comosum* shrub, which is usually rejected by animals, has very similar cover values in the fenced and unfenced sites.

Barth observed that the first obvious sign of overgrazing is the initiation of small wind ripples and the accumulation of sand on the leeward side of small shrubs. Shrub roots then become exposed sometimes at an alarming rate. For example, on the open range the annual plant *Plantago boissieri* was found to have roots exposed to a depth of 5 cm in just four months. Once exposed to continuous strong winds the small ripples evolved into so-called giant ripples with a wavelength of 1 to 2 m and an amplitude of up to 15 cm. In this environment it is virtually impossible for plants to establish themselves successfully. The final state of degradation is the reactivation of fossil dunes and the establishment of new dune fields. In the Jubail area, Barth

Table 9.3 Contrasting plant cover (%) in fenced and open range sites, north Jubail

	Site 1		Site 2		Site 3	
	Fenced	*Open*	*Fenced*	*Open*	*Fenced*	*Open*
Annuals						
Erodium sp.	0.58	0.15	0.40	0.15	0.05	0.05
Erucaria crassifolia	3.80	0.01	1.70	0.00	2.10	0.00
Launea sp.	0.17	0.15	0.40	0.11	0.26	0.08
Perennial Shrub						
Calligonum comosum	3.8	3.6	4.00	4.10	3.90	4.00

Source: Barth (1998).

observed that reactivated dunes easily doubled their extent in twelve to fifteen months in combination with continuous heavy grazing.

The Jubail investigation provides some interesting insights into the relationship between biomass production and optimum stocking rates. Within Barth's study area it was estimated that there was enough annual biomass production, at the very highest stocking rates, for about 600 camels and 1,800 sheep. Amazingly, some 1,050 camels and 4,800 sheep and goats were using the range. The sheep and goats in particular are very harmful to the delicate rangeland ecosystem since they eat the plants down to the roots whereas camels move on after taking one mouthful. The camel's feet are also wonderfully adapted to sandy substrates and have soft pads which hardly disturb the ground. The message from Barth's study, where an ecosystem had been totally degraded, is that the only recovery option is the complete removal of all animals from the area for at least two years. This is more difficult that one might think since the open range is not fenced and where barbed-wire fencing has been tried it is simply pulled down by the *badu* who wish to retain their traditional grazing rights and maintain their cultural behaviour. Clearly there is much scope for environmental education in these circumstances.

The overgrazing of the open range in the Jubail area is typical of many areas in Saudi Arabia. In practically all cases one can trace much of the problem back to an increasingly large, wealthy population in urban areas and the growing demand for meat. This may provide short term financial gains for the *badu* but is leading to a long term degradation of the desert ecosystems. Indeed, it is not just the *badu* who benefit. Many flocks the authors as observed were tended by immigrants from South Asia the owners having migrated to the city.

One relatively long term study of the impacts over overgrazing has been carried out by Shaltout *et al.* (1996) who investigated the differences in vegetation between open range conditions and a 2.5 × 2.5 km fenced area south-east of Hufuf. The protected area was established in 1980 as a research facility for King Faisal University and the authors examined the vegetation and soil differences after 14 years of control. Their results are summarized in Table 9.4. Not only was the vegetation more species rich inside the protected area but the sizes of the dominant species were significantly greater. Many of the species found to be significantly more abundant inside the protected area were important forage and/or fuel plants including: *Anabasis setifera,*

Table 9.4 Impact of range protection on vegetation and soil attributes – Hufuf Research Station

	Protected	Unprotected
Vegetation		
Total Number of Species	68	61
Species per 20 m² stand	18.90	14.20
Total Cover (m 100 m⁻¹)	66.50	39.70
Soil		
Na (mg g⁻¹)	0.10	0.18
P (mg g⁻¹)	0.06	0.14
K (mg g⁻¹)	0.38	0.74
Ca (mg g⁻¹)	5.07	8.52
pH	7.40	6.97
Conductivity (mmhos cm⁻¹)	1.23	2.60

Source: Shaltout *et al.* (1996).

Centaurea sinaica, Cornulaca monacantha and *Panicum turgidum.* Outside the protected area weeds of disturbed habitats and halophytes were significantly more abundant.

The higher soil salinity recorded outside the protected area (Table 9.4). was thought to be due to higher evaporation and the direct heating of the soil by the sun on bare soil exposed by the wind. In grazed areas soil nitrogen levels generally increase because of the passage of herbage through the guts of grazing animals; other soil nutrients were also higher as a result of the redistribution of nutrients by grazing animals. The message here is that if a grazed range can be rested the level of soil nutrients is probably not a limiting factor in its recovery. Even salinity levels can be eventually lowered if soil evaporation is reduced allowing the vegetation to recover.

Hajara and Batanouny (1977) make a number of interesting points concerning desertization (*sic*) in the Asir – an area not often mentioned in this regard. As they note, anyone traveling south from Taif to Abha will be acutely aware of the vast number of abandoned cultivation terraces. The terrace system of agriculture in this region is probably pre-Islamic and if well managed can provide both summer and winter crops. Hajara and Batanouny illustrate the scale of the abandonment with some survey data although it is not known how completely reliable the data are (Table 9.5). The situation is almost certainly much worse today.

The reasons for the deterioration of the terraces include: a) over cutting and/or overgrazing of the natural plant cover leading to increase runoff and soil erosion;

Table 9.5 Terrace survey – Asir and Tihamah

Region	No. of Villages	Population	Area donum ('000 m²)		
			In use	Destroyed	Abandoned
Tihamah	1257	143,163	365,000	250,000	615,000
Asir	1483	161,222	575,000	245,000	532,000

Source: Hajara and Batanouny (1977).

b) lack of labour due to migration to large urban centres; c) lack of suitable draught animals and inaccessibility of the terraces to tractors; d) the high costs of production as compared with market prices, especially for wheat.

9.7 GULF WAR HAZARDS

In August 1990, Iraqi troops occupied Kuwait. The United Nations called upon Iraq to withdraw its troops but Iraq refused and on the 16th January 1991 the Gulf War broke out. Over the next two months some 6 million barrels of oil were released into the coastal waters of Kuwait. The counter-clockwise currents of the Arabian Gulf transported the oil slicks south and the prevailing north-easterly winds kept the slicks close to the shore resulting in some 650 km of the Saudi Arabian coastline being severely contaminated. The oil was largely prevented from traveling further south than Jubail by the protruding islands of Batina and Abu Ali which caused the sheltered embayments of Dauhat ad Dafi and Dauhat al Musallamiya to be severely polluted (Krupp *et al.*, 1996; Smith, 1996). The sheltered embayments contained saltmarsh and mangrove ecosystems and the shallow waters provided extensive feed areas for birds and nursery areas for fish and shrimp. Much of the oil was originally deposited between the high water spring and high water neap tidal levels but soon spread to lower shore levels. The embayments developed extensive areas of tar 'pavements' and oil concentrations were reported as high as 139 g/kg dry sediment with oil burdens as high as 20 kg/m^2.

In an interesting study Vazques *et al.* (2000) compared the hydrocarbon levels in the clam *Meretrix meretrix* at nine locations on a north-south transect along the Gulf coastline during the period 1981–1990 (prior to the 1991 Gulf War) to those of the war and post-war periods. The five northern sites in the study were affected by the war oil spill. There was no impact in *n*-alkane levels but dibensothiophenes (DNTs) and phenanthrenes (PHEN) increased significantly. The DBTs returned to pre-war levels within two years in contrast to the PHENs which, at the five sites affected by the spill, had remained above the pre-war values. Several explanations are thought to account for this difference. Once accumulated, the PHENs, being less soluble in water than the DBTs might be more difficult to remove from the clam's tissue. Or perhaps the clams might more easily metabolize DBTs. The authors make the point that at the four sample sites south of the oil spill impact DBTs and PHENs have, at times, been higher than those from the north. These southern sites, between Jubail and Aziziyah are located in areas related to oil production and/or increased industrial and human disturbance.

Chapter 10

Environmental protection, regulation and policy

Contents

10.1 INTRODUCTION

Although some traditions of environmental management, such as the *hima* system of range management, go back at least to the time of the Prophet Muhammad it is only in the last thirty years or so that formal institutional structures have been developed in the Kingdom. The need to conserve, protect and regulate all aspects of the environment was recognized in the 1970s after a period of rapid industrial and urban growth. Also, at about this time, a steady stream of science graduates returning from studies in Europe and the USA brought with them key scientific skills and awareness of the long term, potentially irreversible, damage of wholesale environmental exploitation.

The first important event in the development of environmental protection and regulation in Saudi Arabia was the creation of the Meteorology and Environmental Protection Administration (MEPA) and the Environmental Protection Coordinating Committee (EPCCOM) in 1981. MEPA, created by Royal Decree No. 7/M/8903, was assigned the responsibility for pollution control, protection of the environment and meteorology. In 1982 MEPA issued the first set of standards designed to protect air and water by limiting the concentration of pollutants. Such standards are vital in order to protect the health, safety and welfare of the population, although there is a good deal of evidence to indicate that the standards are widely ignored. EPCCOM was re-designated the Ministerial Committee on the Environment (MCE) in 1990. The role of the MCE is to coordinate the activities of government bodies involved in environmental protection and particularly to evaluate measures submitted by MEPA and forward them to the Council of Ministers for approval.

From its inception MEPA had no direct role in the protection of wildlife but it was recognized that industrial pollution and uncontrolled hunting were doing great damage to the fauna and flora. The recognition of this state of affairs led to the establishment of the National Commission for Wildlife Conservation and Development (NCWCD) on the 12 May 1986 to: 'develop and implement plans to preserve wildlife in its natural ecology and to propose the establishment of proper protected areas and reserves for wildlife in the Kingdom' (Article 394: Royal Decree No. M/22). By necessity the NCWCD initiated projects for the captive breading and reintroduction of 'flagship' quarry species such as the houbara bustard and has gradually sought to gain popular support for other equally important, but less spectacular, conservation initiatives.

The NCWCD is relatively poorly funded and the relevance of its work is still not appreciated. This is probably due to the very positive attitude towards hunting by many in the Kingdom and this might lie behind the NCWCD's failure to get approval from the then Ministry of Agriculture and Water (MAW) which had overall responsibility for the conservation of natural resources.

The Ministry created a separate Department of National Parks in 1983 and established and administers the Asir National Park – the Kingdom's first national park. In practice is seems that the responsibilities and duties of MEPA, the NCWCD and MAW are quite blurred at the margins but doubtless will become better defined as problems arise. In 2003 the Ministry of Agriculture and Water was split into a Ministry of Agriculture and a Ministry of Water and Electricity.

PERSGA, the Regional Organization for the Conservation of the Environment of the Red Sea and Gulf of Aden, is an intergovernmental body dedicated to the conservation of the coastal and marine environments in the region [http://www.persga.org/]. Operating from Jiddah on the Red Sea, PERSGA is responsible for the development and implementation of regional programmes for the protection and preservation of the unique ecosystem and high biological diversity of this region.

Recent National Development Plans have specifically included environmental strategies. In the Sixth Plan (1995–2000), for example, there were targets to attain the highest possible level of food production within the limit of available natural resources, particularly water, without damaging or severely depleting the existing nonrenewable resource base. Given the rate of fossil water depletion it is pretty clear that this objective has not been met. A particularly ambitious objective was the proposal to use the most advanced and environmentally sound technology in the field of industrial development so as to avoid pollution and rationalize the use of resources and raw materials at all stages of the production process, i.e. in design, construction and operation. Alongside this plan was the suggestion that there should be a national system for environmental impact assessments to be adopted by various development sectors throughout the Kingdom and particularly for industrial, agricultural and urban projects. Some support for conservation came through the proposal that the breeding of wildlife species in the Kingdom, and their re-settlement in their natural habitats, should continue.

In the Seventh Plan (2000–2005) there were targets to link the agencies responsible for environmental management to an information network, and also to update information on natural resources and the environment. The Plan also suggests establishing "Friends of the Environment" societies with branches throughout the Kingdom. This is potentially very interesting and obviously has the concept of stakeholder running through its thinking. Quite how this will work out is not clear and getting environmental information across to a burgeoning urban population should not be neglected. More general targets include increasing the area of protected zones to 6.8 percent of the Kingdom's total area, approving the government's General Environmental Code and preparation of a national plan for wildlife conservation and development.

10.2   ISLAM AND THE ENVIRONMENT

It is important to emphasize that *Islam* (Arabic: '*submission*' – to God) and its teachings pervades, or should pervade, all aspects of environmental planning, protection and policy in Saudi Arabia, from Government decisions on the location and management of protected areas to the action of individuals and their relationship with their fields and animals (Bakader *et al.*, 1983; Bakader *et al.*, 1993; Foltz *et al.*, 2003; Izzi Dien, 1990). Protection, conservation and development of the environment and natural resources, is a mandatory religious duty to which every Muslim should be committed (Ministry of Defence and Aviation, 1989).

In this sense, a code of environmental ethics can be derived directly from interpretations of the *Qur'an*. (Arabic: 'recitation') which is divided up into 114 *surah* (Arabic: 'units') of increasing length. Where there are no relevant *surah* in the *Qur'an* reference can then be made to the multi-volume collections of accounts, called *hadiths* (Arabic: 'story or tradition'), that report the *sunna* (Arabic: 'sayings and deeds' of the Prophet Muhammad). Credence is generally given to the *hadith* collections of al-Bukhari who died in 870 AD and Muslim (Abul Husain Muslim b. al-Hajjaj al-Nisaburi) who died in 875 AD. The *Qur'an* and the *hadiths*, together with the *ijma*' (Arabic: 'consensus of opinion') of the classical jurists, and the systematic *ijtihad* (Arabic: 'thinking by analogy') of founders of the legal schools, were drawn together by Muhammad ibn-Idris al-Shafti'i who died in 898 AD and form the four components of classical Islamic law – the *shariah* (Arabic: 'clear path') – the legal system operating in Saudi Arabia.

Izzi Dien (1990) has summarized some of the most important relationships between the environment and Islam as follows:

The environment is God's creation and to protect it is to preserve its values as a sign of the Creator. To assume that the environment's benefits to human beings are the sole reason for its protection may lead to environmental misuse or destruction.

Izzi Deen is here indicating that Islam recognizes that human beings are part of a complex, integrated ecological system.

> Although human beings cannot totally understand all the complexities of the natural environment the fact that the *Qur'an* describes it is reason enough for its preservation. (*surah* 17:44)

This is taken to mean that human ignorance is no justification for environmental destruction.

> All the laws of nature are the laws of God based on the absolute continuity of existence. Human beings must accept such laws as the will of God (*surah* 22:18). The *Qur'an* acknowledges that humankind is not the only community to live in this world. Although human beings have control over other creatures all living things are worthy of respect and protection. (*surah* 6:38). Islamic environmental ethics is based on the concept that all human relationships are established on justice (Arabic: *'adl*) and equity (Arabic: *ihsan*). (*surah* 16:90)

This *surah* points out that cruelty, even when slaughtering an animal for food, is anti-Islamic. The twin principles of justice and equity are also central to understanding the nature of land use practices such as *hima*, as we shall see below.

> The balance of the universe created by God must also be preserved. For 'everything with him is measured. (*surah* 13:8)

In the above *surah* there is a direct allusion to the balance of nature. In this sense, Islamic teaching is at one with what we now know about the population dynamics and trophic/energy structures of ecosystems. And lastly:

> The environment is the gift of God to all ages, past, present and future (*surah* 2:29). This principle is actually shared by all.

> Man, and no other creature, is entrusted by God with the protection of the environment. (*surah* 33:72)

But more than that, Izzi Deen indicates that God has placed a duty on human beings to take on this task.

Islam also encourages individuals to participate in environmental conservation through gifts, bequests and loans. Of particular importance is the charitable endowment known as *waqf*. From an environmental perspective a *waqf* often takes the form of a land trust granted in perpetuity for charitable purposes, such as agricultural research, wildlife propagation or the construction of a garden (Bakader, 1983; Gulaid, 1990).

To provide some additional perspective on the nature of environmental protection, and particularly the problems of rangeland deterioration it, is also useful to understand a little about Islamic law and land ownership. Three types of land are recognized: developed (*amir*) land, undeveloped (*mawat*) land and 'protective zones' (*harim*). As an interesting aside, Dutton (1992) notes that the word *amir* comes from the Arabic root meaning 'alive', *mawat*, comes from the Arabic root meaning 'dead' and *harim*, from a root meaning 'forbidden'. This means that the use of *harim* land is forbidden to anyone but the owner(s). The *harim* land acts as a protective buffer around the owned land. For example, around a town the *harim* land is traditionally defined as that area which can be reached, and returned from, in the same day for the purposes of collecting fuel and/or pasturing livestock.

Developed land comprises the built environment of towns and villages, and agricultural lands under crops. Undeveloped lands, the natural environment, comprises rough grazing and pasture for use by livestock beyond the *harim* land, and lastly 'wilderness' areas are primarily the domain of wild animals. Developed land is owned by individuals and can be bought, sold and inherited. Rough grazing and pasture is owned in common and everyone has equal rights to its use. Wilderness is owned by the state and is open to all users except where the land is designated as protected and a specific use, such as hunting, is forbidden.

Both in areas of undeveloped land and also in those *harim* under some sort of common ownership, all users have equal rights to water, firewood and fuel, and pasture for livestock. Away from towns and villages, in the so-called wilderness areas, grazing rights were part and parcel of the tribal territory (*dirha*). However, with the unification of the Kingdom, the establishment of a central government and the relocation of many rural communities, the tribal system has been generally abolished and uncontrolled grazing is now common.

10.3   THE LEGAL BACKGROUND

The Kingdom has a suite of laws which, in theory, protect the environment, fauna and flora from willful damage or destruction though it is uncertain if they are ever enforced. Some carry custodial sentences, other a fine. Several of these laws are embodied in the so-called Basic Law of 1992 which is commonly referred to as the '*constitution*' of Saudi Arabia.

Article 1 of the Basic Law defines "environment" as man's surroundings including water, air, land and outer space together with all matter, fauna and flora, different forms of energy, physical systems and operations and human activities. "Environmental protection" is taken mean the preservation of the environment and the prevention and curbing of environmental pollution and degradation.

Article 2 defines the Law's objectives as:

1. Preserve, protect and ameliorate the environment and prevent pollution.
2. Protect public hygiene against the dangers of activities deleterious to environment.
3. Conserve, develop and rationalize the use of natural resources.
4. Make environmental planning an integral part of comprehensive development planning in all industrial, agricultural, and urban fields, etc.
5. Enhance environmental awareness, instill a sense of individual and collective responsibility for environmental protection and improve and encourage national voluntary efforts in this respect.

Article 2 also requires the Concerned Authority (MEPA) to undertake such tasks as may protect and prevent degradation of the environment, and in particular to:

6. Review and assess the state of the environment, upgrade monitoring techniques and tools, collect information and conduct environmental studies
7. Document and publish environmental information
8. Prepare, issue, review, develop and interpret environmental protection standards. Draft environmental laws relevant to its responsibilities
9. Ensure compliance by Public Authorities and individuals with environmental laws, standards and criteria, and take necessary measure to this end in co-operation and co-ordination with the Competent and Licensing Authorities
10. Monitor new developments in the domains of environment and environmental management on regional and international levels
11. Promote environmental awareness on all levels.

According to **Article 5**, the Licensing Authorities are required to ensure that environmental assessment studies are made as part of feasibility studies for any project that may have a negative effect on the environment. The party in charge of executing the project shall be responsible for conducting environmental impact assessment studies in accordance with such environmental principles and standards as may be determined by the Concerned Authority in the Implementing Regulations. **Article 6** also notes that the party in charge of executing new or upgraded projects is required to use the best possible technologies congenial to the local environment as well as the least environment-polluting materials.

Article 7 specifies the efforts that shall be exerted with the object of disseminating environmental awareness in the education, media sectors, etc. In **Article 14** the entry of hazardous, toxic or radioactive waste into the Kingdom, including its territorial waters and its exclusive economic zone is banned. It also obligates those in charge of manufacturing, transporting, recycling, treating or finally disposing of hazardous, toxic or radioactive waste to comply with procedures and controls established by the Implementing Regulations of this Law. In addition, the Article forbids the dumping or disposal of hazardous, toxic or radioactive waste by ships or the like in the territorial waters or the exclusive economic zone of the Kingdom.

Finally, **Article 32** of the Basic Law states that, '*The state works for the preservation, protection, and improvement of the environment, and for the prevention of pollution.*' Theoretically, then, most aspects of the Kingdom's environment are legally protected in one way or another. Enforcement is, however, difficult away from industrial sites which are inspected regularly. Indeed, it is probable that most of the laws protecting the environment are poorly known by the public at large and many are simply flouted.

Much of the Kingdom's recent environmental legislation supports its obligation as a signatory to the United Nations Rio Declaration on Environment and Development, (1992) – known as Agenda 21 – an Action Plan for the 21st Century. In December 1994, the Council of Ministers approved the Kingdom of Saudi Arabia's National Agenda 21. Saudi Arabia became a signatory to the Convention on Biological Diversity on the 3rd October 2001. This convention was originally signed by one hundred and fifty government leaders at the 1992 Rio Earth Summit and was conceived as a practical tool for translating the principles of Agenda 21 into reality.

In addition, Saudi Arabia has signed a number of regional and international agreements, protocols and conventions dealing with various aspects of sustainable development. As the world's largest producer of oil it is natural that the Kingdom takes a fairly protective position vis à vis atmospheric pollution. Saudi Arabia is a signatory to the Montreal Protocol, which calls for phasing out harmful chlorofluorocarbon gases by 2010 but under the United Nations Framework Convention on Climate Change, Saudi Arabia, as a non-Annex I country, is not required to reduce its emissions below 1990 levels. Saudi Arabia ratified the Kyoto Protocol as a "non-Annex I state" in December 2004 but has no specific commitments for greenhouse gas reductions.

Of some general importance is the Public Environmental Law which was enacted by Royal Decree (No. M/34 dated 16 October 2001). The Public Environmental Law creates a general regulatory framework for the development and enforcement environmental rules and regulations, and assigns general responsibility for this to MEPA (now called the Presidency of Meteorology and Environmental Protection).

Under the regulations MEPA is responsible for the following:

1. conducting environmental studies
2. documenting and publishing the results of any environmental studies
3. preparing, issuing and reviewing relevant environmental standards
4. ensuring compliance with relevant environmental standards
5. working in conjunction with other government agencies, establishing plans to deal with environmental catastrophes
6. promoting general awareness for protecting the environment
7. working in conjunction with other government agencies, to deal with violations of applicable environmental standards.

There are specific legal instruments designed to conserve biodiversity (Table 10.1).

In practice it is not clear how well these various laws are enforced, if at all. And in such a vast country it is unlikely that violations of the laws are enforceable. Details

Table 10.1 Legislation concerning the use and control of biodiversity in Saudi Arabia

Regulation/Act	Date
Agriculture and Veterinary Quarantine Regulations	1975
The Uncultivated Land Act	1978
The Forest and Rangelands Act	1979
The Water Resources Conservation Act	1980
The National Commission for Wildlife Conservation and Development Act	1986
The Fishing Exploitation and Protection of Live Aquatic Resources Act	1987
The Wildlife Protected Areas Act	1995
The Wild Animals and Birds Hunting Act	1999
Trade in Endangered Wildlife Species Act	2000
Environmental Code	2002

of all the above Acts are given in the *First Saudi Arabian National Report on the Convention of Biological Diversity* (AbuZinada , no date).

10.4   METEOROLOGY AND ENVIRONMENTAL PROTECTION ADMINISTRATION (MEPA)

The General Directorate of Meteorology, as it was then known, became the Meteorological and Environmental Protection Administration by Royal Decree No. 7/M/8903 in 1981 with responsibility for the control of pollution and protection of the environment as well as its meteorological functions. Under MEPA's charter an Environmental Protection General Directorate (EPGD) was established with responsibility for environmental quality standards and environmental impact assessment. MEPA is also technically responsible for the conservation of the flora of Saudi Arabia although not much seems to have been accomplished. MEPA is now part of the Presidency of Meteorology and Environment (PME).

10.4.1   MEPA and conservation of the flora

There are estimated to be about 2,000 flowering plant species in Saudi Arabia of which about 25 are thought to be endemic. Unlike game, which can be hunted, the conservation effort regarding rare plant species is almost negligible. There are no distribution records, and most descriptions of floral distributions and rarity are at best anecdotal. The following information is from MEPA (1986) who provide some data but its reliability is unknown – the classification is based on the IUCN Red Data Book system:

1.   Extinct species

Some four species are thought now to be extinct including: *Crataegus sinaica*. A single specimen was recorded near An Numas in the Asir but the land has now been cleared for building.

2.   Endangered species

MEPA estimate 54 species to be in imminent danger of extinction including:

> *Ceropegia mansouriana* is an endemic found in foothills of the Asir and threatened by land clearance. (= C. botrys K. Schum)
> *Commiphora erythraea* is represented by a single population on Dumsak Island (Farazan Archipelago). All species of *Commiphora* contain potentially useful resins, some aromatic. The site is threatened by felling.
> *Euphorbia* sp. aff. *parciramulosa* Schweinf – small populations of this short-spined succulent near Abu as Sila, north of Jizan are threatened by clearing and rubbish dumping.

Iris albicans Lange – exists as a reduced population near Baljurashi, Al Bahah district and is threatened by the encroachment of agriculture.

Iris postii Mouterde – has a very limited distribution near Turayf and is threatened by intense grazing.

Pancratium tenuifolium – has a localized distribution on coastal basalts of Harrat al Birk overlying fossil coral reefs. The basalt is used for road fill and the habitat threaten by destruction.

Mimusops laurifolia (Forssk.) Friss – this is the largest tree species in Saudi Arabia, growing to 30 m. It is now reduced to 10 individuals in the Jabal Fayfa area north east of Jizan. There is apparent regeneration but the site is now threatened by development.

3. Rare species

About 100 species could be at risk including: *Ochna inermis, Scadoxus multifloris, Gladiolus dalenii.*

Table 10. 2 Ambient Air Quality Standards

Pollutant	Averaging time period	Acceptable maximum µg/m³ (ppm)	Number of allowable exceedance	Number of violations
Sulfur Dioxide (SO₂)	1-hour	730	2 per any 30 days	0
	24-hours	365	1 per year	0
	Yearly	80	0	0
Inhalable Particulate PM₁₀ Matter	24-hours (15-micron)	340	1 per year	0
	Yearly (15-micron)	80	0	0
Photochemical Oxidants (O₃)	1-hour	295	2 per any 30 days	0
Oxides of Nitrogen (NO₂)	1-hour	660	2 per any 30 days	0
	Yearly	100	0	0
Carbon Monoxide (CO)	1-hour	40,000	2 per any 30 days	0
	8-hours	10,000	2 per any 30 days	0
Hydrogen Sulfide (H₂S)	1-hour	200	1 per year	0
	24-hours	40	1 per year	0
Ammonia (NH₃)	1-hour	1,800	1 per year	0
Non-Methane Hydrocarbons (NMHC)	3-hours	160	0	0
Lead (Pb)	3-month	1.5	0	0

Source: MEPA.

10.4.2 Air quality

Air quality monitoring stations are located at Dammam, Jiddah, as Soudah, Riyadh, Yanbu, Makkah and Hufuf. Daily data are published on the on the Internet [http://www.pme.gov.sa/]. The ambient air quality standards adopted by MEPA are shown in Table 10.2. Field experience suggests that the exceedance levels have to be treated with some skepticism especially in downtown Jiddah and Riyadh, around sea water distillation plants, cement works and the industrial complexes at Yanbu and Jubail.

10.5 THE MINISTRY OF AGRICULTURE

The Ministry of Agriculture (MA) – until 2001 part of the Ministry of Agriculture and Water – (MAW) is responsible for the implementation of economic plans and programmes for agriculture, fisheries, forestry, animal resources, national parks and locust control. In addition to its responsibilities for National Parks the MA has also established some twenty-four enclosures varying in area from 250 and 87,000 donums (1 *donum* = 1,000 m²). Some of these are designed for rangeland and environmental studies and others as reserves for natural fodder to be opened up for grazing in years of drought. This is akin to the *hima* system. Conservation of these rangelands has led to noticeable improvement in vegetation cover and pasture productivity. The MA also imports seeds of trees, shrubs, and perennial grass from Australia, USA, Chile, Pakistan, Syria, Egypt and Tunisia to improve the quality of degraded rangeland and a seed production station has been established at Buseita in the northern part of the Kingdom, where 22 species of perennials have been planted and produce about 4 tons of range seeds annually. In addition the National Centre for Agriculture and Water Research, in collaboration with departments and research centres of the Ministry and other scientific institutions in the Kingdom, have established a bank for collecting and keeping seeds and genetic plant strains with a view to utilizing them in the development of new varieties.

In 1995 the MAW completed a study of the land resources of the Kingdom which defined and identified various agricultural climatic regions and land resources units, and assessed the vulnerability of these units to the risks of degradation, erosion, salinization, and inundation. In this study, the Kingdom was divided into 3,176 terrestrial units shown on maps of a scale of 1:500,000 all the details of which are contained in an *Atlas of Land Resources* and various GIS data bases.

The MA also plans to conduct detailed studies of cultivated areas with the aim of establishing the negative factors which limit the production of crops, identifying the environmental hazards, and monitoring land degradation with a view to making recommendations which include appropriate farming methods and standards that serve optimal and sustainable utilization of soil and water resources. This programme includes identification of areas affected by degradation using maps at a scale of 1:50,000. The detailed study of cultivated land aims at establishing the suitability of various locations for different types of utilization. This in turn will assist in making recommendations on appropriate sustainable farming methods.

10.5.1   The Ministry of Agriculture – forest management

Forests in Saudi Arabia are considered one of the Kingdom's most important renewable natural resources. They provide environmental protection by preserving the soil from water and wind erosion. They also help in the distribution of water and control of its flow, and consequently the increased moisture in the soil. In addition, forests have valuable economic, recreational, scenic and tourist roles. According to the MA the most important problems facing the development and protection of this resource are as follows:

A.   Harsh environmental conditions

Among the constraints and determining factors which restrict expansion in the forestry development programme, particularly in increasing the afforested areas, are the location of the Kingdom in the dry desert belt whose climate is characterized by scarcity of rain, dominance of drought throughout the year, high temperatures especially in summer, and lack of adequate quantities of water or rivers.

B.   Felling of trees and shrubs

People in the Kingdom traditionally use wood and charcoal for heating in winter and for cooking on special occasions. This is still the custom in spite of the availability of electricity, (butane) gas and other petroleum derivatives at token prices. Felling living trees has caused shrinkage of the area covered by natural trees and shrubs. To regulate or restrict this process, the Ministry of Agriculture introduced a licensing system for utilization of dry (dead) plants used for firewood, producing charcoal, or for transporting either of these. The positive impact of these new regulations is evident, and are the results of public awareness campaigns on the importance of maintaining trees and shrubs. There are, however, still unlicensed operations of felling and transporting trees and shrubs and one only has to drive out of any sizeable town, particularly in the Hijaz and Asir to see piles of firewood for sale on the roadsides (Figure 10.1).

C.   Urban expansion into forest areas

Expansion of residential master plans due to the population boom in the Kingdom, particularly in the south-west of the Kingdom, has caused a conflict of interest between the need to protect forested areas and the growth of towns, villages and residential centres. As a result extensive areas of wooded terrain have been removed.

D.   Costs of re-afforestation

The high cost of re-afforestation of areas which have lost their natural vegetation cover, compounded with the scarcity of water, low soil fertility, high temperatures, and low rainfall have all contributed to shrinkage of forested areas, especially in the south-west of the Kingdom. Suitable trees either have to be imported or similar local species propagated. Re-afforestation is not cheap and needs to be adequately funded.

Figure 10.1 Sale of wood used for picnic fires during the *hajj* holidays (Photo: author).

E.   *Shortage of forestry specialists*

The number of forestry specialists is considered very low relative to the number of Government initiatives. The universities might well be encouraged to develop more courses in this subject area, though it is unlikely to be all that popular given that many graduates prefer life in the city and have little affinity with fieldwork. In the mid-1990s MAW initiated a survey and inventory of forests by recruiting a private company to carry out an aerial survey of 70,000 km² of the south-western region of the Kingdom. Aerial photography and interpretation will shortly be completed for the remaining natural forest areas. More than 150 forest wardens have now been appointed to manage designated forest areas though there is little information on their effectiveness.

Trees have been planted in fifty-three locations of degraded forest in various areas of the Kingdom in addition to particular afforestation sites for sand dune stabilisation. Treated sewage water (wastewater) has been used for irrigation of afforested locations in Taif and is also being utilized in the Riyadh area. Quite what the long term implications of this process is in terms of eutrophication remains to be seen. The MA has also established thirty forestry nurseries in various parts of the Kingdom, with an annual production capacity of about 1,000,000 saplings. The capacity of these nurseries can be increased as needed to produce saplings suitable for the Kingdom's specific environments.

Table 10.3 National Parks

Name	Size (ha)	Location
Asir National Park	450,000	Asir province
as Soudah Park	883	25 km from Abha
Al Dalgan Park	440	27 km south-west of Abha
Al Gar'aa Park	420	4 km from Dalgan Park
Al Hadaba Park	10	15 km south-east of Abha
Asir – Visitors Park	–	2 km from Abha
Tour Al Maska Park	27	South from Al Gar'aa
HRH Prince Sultan Park	268	4 km from Dalgan Park
Al Hasa National Park	–	20 km from Hufuf
Suwaidra Park	500	Part of Al Hasa National Park
Jawatha Park	100	Part of Al Hasa National Park
Al Shaybani Park	350	Near Akeer Port, Al Hasa
Sa'ad National Park	300	110 km east of Riyadh
Huraimla National Park	–	80 km north-west of Riyadh
Al Baha National Park	–	Al Baha region

Notes: National Parks have also been proposed for Madinah, Qassim, Ha'il, Tabuk and Jawf. Spellings are somewhat variable e.g. Soudah – also Souda; Hadaba – also Hadba.

10.5.2   The MA and National Parks

National Parks in the Kingdom are managed by the National Parks Department which is part of the Ministry of Agriculture and their establishment is part and parcel of a more general government conservation strategy (Ady and Walter, 1991). In theory, parks have been located so as to represent a variety of ecological niches where biodiversity is high, but in practice very little conservation work is actually undertaken by the Park authorities and their emphasis now seems to be more related to general tourism such as camping and the establishment of picnic sites. Indeed, the term National Parks is directly translated from the Arabic as *National Recreation Areas,* along the lines of parks in America.

10.5.3   The Asir National Park

This is the largest park (450,000 ha) and by far the most important. Its ecosystems range from coral reef communities up to mountain top juniper forests bathed frequently in cloud. The park is formed from four linked units.

1. Dolgen is an area of 440 ha dominated by *Acacia* spp. and grasses.
2. Al Gar'aa is a juniper forests of about 410 ha.
3. As Soudah (the Black Mountain) has an area of about 833 ha and an average elevation of 3,000 m asl. Jabal as Soudah is the Kingdom's highest mountain. The climate here is rather cool and fogs are often present due to air being forced up the escarpment. This area is a favourite place with summer visitors seeking to escape the heat of the plains.
4. Ad Hadaba (plateau) unit is an area with rugged sandstone outcrops on the crest of the escarpment.

10.6   NATIONAL COMMISSION FOR WILDLIFE CONSERVATION AND DEVELOPMENT (NCWCD)

In 1968 David Harrison published an important paper in the journal *Oryx* which reviewed the status and conservation needs of the larger mammals in the Arabian Peninsula. In a way, this paper was instrumental in bringing to the fore the threat of extinction by motorized hunting parties of several species of gazelles especially the dorcas (*Gazella dorcas*) which inhabits open gravel plains in the eastern Hijaz. Conservation and the maintenance of biodiversity was now firmly on the agenda.

At this time biological conservation was the responsibility of two government agencies, the Ministry of Agriculture and the Meteorological and Environmental Protection Agency (MEPA). Although the Ministry of Agricultural and Water established the Asir National Park in 1981 and had given the island of Umm Al-Qamari over to conservation in 1977, it was felt that this dual administrative structure was rather cumbersome and there was a need for a single authority to be responsible for all conservation issues.

On the 12th May 1986 the NCWCD was established by Royal Decree M/22 with the responsibilities for conserving the Kingdom's wild animal and plant resources, and of establishing and managing protected areas for wildlife conservation. As far as conservation of the flora is concerned there was, and still is, some confusion as to whether this is the responsibility of the MA or the NCWCD. In any case, very little flora conservation seems to be conducted.

The long term goal of conservation management efforts by the NCWCD is sustainable utilization. This strategy is totally in line with the resolutions of the IV Global Congress on National Parks and Protected Areas held in Caracus in 1992 which buried the old concept of protected areas preserved as islands outside the local communities and alien to human events. In its place a more up-to-date approach was adopted which makes protected areas the centre of sustainable development strategies, stressing the interrelation between protected areas and their surroundings and paying particular attention to the benefits each area can provide for local communities.

The NCWCD has a powerful Board of Directors, chaired by the Second Deputy Prime Minister and consisting of the Minister of the Interior, the Minister of Foreign Affairs, the Minister of Agriculture, the Governor of the Asir region, the President of the King Abdulaziz City of Sciences and Technology, the President of MEPA and the Secretary-General of the NCWCD (until his retirement in 2006, Professor Abdul Aziz Abu Zinada and since then His Highness Prince Bander Bin Saud Bin Mohammad Al

Saud). The NCWCD now employs about 320 people of whom 163 are rangers working in the field. In 1992 the budget totaled US$ 17.2 million.

Following the near extermination of many other wildlife species through excessive hunting, and the severe destruction of habitat through overgrazing by domestic livestock, major conservation measures have been taken by the NCWCD in an attempt to restore and maintain the biological diversity in Saudi Arabia. These measures include the protection and revitalization of suitable habitats, and the re-establishment of self-sustaining wildlife populations of various species such as gazelles and the Arabian oryx (*Oryx leucoryx*).

The NCWCD has three research centres which concentrate on captive breeding, reintroductions, and related research. Research at the National Wildlife Research Centre (NWRC) in Taif has resulted in the reintroduction to wild conditions of the Arabian oryx. Other native endangered species such as houbara bustard, onager (Asiatic wild ass) and Nubian ibex are also being successfully bred in captivity at the NWRC. The King Khalid Wildlife Research Centre (KKWRC) at Thumamah is another component institution engaged in captive breeding, focusing on native gazelles. A new centre for gazelle research, the Amir Mohammed Al-Sudeiri Gazelle Research Centre at Al Qassim, is beginning to make its own valuable contribution to the captive breeding programme.

10.6.1   System plan for protected areas

The term *protected area* is usually taken to mean a geographically defined area which is designated or regulated and managed to achieve specific conservation objectives. Within any system plan for protection it is important that the choice of protected areas represents the various biotopes in the country (Abuzinada, 1998). In Saudi Arabia, the four following guidelines are used:

1. Representative coverage of the country's physiographic and biogeographic units, on the basis of reliable classifications.
2. Recognition of the need to protect adequate, suitable habitats for a number of high-profile "flagship" species, or those of special significance and to which the NCWCD's species conservation programme is heavily committed.
3. Adequate protection of habitats of key biological importance.
4. An equitable geographical spread of protected areas throughout the Kingdom in order to apportion the benefits of conservation as fairly as possible.

The NCWCD System Plan for Protected Areas (Child and Grainger, 1990; Abuzinada *et al.*, 1991) prioritized protected area establishment by using two sets of criteria, natural values and practical considerations. Natural values include such factors as species value, physiographic considerations, biomes and ecological processes. Practical considerations examined possible conflicts with local grazing practice, any local support, and the actual configuration of the proposed area. Based on these sets of criteria a priority ranking has been allocated to each possible area.

In its original form the System Plan envisaged a network of 103 sites. However, because of the exclusionary nature of the proposed areas the plan ran into both ministerial and public opposition and has still not been formally adopted. Nevertheless, the plan is the main working document at the NCWCD which, in the light of the criticism of the original plan, has modified its approach to protected area management from complete exclusion to the local involvement of people living in the area.

By 2003, 15 protected areas were approved by the Council of Ministers (Figure 10.2). In July 2006 a statement by Prince Bandar ibn Saud, the secretary-general of the National Commission for Wildlife Conservation and Development (NCWCD) indicated that the Kingdom plans to establish sixty more protected areas including two marine reserves at Aqaba and Fatul Wajd on the Red Sea. An updated System Plan is being prepared by Othman Llewellyn – a planner at the NCWCD.

Dr Philip Seddon, formerly Research Coordinator at the NWRC, and now at the University of Otago, has traced the changing management scheme of protected areas at the NCWCD in some detail. He suggests that in the early days of the NCWCD there was a policy of strict protection. This involved the exclusion of all livestock and the restriction of human access. This was supposed to allow the range to recover and also cut down on all illegal poaching. This approach, however, is not terribly successful in a country like Saudi Arabia. The terrain is difficult to monitor and the *badu* regard many

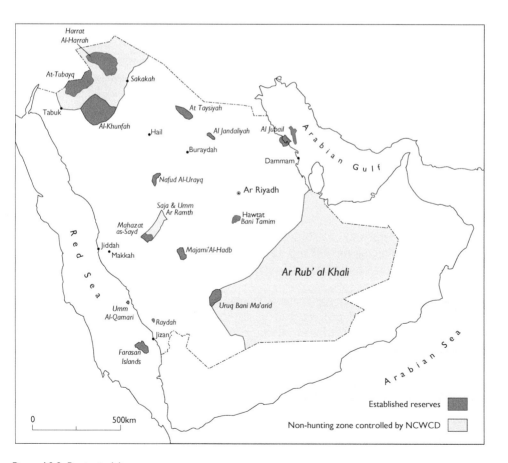

Figure 10.2 Protected Areas.

protected species as good sources of protein. Seddon traces the evolution of policy with regard to four protected areas (Seddon, 2000). His ideas are briefly summarized here:

1. Exclusion Policy

The 20,000 km^2 Al-Khunfah protected area was designated in1988 with little involvement of the surrounding communities. It was chosen to protect the then largest population of sand gazelle in Saudi Arabia and to conserve a large tract of sandy gravel plain bordering the western edge of An Nafud (Child and Grainger, 1990). Only the central core of about 8,000 km^2 are patrolled by rangers, supported by light aircraft. This central zone has Special Nature Reserve status (see Table 10.4) and has the strictest level of protection. All hunting of wildlife, grazing of domestic livestock and settlements of any kind are forbidden. Not surprisingly, the displaced local communities are aggrieved at their loss of traditional grazing grounds, and as a consequence there is heavy poaching of the gazelle made easy by access from major highways. The response of the NCWCD was to build a large ditch and dyke barrier along the southern and part of the western boundary of the core zone but this is hardly thwarts the determined poacher and there are indications that the protected gazelle population is in decline.

2. Fencing

In order to provide a higher level of protection for the Mahazat as-Sayd protected area which was established in 1988 the NCWCD decided to surround this small reserve which has an area a little over 2,200 km^2 with a barbed wire-topped mesh-link fence in 1989. The area was also given Special Nature Reserve Status (Seddon, 1995). As with Al Khunfah, the area is patrolled and public access prohibited. This seems to have worked and there have been no instances of poaching in the last 16 years. On the other hand relationships with the displaced local community are not good and there is no sense in which they can be said to be a part of the scheme.

3. Multiple use resource sites

Fencing large protected areas is extremely costly and the total exclusion of the local population clearly unacceptable especially in a country such as Saudi Arabia with its tradition of nomadism. When the 5,000 km^2 Uruq Bani Ma'arid protected area was designated in 1994 the NCWCD decided on the alternative strategy of incorporating the local community needs into the management plan. The protected area was divided into three zones (Sulayem et al., 1997):

  i) An inner core of 2,400 km^2 was given Natural Reserve status allowing more public access than an SNR. In particular, the transit of livestock through this zone is permitted so that seasonal access to central grazing areas is facilitated.
 ii) A managed grazing zone
iii) A controlled hunting zone

In addition, the NCWCD took the sensible decision to employ local people as rangers and also consult with them regarding the released oryx. As a result the protected area has gained good public support without the need for fences.

Table 10.4 Protected Areas 2004 (categories are defined in Table 10.5)

Protected area	Date established	Area (Km2)	Category	Species in situ	re-introduced	To be introduced
Umm al Qamari	1978	106	NR		none	
Harrat al Harrah	1987	13,775	SNR/RUR	reem; hyaena; wolf; houbara; caracal	none	oryx
Al Khunfah	1987	34,225	SNR/RUR	reem; hyaena; wolf; jackal	none	oryx
Mahazat as-Sayd	1988	2,000	SNR	foxes; cats; houbara	reem;oryx;ostrich; houbara	
Ibex Reserve	1988	2,369	SNR/RUR	ibex; wolf	idmi	ostrich?
At Tubayq	1989	12,200	NR	ibex; caracal; hyaena; wolf; leopard; jackal	none	
Farasan Islands	1989	600	SNR/NR/RUR	idmi	none	ostrich?
Raydah	1989	9	SNR	leopard; caracal	none	none
Majami al Hadb	1993	3,400	SNR/RUR	wolf; hyaena; caracal	idmi	oryx
'Uruq Bani Ma'arid	1994	5,500	SNR/RUR	foxes; caracal; wolf; sand cats	oryx; reem; idmi	ostrich
Gulf Conservation Area	1995	4,262	SNR/RUR	dugong	reem	
At Taysiyah	1995	2,855	RUR	houbara	reem	
Al Jandaliyah	1995	1,160	RUR	houbara		houbara; oryx; reem; ostrich
Umm ar Rimth	1995	5,500	RUR	houbara		houbara
Nafud al 'Urayq	1995	1,900	RUR	houbara; hyaena; wolf		houbara; ostrich

Table 10.5 Protected Area categories

Saudi Arabian category	IUCN equivalent	
Special Nature Reserve (SNR)	I	Strict nature reserve/scientific reserve
	II	National Park
	IV	Nature Conservation Reserve or Managed Reserve/Wildlife Sanctuary
Nature Reserve (NR)	II	National Park
	IV	Nature Conservation Reserve or Managed Reserve/Wildlife Sanctuary
Biological Reserve (BR)	I	Strict Nature Reserve/ Managed Reserve
Resource Use Reserve (RUR)	V	Protected Landscape or Seascape
	VI	Resource Reserve
	VIII	Multiple Use Management Area/ Managed Resource Area
	IX	Biosphere Reserve
Controlled Hunting Reserve (CUR)	VIII	Managed Resource Area

4. Joint management

The Umm ar-Rimth (Arabic: *mother of the rimth*) protected area was declared, or rather imposed, by Royal fiat in 1995 and was not part of the NCWCD's original systems plan. Indeed, it was not at all clear at the time where the exact boundaries were or how they were to be defined. Anyway, this 6,000 km² area was chosen as a new reintroduction site for the Houbara Bustard and the NCWCD is now trying to develop a management plan. The release site is to occupy about 10 percent of the total area and is to be designated an SNR with restricted public access. However, the rest of the area is to be designated as a Resource Use Reserve in recognition that preservation and exclusion are not acceptable or successful in the Saudi Arabian context. The proposed area is, in fact, to be jointly managed by local communities and the NCWCD with the aim of improving the quality of the range by improved grazing practices. Seddon notes that for the first time the NCWCD is taking the advice, co-operation and approval of local sheikhs before formal management decisions are set into place. The central aim of this approach is to develop an empathetic relationship between local stakeholders and the ecological management of the area.

In concluding his analysis of the NCWCD approach to protected area management in the Kingdom, Seddon identified five trends: increased consultation with local communities; increased use of the Resource Use Reserve; increased employment of local people as rangers; increase contact between rangers and local communities; decreased application of strict Special Nature Reserve zones.

The updated Systems Plan also introduces the notion of Key Biological Sites (critical wildlife habitats). The protection of these sites is the single most important action to ensure that biological diversity is preserved for future generations. The NCWCD provides the following general information on such sites (Table 10.6).

Table 10.6 Numbers of sites in different categories identified as key biological sites in the updated system plan

Area	Large sites	Small sites	Major reasons for conservation
Wetlands	16 natural; 10 artificial	22 natural; 3 artificial	Many natural wetlands have been drained by over-use of water; wetlands are important for migrating birds; play important roles in the hydrology of catchments.
Isolated mountain massifs	11	6	Critical centres of biodiversity because they provide a wide array of habitats in close juxtaposition; comparative inaccessibility provides protection; under increasing threat of urban development and recreation use.
Juniper woodlands	10	11	One of the few densely wooded ecosystems of Saudi Arabia; support rich fauna and flora; key areas of soil formation and water conservation.
Marine island	13	13	Isolated areas; high productivity; important breeding sites for birds and marine turtles.
Mangrove swamps	12	14	Major contribution to coastal productivity; breeding and refuge sites for many marine species.
Seagrass beds	16	12	Highly productive areas; feeding grounds for fish, turtles and dugong; stabilise soft sediments and reduce coastal erosion; highly vulnerable to landfilling and other coastal developments.
Coral reefs	15	11	Extremely productive; provide breeding and feeding grounds for many marine species; Red Sea coral reefs amongst most richest in the world; Red Sea reefs of Saudi Arabia still undamaged; highly vulnerable to oil spills and other pollution, physical damage and over-exploitation.
Seed production	12	17	Provide *in situ* seedbanks that can be of crucial importance for re-colonisation of over-grazed and otherwise damaged vegetation.
Other important sites			Woodlands (other than juniper woodlands), saltmarshes, *rawdahs* and algal beds all provide refuges, seedbanks and breeding sites. Identification of sites of particular importance that are in need of conservation is an ongoing activity.

10.6.2   The Jubail Marine Wildlife Sanctuary

The impact of the Gulf War oil pollution was disastrous ecologically. Some fifty percent of the intertidal vegetation of the contaminated shorelines was lost and an estimated 30,000 birds were killed including large numbers of Socotra Cormorants (*Phalacrocorax nigrogularis*) which are endemic to the Gulf and southern Arabia (Symens and Suhaibani, 1994). As a result of cooperation between the European Union and the NCWCD and an extensive research programme which analysed the damage done by the oil spill, the first marine habitat and wildlife sanctuary was established in the western Gulf north of Jubail (Krupp *et al.*, 1996). The sanctuary comprises five offshore coral islands and two large embayments – Dauhat ad-Dafi and Dauhat al-Musallamiya, and covers some 2,400 km². Within the sanctuary there are areas designated Special Nature

Reserves and Resource Use Reserves. The purpose of the Special Nature Reserves is to protect marine life and reduce disturbance to coastal and terrestrial ecosystems after the catastrophe of the war.

10.6.3   NWRC – Captive breeding of flagship species

The National Wildlife Research Centre (NWRC) was established in April 1986 by His Royal Highness Prince Saud al Faisal with the single aim of breeding the houbara bustard in captivity (Seddon, 1996). Today, it is technically a section of the NCWCD and has widened it objectives somewhat to include:

1. Breeding endangered endemic species
2. To reintroduce those species into wild, especially protected areas
3. To undertake scientific research on those species
4. To assist the National Commission for Wildlife Conservation and Development in promoting wildlife conservation by producing films for Saudi Arabian Television

10.6.3.1   houbara bustard

The Asiatic houbara bustard (*Chlamydotis undulata macqueenii*) is the symbol of wildlife in Saudi Arabia and a good choice for a flagship species around which conservation efforts in Saudi Arabia can focus (Figure 10.3). As the favoured quarry for Arab falconers hunting from camel-back, populations of the houbara seemed to have achieved some sort of equilibrium over many hundreds of years. All this changed with the advent of 4WD vehicles and firearms. Huge falconry parties can now hunt in even the most remote regions, and the allure of the houbara is such that even those without a trained falcon may attempt to shoot the birds wherever they are found. Increasing and relentless hunting pressure within Saudi Arabia targeted both migrant and resident houbara. Migrant birds gained respite from hunting when they left the Peninsula in spring each year to return to their breeding grounds. However, the resident houbara remained vulnerable to hunting throughout the year, even during the breeding season when displaying males would advertise their presence to females and to falconers. The end result was an inevitable decline in the numbers and range of resident houbara in Saudi Arabia.

Saudi Arabia contains both breeding populations and migratory populations of the Asiatic houbara bustard (van Heezik and Seddon, 1996). Resident houbara were once widespread throughout the plains of Saudi Arabia (Meinertzhagen, 1954) and their simple nest scrapes containing two to four eggs could be found each spring and provided a seasonal dietary supplement for the *badu*. Migratory houbara enter the Arabian Peninsula each autumn, probably moving south to avoid the freezing winter temperatures of their breeding grounds in central Asia.

The NCWCD, in conjunction with the NWRC, has adopted a two-pronged approach to Houbara bustard conservation in Saudi Arabia (Saint Jalme *et al.*, 1996):

1. The restoration of resident breeding populations, principally through the reintroduction of captive-bred birds.
2. The protection and encouragement of migrant houbara.

Figure 10.3 houbara bustard (Photo P. Seddon).

As a high profile, high budget, operation the captive breeding (Saint Jalme and van Heezik, 1996) and reintroduction (Seddon and Maloney, 1996) of captive-bred houbara has attracted much attention. But equally important from a regional perspective is the protection of migrants in their wintering grounds. Migrant houbara support the falconry that still takes place between December and February each year in Saudi Arabia, and while protection of these birds on their breeding grounds is essential to ensure the maintenance of a hunting resource, it is equally important that hunting on wintering grounds is regulated. Although hunting regulations do exist in Saudi Arabia, setting out hunting seasons and permissible methods by which houbara may be taken, enforcement of these regulations is extremely difficult. In recognition of these difficulties the NCWCD has been active in the creation of refuges within which migrant houbara are completely protected from hunting.

Four new sites have been established by the NCWCD Board of Governors, a committee with representation from the Ministry of Agriculture and Water, The Ministry of Interior, MEPA, and the Emara of Makkah, Al-Qassim and Ha'il. The new sites range in size from 1,150 km² to 5,670 km², and provide a total of 12,650 km² of houbara habitat within which hunting is theoretically prohibited and the grazing by domestic livestock regulated. Up to 10 percent of each site is designated "non-grazing", in order to create a core zone of complete protection from all forms of disturbance (Table 10.7). The location of the no-grazing core zones was determined by NCWCD field survey teams. Core zones are fenced to ensure livestock are kept out, while the entire reserve areas is be patrolled by NCWCD rangers with participation from local tribes. Following establishment of the reserves, monitoring of vegetation recovery is undertaken to assess the potential of these areas possibly to act as reintroduction sites in the future.

The sites are:

1. Harrat al-Harrah, which also protects the last resident houbara population in Saudi Arabia (Seddon and van Heezik, 1996)
2. Al Khunfa
3. Mahazat as-Sayd, also a reintroduction site for captive-bred houbara
4. Uruq Bani Ma'arid on the western edge of the Empty Quarter.

Since their protection, migrant houbara numbers have been recorded seasonally in each of these reserves.

In the last few years, the NCWCD has been working to expand its network of protected sites for houbara, with particular emphasis on the central and eastern regions of Saudi Arabia, areas traditionally known by falconers to contain good

Table 10.7 Newly created NCWCD protected areas for migrant houbara bustards in central and eastern Saudi Arabia

	Core area km² No grazing	Total area km² No hunting
At-Taysiyah	393	3,937
Al-Jandaliya	115	1,150

numbers of migrant houbara in winter (Table 10.7). It is important to provide sites within these hunting grounds, inside which houbara can rest and feed undisturbed. The houbara has hung on and is now being helped to recover. Other species have not been so fortunate. Kingdon (1991) observed that visitors to Arabia in the 1840s described herds of gazelles, oryx and ostrich and numerous ibex, hyenas, wolves. This was just before the mass export of cheap fire arms from Europe and their replication in markets all over the Peninsula. By the early 1900s carrying guns was commonplace among young adult men and the Arabian wild ass (*Equus hemionus hemipus*) and lion (*Panthera leo*) soon became extinct.

Indiscriminate hunting over the last 60 years finally put paid to the Arabian sub-species of the ostrich (*struthio camelus syriacus*) in the 1930s, and more recently the Saudi Arabian population of the Arabian oryx (*Oryx leucoryx*), the Saudi gazelle (*Gazella saudiya*) and the cheetah (*Acinonyx jubatus*). Thouless (1991) highlights the bit-ter irony that while oil discovery brought wealth, it also brought foreign oil-workers whose idea of a weekend's sport was a motorized hunting expedition. The situation is not completely hopeless and a number of captive breeding programs are underway with a view to building viable populations that can eventually populate more remote parts of the country, particularly the protected areas set up by the NCWRC.

10.6.3.2   Arabian oryx

This beautiful animal with its silvery white coat and long tapered horns was once widespread in the Arabian Peninsula but by 1960 had dwindled to about 200 indi-viduals living in the Empty Quarter (Figure 10.4).

Figure 10.4 Arabian oryx (Photo: NCWCD).

As Kingdon (1991) notes, the coloration of the oryx may well have evolved as a useful device enabling these mainly solitary animals to keep in touch with each other. The sight of an oryx standing on the crest of a dune must have been a magnificent spectacle. Unfortunately, it also made it an easy target for a marksman, and yesterday's asset became today's tragedy and the species was declared extinct in the wild in 1972 when the last specimen was shot in Oman.

The Arabian oryx has now become one of the most illustrative symbols of the recovery of a species extinct in the wild by "*ex situ*" conservation measures. Previously, in 1962–63 the Fauna and Flora Preservation Society captured three specimens (2 males and 1 female) in the Yemen. These animals were sent first to Phoenix Zoo in the USA and then moved on to the San Diego Zoo. To these were added one female from Kuwait and four animals from a private herd established originally by H.R.M. King Saud. These eight animals established was what is now called 'The World Herd'. The herd has bred successfully and now numbers several hundred individuals. In the mean time, H.R.H. King Khaled established a herd on his farm, now the King Khaled Wildlife Research Centre (KKWRC), at Thumama near Riyadh with animals from Saudi Arabia, Qatar and from the World Herd. Some of these animals were transferred to the NWRC in May, 1986. A herd of seventy oryx now breeds at the centre with a smaller herd also firmly established at the KKWRC.

The only protected areas in Saudi Arabia to be fenced by the NCWCD is the Mahazat As Sayd reserve some 180 km east of Taif. It is the major site for all captive breeding animal releases. NCWCD rangers patrol the reserve and liaise with local people to make them aware of the purpose of the fencing and the importance of keeping it intact so as to stop domestic stock entering and grazing vegetation needed by the released animal populations.

Until recent times its 2,190 km² of sand and gravel plateau held relatively large herds of Arabian gazelles – reem, idmi and afri . This protected area was chosen for the release of 15 oryx in 1990. They came from two sources: 9 from San Diego Zoo in the USA and 6 from the Shaumari Wildlife Reserve in Jordan. By 1993 the free-ranging population of oryx in Mahazat grew from 62 to 113 animals. Thirty-one calves were born in the 'wild' and sixteen new animals were released into the reserve. Mortality rates of free-ranging adults and calves are less than 10 percent. 1993 is the first year in which natural reproductive recruitment has accounted for the majority increase in the free-ranging population since the initial re-introduction.

In early May 1994, 13 more oryx were released into the reserve, bringing the estimated total number of animals in the 'wild' to 134. With a target population size of around 300 oryx, and with a good natural rate of increase, no further releases of captive-bred oryx are planned for Mahazat. New release sites are currently being considered in unfenced protected areas in the far north and south of Saudi Arabia.

10.6.3.3   The ostrich

The Arabian subspecies of the ostrich (*Struthio camelus syriacus*) was shot to extinction in the 1950s and re-introduction is therefore not possible. However, the National Commission for Wildlife Conservation and Development took the decision to introduce red-necked ostriches (*Struthio camelus camelus*) from Sudan, the nearest living relative of the extinct Arabian subspecies, in order to develop a breeding stock at the

NWRC. In June 1994 four male and three female captive-bred red-necked ostriches were released into the fenced Mahazat as-Sayd reserve – an event which marked the start of the 're-introduction' of ostriches in Saudi Arabia. In July 1995, a second release of one male and one female ostrich took place, also in Mahazat as-Sayd. Because the released birds were closely related to each other, introductions at this first stage were restricted to the fenced Mahazat reserve. Captive breeding undertaken at the NWRC is attempting to increase the genetic diversity of stock by obtaining new founders from Sudan and from collections in the Middle East. The purpose of the releases was to determine if ostriches can survive and breed without provisions of food and water. All the released birds were radio-tagged and their movements and behavior monitored closely throughout 1994 and 1995.

In April 1995, after good spring rainfall, the first breeding attempt was initiated. A single female paired with one male and laid a total of eleven eggs. Incubation progressed normally until the 32nd day when the nest was uncovered and left for twenty four hours. The nest was finally abandoned after more than sixty days of incubation, and the eggs were collected and examined. Eight of the eleven eggs were fertile, but all the embryos died when approximately four to five weeks old, possibly as a result of the exposure to direct sun for one day.

During August and September 1995, four ostriches died in the reserve, including the breeding pair and female from the second release. Veterinary examination suggested that at least three of the deaths were related to dehydration and progressive decline in body condition, whereas the fourth involved injury after being kicked by a territorial male. It appears that few of the birds were able to cope with the harsh summer conditions. The breeding pair may have been additionally stressed by incubation relatively late in the season. A fifth bird, the male from the second release, was recaptured after being found prone, thin and dehydrated. This left four birds in the reserve, three males and one female. In February 1996, a second nest was recorded, with laying throughout the month resulting in a total of 13 eggs. Incubation proceeded normally, but by mid-April, well after the normal forty-five day incubation span, none of the eggs had hatched. Examination of several eggs showed them to be infertile. The remaining eggs were collected from the nest and found also to be infertile.

The deaths were disappointing but have answered some questions about the ability of the birds to survive. It is clear that some birds can survive, and that the loss of a proportion of ill-adapted ostriches is to be expected. Future releases will attempt to extend the pre-release phase, with a gradual transition from captive-care through to full independence.

A further 9 ostriches are currently being held in a 25 ha enclosure in the reserve. They were provided with food and water throughout the summer of 1996 and, after radio-tagging, were moved into a 200 ha naturally-vegetated enclosure. Food and water continue to be supplied and the birds use of natural plants monitored. A third release took place in the autumn and winter of 1996–7.

10.6.3.4 gazelles

Nader (1990) names three species of gazelle in his checklist of Saudi Arabian mammals: the sand gazelle or *reem* (*Gazella subgutturosa*), dorcas gazelle or *afri* (*Gazella dorcas saudiya*) and the mountain gazelle or *idmi* (*Gazella gazella*). In spite of the dire warnings of the impacts of overhunting of these species issued by Harrison in 1968, some twenty years later they faced extinction in the wild (Habibi, 1986). Fortunately, mountain gazelles and sand gazelles formed part of the large collection of ungulates kept at King's Khalid's farm at Thumamah, near Riyadh although they were in poor condition due to lack of management. The wild dorcas gazelle is thought now to be extinct in Saudi Arabia. A large population of about 1,000 mountain gazelles on the Farazan Islands (60 km off shore from Jizan) was discovered in the 1980s. A successful captive breeding programme has allowed re-introduction of populations of gazelle to several protected areas (Table 10.4).

The major threat to gazelle populations in Saudi Arabia is illegal hunting. In principle, gazelle in protected areas should be relatively safe but enforcing 'no hunting' legislation in such large, unfenced areas is impractical. To poor *badu*, the temptation of easily caught fresh meat must be an enormous temptation. Perhaps one of the surest ways of guaranteeing freedom from poaching is to employ local tribesmen as rangers (Thouless, *et al.*, 1991; Magin and Greth, 1994). The success of such operations depends very much on the attitudes of the local people and also on the local Emir who must act in cases of suspected hunting.

10.6.4   Potential plant flagship species

In the updated System Plan it has been suggested that several plants might serve as flagship species. Many are relict plants such as the almond *Prunus korshinskyii*, the tulip *Tulipa biflora* and the heather *Erica arborea*. Some of the more spectacular species might also come into this category. For example, the dragon tree *Dracaena ombet* and the lote tree *Ziziphus spina-Chrisi*. This species of Sudanese origin has many uses in Arabian culture and is used in the production of soap, timber, forage, honey, fruit and various medicines. It is also mentioned in the *Qur'an* (*surah* 56:28 – Al-Waaqe`ah – the Inevitable) "*Companions with beautiful, big, and lustrous eyes, – Like unto Pearls well-guarded. All of this among Lote-trees without thorns.*"

10.7   TRADITIONAL LAND MANAGEMENT – *HIMA* (SOMETIMES *HEMA*)

A *hima* (Arabic: pl. *ahmyah* or *ahmia*) is a protected grazing reserve belonging to a tribal group or village and is one of the oldest systems of rangeland management in the world (Draz, 1985). In Arabic, the term translates as 'to bring under protection'. Less common is the family-owned *hima* known as a *hojjra* (sometimes spelled *hodjra*) which is often used for bee keeping and the production of honey – a much sought after, and expensive, item in the souks.

The *hima* system is actually pre-Islamic in origin but was adopted by the Muslim community because the benefits of this approach to range management do not contradict any fundamental Islamic principles (Dutton, 1992). The *hima* system or variants of it was once widely practiced in the Arabian Peninsula as a whole (Grainger and Llewellyn, 1994) but has now all but died out. In Oman, Wilkinson (1978) describes the practice of communal range control by villages of eastern oases and Thesiger (1959) remarks on the areas called *hawtah* where hunting, cutting or grazing were proscribed. Similar regulations governing rangeland have been recorded in Syria (locally called *mahmia* or *ma'a*), from the Kurdish areas of Iraq and Turkey where they are referred to as *koze* (Draz, 1978) and from Tunisia where they a referred to as *ghidal* or *zenakah* (Hobbs, 1985).

In pre-Islamic times when a tribal chief settled on a piece of land he had a dog bark and then made the area, within the bark of the dog, exclusive for his own horses, camels and beasts, and no one else could share it. Thus public *mawat* (undeveloped) land was converted into a *hima* for his own personal use. The Prophet Muhammad actually forbade the appropriation of land as a private *hima* and in Islam it is only possible to declare land as a *hima* for the general benefit of the community or specifically for those in need within the community. Dutton (*op. cit.*) cites the example of the Prophet Muhammad who set aside an area around Madinah 16 km in radius as a *hima* for those horses which have been used in war. The practice was continued by the caliphs immediately after him. Abu Bakr, the first caliph created a *hima* near ar Rabdah for the animals collected as *zakat* (Arabic: annual wealth tax). Omar Ibn al Khattab, the second caliph, created the Hima al Rabza, near Dari'ya, a grazing reserve nearly 250 km long and Othman, the third caliph, extended it so as to provide grazing for up to 40,000 camels and horses.

The practice of *hima* creation seems to have spread with Islam and some *ahmyah* are even found in northern Nigeria where 'Uthman Ibn Fudi, a Fulani mystic, philosopher, teacher and revolutionary reformer created a new Muslim state, the Fulani empire, in the early 1800s. 'Uthman ibn Fudi wrote excellent Islamic legal texts, indicating one on the importance of the *hima* and the regulations pertaining to it. Range reserves, similar to *ahmyah*, and called *mahram*, are described in legal documents from the same period in the state of Bornu, the present-day Lake Chad region.

The only conditions for a *hima* to be valid are: a) that it is constituted by the legitimate authority; b) that it is for the public good; c) and that it avoids causing undue hardship to the local people by depriving them of resources they needed to survive, and d) actually provides more benefits than costs to society. The whole system is designed to be extremely flexible and easily adapted to the needs of the local population.

According to Batanouny (1998) various *ahmyah* are endowed with written statute whereas others have no documentation and are authority handed down from generation to generation. Some documentation might also exist in the Islamic courts, particularly where there has been a legal dispute, but this has not so far been researched.

The actual administration of the *hima* varied considerably from place to place but was usually the responsibility of the local village, tribal head or sheikh who decided when the *hima* was to be opened for grazing, for how long, and for what quantity of livestock. In times of special need the sheikh also determined if the *hima* could be used by other tribes. In the large *hima* at Bilad Zahran, south of Taif, Batanouny (*op. cit.*) describes the situation as follows. No direct grazing is allowed and tribal members may only cut the grass for hay making once it has grown to arms length. Adult members of the clan are allowed to cut the hay one day per week and women on another day. Those authorized may cut from dawn to dusk and tribal guards (*hurras*) watch their work. These guards supervise the *hima* and depend on a sort of administrative council whose members are called *muwamin* – they are elected for life by the male members of the clan. The *muwamin* decide if and when the hay can be cut and what maintenance is to be done, such as keeping the border in order and making paths. They also decide the guard rota and how trespassers should be penalized. At Hima Quraish, which is located on a hill top between Wadi Ghadirain and Wadi Dhahiyya near Taif, the grazing is for cattle only which are still used to plough small fields. In the past, 5 percent of any trespassing sheep were confiscated but this is now illegal and protection difficult (Ady, 1995).

Draz (1978) recorded five main types of *hima*:

a)   animal grazing is prohibited, but grass cutting for fodder is permitted in years of drought and special privileges may be granted to needy families to cut mature grass e.g. Hima Bani Sar near Al Baha in the Asir.

b)   where grazing and cutting are permitted on a seasonal basis as in Hima al Azahra and Hima Hameed around Baljurashi about 40 km south of Al Baha.

c)   where grazing is allowed but restricted to certain numbers and or kinds of stock e.g. Hima Thamalah near Taif.

d)   where the area is kept primarily for bee-keeping with no grazing allowed during the flowering season. These are primarily private *ahmyah* or *hodjra* (also transliterated *hojjra*) which are found particularly in the Bani Malik region of the Asir.

e)   where the area is primarily to protect forest trees such as juniper, acacia or ghada (*Haloxylon persicum*). These himas are usually the common property of a village or a tribe. Cutting of trees is prohibited except in great emergencies or needs, such as rebuilding a house destroyed by a calamity. Sometimes wood is also sold for the benefit of the village or tribe e.g. *ahmyah* Huraymila in Jabal Tuwaiq.

From a range protection point of view *hima* have particular rôles in allowing the rehabilitation of degraded rangeland or its maintenance even where not degraded. Hima also help preserve plant species diversity and improve pasture quality. They also protect watersheds and water catchment areas and thus reduce erosion. Lastly they provide wildlife refuges.

In practice, the *hima* system is a very sensitive ecological and range management instrument and can readily be adapted to local needs and conditions. An interesting example of this is the Hima Unayzah, in the central Najd about 200 km north-west of Riyadh. This 70 km by 40 km *hima* has the unique objective of protecting *Holyxylon persicum* trees which provide effective stabilization of the sand dunes of the Nafud ash Shuqayyiqah to the west of Unayzah.

Available descriptions of *ahymah* indicate that they were mostly located in the Hijaz and Asir mountain region where there are many villages. Draz (1969) indicates that there might have been about some 3,000 in existence in 1965 although this figure could be wildly wrong. No proper survey was actually made and he probably came to this conclusion by assuming that most villages in the Al Baha region of the Asir had a *hima*. He suggests that there were thirty *ahymah* in the Taif area and by 1981 only three of these were actively maintained. Eighmy and Ghanem (1982) estimated that the number of active *ahymah* in a 200 km belt between Taif and Al Baha approached 200. In Saudi Arabia, there are no reports of *ahymah* in the eastern deserts used by nomadic pastoralist and *ahymah* seem to be developed among the village country of the Asir and Hijaz. *Ahymah* can vary enormously in size, depending on their purpose and the space available. An incomplete study by Grainger and Ganadelly (1984) in the northern Asir revealed 71 separate *ahymah* which varied in size from 10 ha to over 1,000 ha. The larger *ahymah* were located away from the villages and most were in areas unsuited to cultivation.

A few *ahymah* have been studied in detail. Draz (1985) examined the Hima Beni Sar located north of Baljurashi in the Al Bahah district. This *hima*, of about 1,000 hectares, has been protected by the Beni Sar tribe for a long time and soil sections reveal a good build up of organic matter under the grass cover. The lack of grazing stress has allowed palatable grass species such as *Themeda triandra, Aristida spp. Andropogon spp.* and *Stipa spp.* to thrive. No year long grazing is allowed but cutting of grasses is permissible during scarcity or late in the summer season when the grass seeds have matured. Permits for cutting or collection of grass are granted by the tribal head. No more than a specified number of persons from each family are allowed to cut mature grass and only on certain days of the week. In Hima Hureimla, in the floor of a *wadi* some 80 km north of Riyadh, Draz counted some 28,000 individual *Acacia* plants in the 4 km × 1 km protected area, as compared with none along the upper and lower parts of the same *wadi*.

Ghanem and Eighmy (1984) have probably conducted the most detail fieldwork of how a *hima* actually functions in relation to other farming activities. They carried out two case studies one in the villages belonging to the Thalama tribe 25 km southeast of Taif and the other belonging the Bani Sar tribe in villages north of Al Baha. The authors note that almost every Thalama farmer has expropriated a portion of tribal/village land next to his fields and/or house as his private *hodjra*. The legality of this is dubious but seems to be locally respected when used for domestic crops. Many *hodjra* have been sown with a boundary of prickly-pear cactus which saves the villager the trouble of building an enclosing wall to restrict grazing and to protect any beehives. The fruits of this cactus are also edible and rich in vitamin C. They are sometimes collected and sold at local *souks*.

The demise of the *hima* system over the last seventy years or so appears to have come about for several reasons. Ady and Walter (1991) note that they were formally abolished in 1932 at the time when tribal boundaries were legally abolished. According to Draz (1985) a key factor in those remaining seems to have been the widespread misunderstanding of a Royal decree in 1953 which was interpreted by local Emirs as the withdrawal of all controlled grazing measures and the opening up of all rangeland for general use. As a result, *ahmyah* rapidly became overgrazed and the reduction of vegetation cover led to serious soil erosion, especially in the Asir. Many ancient dams and water conservations systems in *widyan* then failed and became silted up or were swept away. The collection of water behind dams across *widyan* has been practiced since pre-Islamic times and at a local scale can be important in helping groundwater recharge.

Draz's comments need further elaboration, however. The Royal Decree actually abolished *ahmyah* that King Abdulaziz had established for the cavalry and other government purposes. It was not aimed at the hundreds of traditional *hima* found in the Kingdom. The Ministry of the Interior made this quite clear in response to questions by some of the provincial governors. The *hima* has not disappeared entirely from Saudi Arabia but since they appear now to have no legal support from the Government their protection is very much up to the will and consensus of the local community. In fact, many *ahmyah* have simply stopped functioning because of disputes arising over grazing rights or the requirement of land for building. A few *ahmyah* are definitely still functioning, among them Hima Bani Sar, Hima al-Fawqah and Hima Quraysh but the system is in serious decline and soon there will be few villagers left who have direct knowledge of the workings of this important range management system. A list of *ahmyah* thought to still to be functioning or at least be functioning until very recently has been collected by the NCWCD and is shown in Table 10.8. Most are located in the northern Asir.

The value of the *hima* system of range management has not gone unnoticed by the NCWCD and Gladstone (2000) refers to this concept in his study of the Farazan Islands Marine Protected Area in which he refers to five *hima* planning zones: resource

Table 10.8 Existing or Recently Existing *ahyma*

Name	Size (ha)	Name	Size (ha)
Hima Abu Raklah	8,000	Hima as-Sakhayit	600
Hima adh-Dhibah	150	Hima as-Sidad	100
Hima Afar	-	Hima as-Suhdum	600
Hima al-Abal		Hima as-Samur	
Hima al-Abasah		Hima Awf	
Hima al-Abbas		Hima Awn	700
Hima al-Arish		Hima Bani Amr min Sufyan	
Hima al-Fawqah		Hima Bani Sar	
Hima al-Fiqrah	54,000	Hima Bani Umar	350
Hima Unayzah		Hima Duqan	
Hima al-Habl		Hima Humayd	700
Hima al Halaqah	900	Hima Huraymila	
Hima al-Hamdah	300	Hima Jabal Ral	3,600
Hima al-Hayafin	200	Hima Murrah	1,000
Hima al-Huzaym	400	Hima Nifar	300
Hima al-Jawf	2,000	Hima naqib	700
Hima al-Makhadah	500	Hima Quraydah	2,000
Hima al-Mazhar	500	Hima Quraysh	1,500
Hima al-Mindaq		Hima Saysad	420
Hima al-Qadirah	500	Hima Shada	
Hima an-Naqi		Hima Shiban	500
Hima an-Nisayit		Hima Thumala	800
Hima al-Namur	3,000	Hima Tuwayriq	1,000
Hima an-Nasur		Hima Urush	200
Hima ar-Rabdah		Hima Wual	
Hima ar-Rafidah		Hima Jabal Juwwah	100
		Hima Bilhar Zaran	

Source: NCWCD (unpublished); Ady (1995).

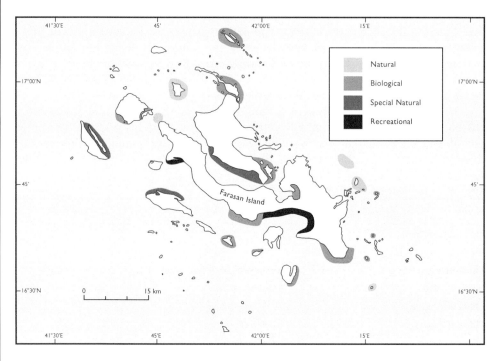

Figure 10.5 Farazan *hima* planning zones. Modified from Gladstone (2000).

use *hima*; biological *hima*; natural *hima*, special natural *hima* and recreational *hima* (Figure 10.5). It seems that this sort of management scheme is empathetic with the traditional way of life of the small fishing villages on the islands. There has, for example, been some form of traditional *hima*-like conservation of the reef fishing stocks practiced by villagers in which fishing grounds are rotated in order to prevent over-fishing.

10.8   ECO- AND GEOTOURISM

In 2000, the Kingdom set up a Supreme Commission for Tourism chaired by Prince Sultan Ibn Abdul Aziz. One market which the new Commission immediately targeted for development is ecotourism within so-called Natural Heritage Areas. These comprise the National Parks operated by the Ministry of Agriculture and the Protected Areas operated by the NCWCD.

A year later, in 2001, the NCWCD informally classified Protected Areas into three groups on the basis of their suitability or otherwise for tourism activities:

1. Strict protection: very low numbers of visitors by permit only, primarily for specialist and research interest; few or no facilities; no interpretation facilities.
2. Ecotourism: restricted to groups of less than 16 people at any one time; low impact activities only, e.g. guided walks, wildlife viewing; very basic facilities; limited interpretation facilities.
3. Nature-based tourism: visitor numbers and activities according to site-specific constraints; many forms of natural area tourism activity possible.

The majority of existing Protected Areas fall into the ecotourism category. The NCWCD has independently identified five existing Protected Areas for high priority development as eco-tourism sites, and a further nine candidate Protected Areas have been given high priority for official protection and tourism development. Seddon (2000) notes that the Kingdom's Protected Areas are inherently sensitive sites and increased visitor numbers will undoubtedly impact in both negative and positive ways.

In 2004 the Saudi Geological Survey initiated a collaborative program with other agencies to develop a series of geoparks in the Kingdom and also to identify important geological and geomorphological geotopes – specific sites of heritage value. Potential parks include the magnificent suite of volcanic landforms along side the Madinah-Jiddah road and those on Harrat Rahat, an hour's drive south of Madinah. Obviously, there needs to be a balance here between conserving sensitive sites, and the provision of environmental information and education. No one wants wholesale damage and this will depend entirely on how the worth of some rock formations is valued by visitors. Already, rock and mineral specimens are widely available in the markets of Jiddah.

Whilst the development of infrastructures capable of handling small groups of tourists might be unproblematic the whole point of ecotourism in the Kingdom is to provide employment and raise funds for environmental projects. This inevitably means coping eventually with quite large numbers of both local and foreign tourists. Issues such as transport, accommodation, safety and so on need to be thought through, as do issues such as waste disposal in pristine desert habitats. Above all the Kingdom, and agencies such as the NCWCD, need to train a workforce specifically for the purpose.

References

Aba-Husayn, M.M., Dixon, J.B. & Lee, S., 1980. Mineralogy of Saudi Arabian soils: southwestern region. *Soil Science of America Journal* 44: 643–649.

Abahussain, A.A., Adu, A.S., Al-Zbari, W.K., El-Deen, N.A. & Abdul-Raheem, M., 2002. Desertification in the Arab Region: analysis of current trends. *Journal of Arid Environments* 51: 521–545.

Abderrahman, W.A. & Rasheeduddin, M., 1994. Future groundwater conditions under long-term water stresses in an arid urban environment. *Water Resources Management* 8: 245–264.

Abderrahman, W.A., 1989. Effect of groundwater use on the chemistry of spring water in Al-Hassa Oasis. *Journal of King Abdulaziz University, Earth Sciences* 3: 259–265.

Abdulrazzak, M.J., 1995. Losses of floodwater from alluvial channels. *Arid Soil Research & Rehabilitation* 9: 15–24.

Abed, A., Carbonel, P., Collina-Girard, J., Fontugne, M., Petit-Maire, N., Reyss, J-C. & Yasin, S., 2000. Un paléolac du dernier interglaciere Pléistocène dans l'Extrême-Sud hyperaride de la Jordanie. *Les Comptes rendus de l'Académie des sciences, Paris, Science de la terre et des planêtes* 300: 259–264.

Abo-Ghobar, H.M. & Mohammad, F.S., 1995. Actual evapotranspiration measurements by lysimeters in a desert climate. *Arab Gulf Journal of Scientific Research* 13: 109–122.

Abu-Jarad, F. & Al-Jarallah, M.I., 1986. Radon in Saudi Houses. *Radiation Protection Dosimetry* 14: 243–249.

Abu-Rizaiza, O.S., 1999. Modification of the standards of wastewater reuse in Saudi Arabia. *Water Research* 33: 2601–2608.

Abuzinada, A.H., Grainger, J. & Child, G., 1991. Planning a system of protected areas in Saudi Arabia. *Parks* 2: 12–17.

Abuzinada, A.H., 1998. Restoration of desert ecosystems through wildlife management – the Saudi Arabian experience. In: S.A.S. Omar, R. Misak, D. Al-Ajmi & N. Al-Awadhi (eds.). *Sustainable Development in Arid Zones. Volume 2: Management & Improvement of Desert Resources*. Rotterdam: Balkema. pp. 641–648.

Ady, J. & Walter, E., 1991. Planning for conservation, recreation and tourism in Saudi Arabia. *Proceeding of the Conference on Landscape Architecture in Developing Countries*. Singapore: International Federation of Landscape Architects. pp. 99–109

Ady, J., 1995. The Taif escarpment, Saudi Arabia: A study for nature conservation and recreational development. *Mountain Research and Development* 15: 101–120.

Ahmad, Z., Allam, I.M. & Abdul Aleem, B.J., 2000. Effect of environmental factors in the atmospheric corrosion of mild steel in aggressive sea coastal environments. *Anti-Corrosion Methods and Materials* 47: 215–225.

Ahmed, S.S., 1972. Geology and petroleum prospects in eastern Red Sea. *Bulletin of the Association of Petroleum Geologists* 56: 707–719.

Al Kolibi, F.M., 2002. Possible effects of global warming on agriculture and water resources in Saudi Arabia: impacts and responses. *Climatic Change* 54: 225–245.

Al Yahya, E., (ed.), 2006. *Travellers in Arabia: British Exploration in Saudi Arabia*. London: Stacey.

Al-Abdela, S.I., Sabrah, R.E.A., Rabie, R.K. & Magid, H.M.A., 1997. Evaluation of groundwater quality for irrigation in Central Saudi Arabia. *Arab Gulf Journal of Scientific Research* 15: 361–377.

Al-Abdul Wahhab, H. & Hasnain, J., 1988. Laboratory study of asphalt concrete durability in Jeddah. *Building and Environment* 33: 319–239.

Al-Ahamadi, M.I., Jones, G. & Dotteridge, J., 1994. Agricultural production or conservation of groundwater? A case study of the Wajid aquifer, Saudi Arabia. *In*: C. Reeves & J. Watts, (eds.). *Groundwater – Drought, Pollution & Management*. Rotterdam: Balkema. pp. 13–21.

Al-Alawi, J. & Abdulrazzak, M., 1994. Water in the Arabian Peninsula: problems. *In*: P. Rogers & P. Lyndon, (eds.). *Water in the Arab World – Perspectives & Prognoses*. Cambridge: Harvard University Press. pp. 171–202.

Al-Dhowelia, K.H. & Shammas, U.K., 1991. Leak detection and quantification of losses in a water network. *Water Resources Development* 7: 30–38.

Al-Ghobari, H.M., 2000. Estimation of reference evapotranspiration for southern region of Saudi Arabia. *Irrigation Science* 19: 81–86.

Al-Harthi, A.A. & Bankher, K.A., 1999. Collapsing loess-like soil in western Saudi Arabia. *Journal of Arid Environments* 41: 383–399.

Al-Hinai, K.G., Moore, J.M. & Bush, P.R., 1987. Landsat image enhancement study of possible submerged sand-dunes in the Arabian Gulf. *International Journal of Remote Sensing* 8: 252–258.

Al-Jerash, M.A., 1985. Climatic subdivisions in Saudi Arabia: an application of principal component analysis. *Journal of Climatology* 5: 307–323.

Al-Jerash, M.A., 1989. Data for climatic water balance in Saudi Arabia: 1970–1986 A.D. Jiddah: Scientific Publishing Centre, King Abdulaziz University.

Alkolibi, F.M., 2002. Possible Effects of Global Warming on Agriculture and Water Resources in Saudi Arabia: Impacts and Responses. *Climatic Change* 54: 225–245.

Allen, R., Jensen, M.E., Wright, J. & Burman, R., 1989. Operational estimates of reference evapotranspiration. *Agronomy Journal* 81: 650–662.

Al-Mutaz, I.S., 1991. Environmental impact of seawater desalination plants. *Environmental Monitoring & Assessment* 16: 75–84.

Al-Rehaili, A.M., 1997. Municipal wastewater treatment and reuse in Saudi Arabia. *The Arabian Journal for Science & Engineering* 22: 143–152.

Al-Rehaili, M. & Roobol, M.J., 2002. Geohazards map for the Kingdom or Saudi Arabia. Scale 1:3,000,000. Open-File Report SGS-OF-2002–9. Jiddah: Saudi Geological Survey.

Al-Saafin, A.K., Bader, T.A., Shehata, W., Höetzl, H., Wöhnlich, S. & Zöetl, J.G., 1990. Groundwater recharge in an arid karst area in Saudi Arabia. *Selected Papers in Hydrology* 1: 29–41.

Al-Saeed, S., 1989. Local Production: Consumption and Imports. Riyadh: Research Department, Ministry of Commerce.

Al-Shaibani, A.M., 2002. Lava fields as potential groundwater sources in western Saudi Arabia: a case study of northern Harrat Rahat. *In*: *Proceedings, International Conference on Water Resource Management in Arid Regions*. Kuwait Institute for Scientific Research.

Alsharhan, A.S. & Nairn, A.E.M., 1997. *Sedimentary Basins & Petroleum Geology of the Middle East*. Amsterdam: Elsevier.

Alsharhan, A.S., Rizk, Z.A., Nairn, A.E.M., Bakhit, D.W. & Alhajari, A.S., 2001. *Hydrogeology of an Arid Region: the Arabian Gulf & Adjoining Areas*. Amsterdam: Elsevier.

Al-Sulaimi, J.S. & Pitty, A.F., 1995. Origin and depositional model of Wadi Al-Batin and its associated alluvial fan, Saudi Arabia and Kuwait. *Sedimentary Geology* 97: 203–229.

Ambraseys, N.N., Melville, C.P. & Adams, R.D., 1994. *The Seismicity of Egypt, Arabia & the Red Sea: A Historical Review.* Cambridge: Cambridge University Press.

Amin, A.A. & Bankher, K.A., 1997. Karst hazard assessment of Eastern Saudi Arabia. *Natural Hazards* 15: 21–30.

Anati, E., 1972. Rock-Art in Central Arabia: Corpus of the Rock-Engravings. Parts I and II. *Publications de l'Institut Orientaliste de Louvain* 3. p. 167.

Anati, E., 1974. Rock-Art in Central Arabia: Corpus of the Rock-Engravings. Parts III and IV. *Publications de l'Institut Orientaliste de Louvain,* 3. p. 262.

Anton, D. & Vincent, P., 1986. Parabolic dunes of the Jafurah Desert, Eastern Province, Saudi Arabia. *Journal of Arid Environments* 11: 187–198.

Anton, D., 1983. Modern eolian deposits of the Eastern Province of Saudi Arabia. *In:* M.E. Brookefield & T.S. Ahlbrandt, (eds.). *Eolian Sediments & Processes.* Amsterdam: Elsevier Science Publishers. pp. 365–378.

Anton, D., 1984. Aspects of geomorphological evolution; paleosols and dunes in Saudi Arabia. *In:* A.R. Jado & J.G. Zötl, (eds.). *Quaternary Period in Saudi Arabia. Volume* 2. Vienna: Springer-Verlag. pp. 273–294.

Bagnold, R.A., 1941. *The Physics of Blown Sand & Desert Dunes.* London: Methuen.

Bagnold, R.A., 1951. Sand formations in southern Arabia. *Geographical Journal* 117: 78–86.

Bahafzullah, A., Fayed, L.A., Kazi, A. & Al-Saify, M., 1993. Classification and distribution of the Red Sea coastal sabkhas near Jeddah – Saudi Arabia. *Carbonates and Evaporites* 8: 23–38.

Baierle, H.U. & Frey, W., 1986. A vegetation transect through central Saudi Arabia (at-Taif – ar-Riyad). *In:* H. Kürschner, (ed.). *Contribution to the Vegetation of Southwest Asia.* Wiesbaden: Dr. Ludwig Reichart Verlag. pp. 111–136.

Bailey, G.N, Flemming, N., King, G.C.P., Lambeck, K., Momber, G., Moran, L., AlSharekh, A., & Vita-Finzi, C., 2007. Coastlines, submerged landscapes, and human evolution: the Red Sea Basin and the Farasan Islands'. *Journal of Island and Coastal Archaeology* 2 (in press).

Bakader, A.H., al-Sabbagh, A.L., Al-Glenid, M.A. & Izzidien, M.Y., 1983. *Islamic Principles for the Conservation of the Natural Environment.* Gland: International Union for Conservation of Nature and Natural Resources (IUCN), and Riyadh: Meteorological & Environmental Protection Administration (MEPA).

Bakader, A.H., Al-Sabbagh, A.L., Al-Glenid, M.A. & Izzidien, M.Y., 1993. *Islamic Principles for the Conservation of the Natural Environment.* Gland: International Union for Conservation of Nature and Natural Resources (IUCN), and Riyadh: Meteorological & Environmental Protection Administration (MEPA). Second Edition online at http://www.islamset.com/env/contenv.html [May 2007].

Baker, J.M. & Dicks, B., 1982. The environmental effects of pollution from the Gulf oil industry. IUCN/MEPA report for the Expert Meeting of the Gulf Co-ordinating Council to review environmental issues.

Bakiewicz, W., Milne, D.M. & Noori, M., 1982. Hydrogeology of the Umm Er Radhuma aquifer, Saudi Arabia, with reference to fossil gradients. *Quarterly Journal of Engineering Geology, London* 15: 105–126.

Bankher, K.A. & Al-Harthi, A.A., 1999. Earth fissuring and land subsidence in western Saudi Arabia. *Natural Hazards,* 20: 21–42.

Bannerman, D.A., 1937. Natural history exploration in Arabia. *Discovery* (September), 260–265.

Barger, T.C., 2000. *Out in the Blue: Letters from Arabia 1837–1940.* Vista: Selwa Press.

Barth, H.-J., 1998. Status of vegetation & an assessment of the impact of overgrazing in an area north of Jubail, Saudi Arabia. *In:* S.A.S. Omar, R. Misak, D. Al-Ajmi & N. Al-Awadhi, (eds.).

Sustainable Development in Arid Zones. Volume 2: Management & Improvement of Desert Resources. Rotterdam: A.A. Balkema, 435–450.

Barth, H.-J., 1999. Desertification in the Eastern Province of Saudi Arabia. *Journal of Arid Environments* 43: 399–410.

Barth, H.-J., 2002. The sabkhat of Saudi Arabia – an Introduction. *In*: H-J. Barth & B. Böer, *Sabkha Ecosystems. Vol I: The Arabian Peninsula & Adjacent Countries.* pp. 37–51.

Barth, H.-K., 1976. Pedimentgenerationen und Reliefentwicklung im Schichtstufenland Saudi-Arabiens. *Zeitschrift für Geomorphologie*, N.F., Supplement band 24: 111–119.

Basamed, A.S., 2001. Hydrochemical Study and Bacteriological Effect on Ground water in the Northern Part of Jeddah District. Unpublished M.Sc. King Abdulaziz University.

Batanouny, K.H., 1978. Natural History of Saudi Arabia: a Bibliography. Publication of King Abdulaziz University, Biology 1. p. 121.

Batanouny, K.H., 1998. Traditional land use in the deserts of the Arab World. *In*: S.A.S. Omar, R. Misak, D. Al-Ajmi & N. Al-Awadhi, (eds.). *Sustainable Development in Arid Zones. Volume 2: Management & Improvement of Desert Resources.* Rotterdam: A.A. Balkema. pp. 697–705.

Bazuhair, A.S.A., 1989. Optimum aquifer yield of four aquifers in Al-Karj area, Saudi Arabia. *Journal of King Abdulaziz University, Earth Science* 2: 37–49.

Bazuhair, A.S.A. & Hussein, M.T., 1990. Springs in Saudi Arabia. *Journal of King Abdulaziz University, Earth Sciences* 3: 259–265.

Bazuhair, A.S.A. & Wood, W.W., 1996. Chloride mass-balance method for estimating ground water recharge in arid areas: examples from western Saudi Arabia. *Journal of Hydrology* 186: 153–159.

Beaumont, P., 1977. Water and development in Saudi Arabia. *Geographical Journal* 143: 41–60.

Behairy, A.K.A, El-Sayed, M. Kh. & Durgaprada Rao, N.V.N., 1985. Eolian dust in the coastal area north of Jeddah, Saudi Arabia. *Journal of Arid Environments* 8: 89–98.

Bensichke, R., Fuchs, G. & Weissensteiner, V., 1987. Speläologische Untersuchungen in Saudi-Arabia (Eastern Province, As-Summan Plateau). Region Ma'aqala. *Die Höhle, Zeitschrift für karst und Höhlenkunde* 3: 61–76.

Beyth, M. & Heimann, A., 1999. The youngest igneous event in the crystalline basement of the Arabian-Nubian Shield, Timna Igneous Complex. *Israel Journal of Earth Science* 48: 113–120.

Bidwell, R., 1976. *Travellers in Arabia.* London: Hamlyn.

Blunt, A., 1881. A Pilgrimage to Nedj. London: John Murray (reprinted 1985, London: Century Publishing).

Bögli, A., 1980. *Karst Hydrology & Physical Speleology.* Berlin: Springer-Verlag.

Braithwaite, C., 1987. Geology and Palaeogeography of the Red Sea Region. *In*: A.J. Edwards & S.M. Head, (eds.). *Red Sea.* London: Pergamon. pp. 22–44.

Breed, C.S., Fryberger, S.G., Andrews, S., McCauley, C.K., Lennartz, F., Gebel, D. & Horstman, K., 1989. Regional Studies of Sand Seas using Landsat (ERTS) Imagery. *In*: E.D., McKee, ed., *A Study of Global Sand Seas.*, Tunbridge Wells: Castle House Publications. pp. 309–381.

Brookes, I., 2003a. Lawrence's Landscapes: "Seven Pillars of Wisdom" as geographer – part I: landscape cognition, topopsychology, regional geography, and application to war. *Arab World Geography* 5: 156–171.

Brookes, I., 2003b. Lawrence's Landscapes: "Seven Pillars of Wisdom" as geographer – part II: regional description and interpretation. *Arab World Geography* 5: 172–192.

Brown, G.F., 1970. Eastern margin of the Red Sea and the coastal structures in Saudi Arabia. *In*: N.L. Falcon, I.G. Gass, R.W. Girdler & A.S. Laughton. A discussion on the structure and evolution of the Red Sea and the nature of the Red Sea, Gulf or Aden and Ethiopian Rift Junction. *Royal Society (London) Philosophical Transactions* 267: 75–87.

Brown, G.F., 1972. Tectonic Map of the Arabian Peninsula. Saudi Arabian Directorate General of Mineral Resources. Arabian Peninsula Map AP-2, scale 1:4 000 000.

Brown, G.F., Schmidt, D.W. & Huffman, A.C., 1989. Geology of the Arabian Peninsula. Shield area of western Saudi Arabia. *United States Geological Survey, Professional Paper*, 560-A.

Burek, P.J., 1969. Structural effects of sea-floor spreading in the Gulf of Aden. *In*: E.T. Degens & D.A. Ross, (eds.). *Brines & Recent Heavy metal Deposits in the Red Sea, a Geochemical and Geophysical Account.* New York: Springer-Verlag. pp. 59–79.

Burkhart, G.E., 1998. National Security and the Internet in the Persian Gulf Region. http//www.georgetown.edu/research/arabtech/pgi98-9.html [05/2007].

Camp, V.E., 1984. Island arcs and their role in the evolution of the western Arabian shield. *Geological Society of America Bulletin* 95: 913–921.

Camp, V.E., 1986. Geological map of the Umm al Birak quadrangle. Sheet 23D, Kingdom of Saudi Arabia Map GM 87. Saudi Arabian Directorate General of Mineral Resources.

Camp, V.E., Hooper, P.R., Roobol, M.J. & White, D.L., 1987. The Madinah eruption, Saudi Arabia: magma mixing and simultaneous extrusion of three basaltic chemical types. *Bulletin of Volcanology* 49: 489–508.

Camp, V.E. & Roobol, M.J., 1992. Upwelling asthenosphere beneath western Arabia and its regional implication. *Journal of Geophysical Research* 97: 255–271.

Cann, R.L., Stoneking, M. & Wilson, A.C., 1987. Mitochrondrial DNA and human evolution. *Nature* 325: 31–36.

Chapman, R.W., 1971. Climatic changes and the evolution of landforms in the Eastern Province of Saudi Arabia. *Geological Society of America Bulletin* 82: 2713–2728.

Chapman, R.W., 1974. Calcareous duricrust in Al-Hasa, Saudi Arabia. *Geological Society of America Bulletin* 85: 119–130.

Chapman, R.W., 1978. General information on the Arabian Peninsula: Geomorphology. *In*: S. Al-Sayari & J.G. Zötl, (eds.). *Quaternary Period in Saudi Arabia*, Vol. 1. Vienna: Springer Verlag. pp. 19–30.

Chaudhary, S.A., 1983. Vegetation of the Great Nefud. *Journal of the Saudi Arabian Natural History Society* 2: 32–7.

Chaudhary, S.A., 1989. Understanding the desert range of plants of Saudi Arabia. *In*: A.H. Abuzinada, P. Goriup & I.A. Nader, (eds.). *Wildlife Conservation & Development in Saudi Arabia*. Riyadh: NCWCD Publication No. 3. pp. 156–164.

Child, G. & Grainger, J., 1990. *A System Plan for Protected Areas for Wildlife Conservation & Sustainable Development in Saudi Arabia*. IUCN, Gland, Switzerland and National Commission for Wildlife Conservation and Development, Riyadh, Saudi Arabia.

Clapp, N., 1998. *The Road to Ubar: Finding the Atlantis of the Sands*. Houghton Mifflin Company.

Clemens, S. & Prell, W.L., 1990. Late Pleistocene variability of Arabian Sea summer monsoon winds and continental aridity: eolian records from the lithogenic component of deep-sea sediments. *Paleoceanogaphy* 5: 109–146.

Cleuziou, S. & Tosi, M., 1998. Hommes, climates et environnements de la Péninsule Arabique à l'Holocène. *Paléorient* 23: 121–135.

Cochran, J.R., 1983. Model for development of the Red Sea. *American Association of Petroleum Geologists Bulletin* 76: 41–69.

Collenette, P. & Grainger, D.J., 1994. *Mineral Resources of Saudi Arabia*. Directorate General of Mineral Resources, *Special Publication*, SP–2.

Collenette, S., 1988. *Checklist of the Botanical Species in Saudi Arabia*. Burgess Hill: International Asclepiad Society.

Collenette, S., 1999. *Wildflowers of Saudi Arabia*. Riyadh: National Commission for Wildlife Conservation and Development.

Cooke, R.U., Brunsden, D., Doornkamp, J.C. & Jones, D.K.C., 1982. *Urban Geomorphology in Drylands.* Oxford University Press.

Courtney-Thompson, E.C.W., 1975. Rock engravings near Medinah, Saudi Arabia. *Seminar for Arabian Studies* 5:22–32.

Cressman, K., 1998. A detailed analysis of a desert locust upsurge in Saudi Arabia (November 1996–May 1997). Rome: FAO.

Dabbagh, A.E. & Abderrahman, W.A., 1997. Management of groundwater resources under various irrigation water use scenarios in Saudi Arabia. *The Arabian Journal for Science & Engineering* 22: 47–64.

Dabbagh, A.R., al-Hinai, K.G. & Asif Khan, A., 1998. Evaluation of the Shuttle Imaging Radar (SIR-C/X-SAR) data for mapping paleo-drainage systems in the Kingdom of Saudi Arabia. *In*: A.S. Alsharhan, K.W. Glennie, G.L. Whittle & C. Kendall, (eds.). *Quaternary Deserts & Climatic Change.* A.A. Balkema: Rotterdam. pp. 483–493.

Davies, F.B. & Grainger, D.J., 1985. Geologic map of the Al Muwaylih quadrangle, sheet 27A, Kingdom of Saudi Arabia. Saudi Arabian Directorate General of Mineral Resources Geological Map GM-82, scale 1:250,000.

Delany, M.J., 1989. The zoogeography of the mammal fauna of southern Arabia. *Mammalian Review* 19:133–152.

DeMenocal, P.B., 1993. Sensitivity of Asian and African climate to variations in seasonal insolation, glacial ice cover, sea surface temperature, and Asian orography. *Journal of Geophysical Research* 98: 7265–7287.

Dincer, T., Al-Mugrin, A. & Zimmerman, U., 1974. Study of the infiltration and recharge through the sand dunes in arid zones with special reference to the stable isotopes and thermonuclear tritium. *Journal of Hydrolology* 23: 79109.

Dorman, A.U. & Abdulrazzak, M.J., 1993. Flood hydrograph estimation for ungaged wadis in Saudi Arabia. *Journal of Water Resources Planning and Management* 119:45–63.

Draz, O., 1969. The hema system of range reserves in the Arabian peninsula. Its possibilities in range improvement conservation projects in the Middle East. FAO/PL: PFC/13.11 Rome: FAO.

Draz, O., 1978. Revival of the hema system of range reserves as a basis for the Syrian range management programme. Proceedings, 1st International Rangeland Congress, Denver, Colorado. pp. 100–103.

Draz, O., 1985. The hema system of range reserves in the Arabian Peninsula. *In*: J.A. McNeely & D. Pitt, *Culture & Conservation: The Human Dimension in Environmental Planning.* Croom Helm, London. pp. 109–121.

Drysdall, A.R., Jackson, N.J., Ramsay, C.R., Douch, C.J. & Hackett, D., 1984. Rare element mineralization related to Precambrian alkali granites in the Arabian Shield. *Economic Geology* 79: 1366–1377.

Dullo, W.-C., 1990. Facies, fossil record and age of Pleistocene reefs from the Red Sea (Saudi Arabia). *Facies 22*: 1–46.

Dutton, Y., 1992. Natural Resources in Islam. *In*: F. Khalid & J. O'Brien. *Islam & Ecology.* Cassell:London. pp. 51–69.

Dzerdzeevskii, B.L., 1958. On some climatological problems and microclimatological studies of arid ad semi-arid lands in the U.S.S.R. *In*: Climatology and Microclimatology, Vol. 11, Proceedings Canberra Symposium, UNESCO, Paris. pp. 315–25

Edgell, H.S., 1990a. Evolution of the Rub' al Khali desert. *Journal of King Abdulaziz University, Earth Sciences* 3:109–126.

Edgell, H.S., 1990b. Karst in northeastern Saudi Arabia. *Journal of King Abdulaziz University, Earth Sciences* 3: 81–94.

Edgell, H.S., 1990c. Geological framework of Saudi Arabia groundwater resources. *Journal of King Abdulaziz University, Earth Sciences* 3: 267–286.

Edwards, A.L. & Head, S.M., 1987. *Red Sea*. Oxford: Pergamon.

Eighmy, J. & Ghanem, Y., 1982. The hima system. Prospects for traditional subsistence systems in the Arabian Peninsula. Working paper II CID/FMES Project. School of Renewable Resources, University of Arizona.

El Din, M.N.A., Madany, I.M., Al-Tayaran, A., Al-Jubair, H.A. & Gomaa, A., 1993. Trends in water quality of some wells in Saudi Arabia. *Science of the Total Environment* 154: 110–122.

El Prince, A.M., Mashhady, A.S. & aba-Husayn, M.M., 1979. The occurrence of pedogenic palygorskite (attapulgite) in Saudi Arabia. *Soil Science* 128: 211–218.

El-Demerdash, M.A. & Zilay, A.M., 1994. An introduction to the plant ecology of the Tihama plains of the Jizan region, Saudi Arabia. *Arab Gulf Journal of Scientific Research* 12: 286–299.

El-Ghani, M.M.A., 1996. Vegetation along a transect in the Hijaz mountains (Saudi Arabia). *Journal of Arid Environments* 32: 289, 304.

Elhadj, E., 2004. Household water and sanitation services in Saudi Arabia: an analysis of economic, political and ecological issues. SOAS/KCL Water Research Group, University of London. Occasional Paper 56.

El-Hage, B., 1997. *Saudi Arabia: Caught in Time 1861–1939*. Reading: Garnet.

El-Khatib, A.B., 1980. *Seven Green Spikes – Water & Agricultural Development*. 2nd Edition. Riyadh: Ministry of Agriculture & Water.

Erol, A., 1989. Engineering geological considerations in a salt dome region surrounded by sabkha sediments, Saudi Arabia. *Engineering Geology* 26: 215–232.

Euting, J., 1896. Tagebuch einer Reise in Inner-Arabien" ("Diary of a Journey to Inner Arabia"); Part I, Leiden 1896; Part II, published by Enno Littmann, Leiden 1914. (Reprinted as a single volume: Hildesheim 2004).

Field, H., 1960. Carbon-14 date for 'Neolithic' site in the Rub' al Khali. *Man* 58: 162.

Fisher, M., 1997. Decline in the juniper woodlands of Raydah reserve in southwestern Saudi Arabia: a response to climatic changes? *Global Ecology & Biogeography Letters* 6: 379–386.

Fleitmann, D., Matter, A., Pint, J. & Al-Shanti, M.A., 2004. The speleothem record of climate change in Saudi Arabia: Saudi Geological Survey Open-File Report SGS-OF-2004-8, p. 40.

Folk, R.L., Roberts, H.H. & Moore, C.H., 1973. Black phytokarst from Hell, Cayman Islands, British West Indies. *Geological Society of America Bulletin* 84: 2351–2360.

Foltz, R.C., Denny, F.M & Baharuddin, A., 2003. *Islam & Ecology*. Cambridge: Harvard University Press.

Förstner, U. & Wittman, G.T.W., 1979. *Metal Pollution in the Aquatic Environment*. Berlin: Springer Verlag.

Fourniguet, J., Alabouvette, B., Kluyver, H.M., Ledru, P. & Robelin, C., 1985 (1405.a.h). Evolution of Western and Central Saudi Arabia during Late Tertiary and Quaternary. A Bibliographic Review. Open-File Report BRGM-OF-05–10. Jiddah, Ministry of Petroleum and Mineral Resources.

Freeth, Z. & Winstone, V., 1978. *Explorers of Arabia*. London: George Allen & Unwin.

Frey, W. & Kurschner, H., 1989. Die Vegetation im Vorderen Orient. Erläuterungen zur Karte A VI 1 Vorderer Orient. Vegetation des Tübinger Atlas des Vorderen Orients. *Beihefte zum Tübinger Atlas des Vorderen Orients, Reihe A (Naturwissenschaften)* Nr. 30. Wiesbaden: Dr. Ludwig Reichert Verlag.

Fricke, H.W. & Landmann, G., 1983. Origin of the Red Sea submarine canyons: observations by submersibles. *Naturwissenschten* 70: 195–197.

Fryberger, S.G., 1980. Dune forms and wind regimes. *In*: E.D. McKee. ed., *A Study of Global Sand Seas*. Tunbridge Wells: Castle House Publications. pp.137–169.

Fryberger, S.G. & Ahlbrandt, T.S., 1979. Mechanisms for the formation of eolian sand seas. *Zeitschrift für Geomorphologie* 23: 440–460.

Fryberger, S.G., Al-Sari, A.M. & Clisham. T.J., 1983. Eolian dune, interdune, sand sheet, and siliciclastic sabkha sediments of an offshore prograding sand sea, Dhahran area, Saudi Arabia. *American Association of Petroleum Geologists Bulletin* 67:280–312.

Fryberger, S.G., Al-Sari, A.M., Clisham, T.J., Rizvi, S.A.R. & Al-Hinai, K.G., 1984. Wind sedimentation in the Jafurah sand sea, Saudi Arabia. *Sedimentology* 31: 413–431.

Garfunkel, Z., 1999. History and paleogeography during the Pan-African orogen to stable platform transition: reappraisal of the evidence from the Elat area and the northern Arabian-Nubian Shield. *Israel Journal of Earth Science* 48: 135–157.

Garrard, A.N. & Harvey, C.P.D., 1981. Environment and settlement during the Upper Pleistocene and Holocene at Jubba in the Great Nafud, northern Arabia. *Atlai* 5: 137–148.

Gass, I.G., 1981. Pan-African (Upper Proterozoic) plate tectonics of the Arabian-Nubian Shield. *In*: A. Kroner, ed., *Precambrian Plate Tectonics*. Amsterdam: Elsevier. pp. 387–405.

Ghanem, Y. & Eighmy, J., 1984. Hema and traditional land use management among arid zone villagers in Saudi Arabia. *Journal of Arid Environments* 7: 287–297.

Ghazanfar, S.A. & Fisher, M., (eds.), 1998. *Vegetation of the Arabian Peninsula*. Dordrecht: Kluwer Academic Publishers.

Ghebreab, W., 1998. Tectonics of the Red Sea region reassessed. *Earth-Science Reviews* 45: 1–44.

Girdler, R.W. & Southern, R.C., 1987. Structure and evolution of the northern Red Sea. *Nature* 330: 716–721.

Gladstone, W., 2000. The ecological and social basis for management of a Red Sea marine-protected area. *Ocean & Coastal Management* 43: 1015–1032.

Glennie, K.W., 1970. *Desert Sedimentary Environments*. Volume 11: *Developments in Sedimentology*. Amsterdam: Elsevier Scientific Publishing Company.

Glennie, K.W., Singhvi, A.K., Lancaster, N. & Teller, J.T., 2002. Quaternary climate changes over Southern Arabia and the Thar Desert, India. *In*: P.D. Cliff, D. Kroon, C. Gaedicke & J. Craig (eds.). *The Tectonic & Climatic Evolution of the Arabian Sea Region. Geological Society, London. Special Publication* 195: 301–316.

Gorin, G.E., Racz, L.G. & Walter, M.R., 1982. Late Precambrian-Cambrian sediments of the Huqf Group, Sultanate of Oman. *American Association of Petroleum Geologists Bulletin* 66: 2609–2627.

Goudie, A., 1999. Wind erosional landforms: yardangs and pans. *In*: A.S. Goudie, I.I. Livingstone, & S. Stokes, (eds.). *Aeolian Environments, Sediments & Landforms*: Chichester: Wiley. pp. 167–181.

Goudie, A. & Viles, H., 1997. *Salt Weathering Hazards*. Chichester: Wiley.

Grainger, J. & Ganadelly, A., 1984. Himas: an investigation into a traditional conservation ethic in Saudi Arabia. *Journal of Saudi Arabian Natural History Society* 2: 28–32.

Grainger, J. & Llewellyn, O., 1994. Sustainable use: lessons from a cultural tradition in Saudi Arabia. *Parks* 4: 8–16.

Greenwood, W.R., Anderson, R.E., Fleck, R.J. & Roberts, R.J., 1980. Precambrian geologic history and plate tectonic evolution of the Arabian Shield. *Mineral Resources Bulletin*, 24. Directorate General of Mineral Resources Jiddah.

Greenwood, W.R., 1973. The Ha'il Arch – A key to deformation of the Arabian Shield during the evolution of the Red Sea rift. *Mineral Resources Bulletin*, 7. Directorate General of Mineral Resources Jiddah.

Greth, A., Magin, C. & Ancrenaz, M., 1996. *Conservation of Arabian Gazelles*. Riyadh, National Commission for Wildlife Conservation and Development.

Guarmani, C., 1866. *Il Neged settentrionale: Itinerario da Gerusalemme a Aneizeh nel Cassim*. Jerusalem: Press of the Franciscan Fathers. Edited version prepared for the Arab Bureau, Cairo, 1917 and translated version Northern Nejd: journey from Jerusalem to Anaiza in Kasim. Translated by Lady Capel-Cure and introduced by Douglas Carruthers, London: Argonaut Press, 1938. (reprinted 1971).

Guilcher, A., 1954. Structure et relief de l'Arabie, du Sinaï et de la mer Rouge. *L'information Geographique* 2: 56–63.

Guilcher, A., 1955. Géomorphologie del l'extrémité septentrionale du banc Farsan (Mer Rouge). *Annals, Institut d' Océanographie* 30: 55–100.

Gulaid, M.A., 1990. *Effect of Islamic Law & Institutions on Land Tenure with Special Reference to some Muslim Countries.* Research Paper No. 8. Jiddah: Islamic Research and Training Institute/Islamic Development Bank.

Gvirtzman, G., Buchbinder, B., Sneh, A., Nir, Y. & Friedman, G., 1977. Morphology of the Red Sea fringing reefs: a result of erosional pattern in the last-glacial low-stand sea level and the following Holocene recolonisation. *Memoirs du Bureau des Reserches Geologiques et Minières* 89: 480–491.

Habibi, K., 1986. Arabian ungulates – their status and future protection. *Oryx* 20: 100–103.

Hajar, A.S.M., 1993. A comparative ecological study of the vegetation of the protected and grazed parts of Hema Sabihah, in Al-Bahah Region, South Western Saudi Arabia. *Arab Gulf Journal of Scientific Research* 11: 259–280.

Hajara, H.H. & Batanouny, K.H., 1977. Desertization in Saudi Arabia, *Proceedings of the Saudi Biological Society* 1: 34–52.

Harrigan, P., 2002. The captain and the king. Saudi Aramco World, 53 (5) Sept/October.

Harrison, D.L. & Bates, P.J.J., 1991. *The Mammals of Arabia.* 2nd ed. Sevenoaks: Harrison Zoological Museum.

Harrison, D.L., 1968. The large mammals of Arabia. *Oryx* 6: 357–363.

Hattstein, M. & Delius, P. (eds.), 2000. *Islam Art & Architecture.* Cologne: Könemann.

Herbert, T., Geraads, D., Janjou, D., Vaslet, D., Memesh, A., Billiou, D., Bocherens, H., Dobigny, G., Eisenmann, V., Gayet, M., Broin, F., Petter, G., & Halawani, M., 1998. First Pleistocene faunas from the Arabian Peninsula: An Nafud desert, Saudi Arabia. *Comptes rendus de l'Académie des Sciences – Series IIA – Earth & Planetary Science* 326: 145–152.

Hobbs, J., 1985. Bedouin Conservation of Plants and Animals in the Eastern Desert, Egypt. *In: Conference on Arid Lands: Today & Tomorrow.* Westview Press Boulder, Colorado. pp. 997–1005.

Hoelzmann, P., Gasse, F., Benkaddour, A., Dupont, L., Leuschner, D.C., Salzmann, U., Sirocko, F. & Staubwasser, M., 2002. Palaeoenvironmental changes in the arid and subarid belt (Sahara-Sahel Arabian Peninsula) from 150 ka to present – *In:* R. Batterbee, F. Gasse & C. Stickely, (eds.). *Past Climate Variability Through Europe & Africa.* Amsterdam: Kluwer.

Holm, D.A., 1960. Desert geomorphology of the Arabian Peninsula. *Science* 132: 1369–1379.

Hötzl, H., Maurin, V. & Zötl, J.G., 1978. Geologic history of the Al Hasa area since the Pliocene. *In:* S.S. Al-Sayari & J.G. Zötl, (eds.). *Quaternary Period in Saudi Arabia.* Vienna: Springer-Verlag. pp. 58–74.

Hötzl, H., Maurin, V., Moser, H. & Rauert, W., 1978. Quaternary studies on the recharge area situated in crystalline rock regions. *In:* S.S. Al-Sayari & J.G. Zötl, (eds.). *Quaternary Period in Saudi Arabia.* Vienna: Springer-Verlag pp. 230–239.

Hötzl, H. & Zötl, J.G., 1978. Climatic changes during the Quaternary Period. *In:* S.S. Al-Sayari & J.G. Zötl, (eds.). *Quaternary Period in Saudi Arabia.* Vienna: Springer-Verlag. pp. 310–311.

Hötzl, H., Wohnlich, S., Zötl, J.G. & Beischke, R., 1993. Verkarstung und Grundwasser im As Summan Plateau (Saudi Arabien). *Steirische Beiträge zur Hydrogeologie* 44: 5–157.

Huber, C., 1891. Journal d'un Voyage en Arabie (1883–1884). Paris:Ernest Lerroux. p. 778.

Hudson, N.W., 1981. *Soil Conservation.* London: Batsford.

Hussain, G. & Al-Saati, A.J., 1999. Wastewater quality and its reuse in agriculture in Saudi Arabia. *Desalination* 123: 241–251.

Imcos Marine Ltd., 1974a. Handbook of Weather in the Gulf – Surface Wind Data. London: Austral Press.

Imcos Marine Ltd., 1974b. Handbook of Weather in the Gulf – Bibliography. London: Austral Press.

Imcos Marine Ltd., 1976. Handbook of Weather in the Gulf – General Climate Data. London: Austral Press.

IPCC., 2001. *Climate Change 2001: The Scientific Basis.* Contribution of Working Group I to the Third Assessment Report of the Intergovernmental Panel on Climate Change. Houghton, J.T., Y. Ding, D.J. Griggs, M. Noguer, P. van der Linden, X. Dai & K. Maskell, (eds.). Cambridge: Cambridge University Press.

Ishaq, A.M. & Alassar, R.S., 1999. Characterizing urban storm runoff quality in Dharhan City in Saudi Arabia. *International Water Resources Association*, 24: 53–58.

Ishaq, A.M. & Khan, A.A., 1997. Recharge of aquifers with reclaimed water: a case study for Saudi Arabia. *The Arabian Journal for Science and Engineering* 22: 134–141.

Italconsult, 1967. Water supply surveys for Jeddah-Makkah-Taif areas. Report No. 2. Ministry of Agriculture and Water, Saudi Arabia.

Italconsult, 1969. Water and agricultural development surveys for Areas I & II: Final Report. Ministry of Agriculture and Water. Saudi Arabia.

Italconsult, 1979. Water and Agricultural Development Studies. Arabian Shield – South. Climate & Surface Hydrology, Annex 8. Ministry of Agriculture and Water. Saudi Arabia.

Izzi Deen, M.Y., 1990. Islamic environmental ethics, law, and society. *In*: J.R. Engel & J.G. Engle, (eds.). *Ethics of Environmental & Development.* London: Belhaven Press. pp. 189–198.

Jado, A.R. & Zötl, J.J., 1984. *Quaternary Period in Saudi Arabia. Volume 2.* Vienna: Springer-Verlag.

Jellicoe, P., 1989. Forward. *Arabia Deserta.* 2nd Edition. London: Bloomsbury.

Jennings, M.C., 1986. The distribution of the extinct Arabian ostrich *Struthio camelus syriacus* Rothchild 1919. *Fauna of Saudi Arabia* 8: 447–461.

Jennings, M.C., 1995. *An Interim Atlas of the Breeding Birds of Arabia.* Riyadh: National Commission for Wildlife Conservation and Development.

Johnson, P.R. & Jastaniah, H., 1993. Geomorphological studies in western Saudi Arabia – a progress report, in Annual Report of the USGS Mission, Kingdom of Saudi Arabia, for the Fiscal Year 1992. Saudi Arabian Directorate General of Mineral Resources. *Technical Report USGS-Tr-93-1*, pp. 19–29.

Johnson, P.R. & Kattan, F., 1999. The timing and kinematics of a suturing event in the northeastern part of the Arabian shield, Kingdom of Saudi Arabia: Saudi Arabian Deputy ministry for Mineral Resources Open-File Report USGS-OF-99–3, p. 29.

Johnson, P.R. & Woldehaimanot, B., 2003. Development of the Arabian-Nubian Shield: perspectives on accretion and deformation in the northern East African Orogen and the assembly of Gondwana. *In*: M. Yoshida, B.F. Windley, & S. Dusgupta, (eds.). *Proterozoic East Gondwana: Supercontinent Assembly & Breakup.* Geological Society, London: Special Publication 206: 289–325.

Jumgus, H., 1983. The role of indigenous flora and fauna in rangeland management systems of the arid zones in western Asia. *Journal of Arid Environments* 6: 76–86.

Kerr, R.O. & Nigra, J.O., 1952. Eolian sand control. *Bulletin of the American Association of Petroleum Geologists* 36: 1541–1573.

Khalid, F. & O'Brien, J., 1992. *Islam & Ecology.* London: Cassell.

Kiernan, R.H., 1937. *The Unveiling of Arabia.* London: George H. Harrap.

Kingdon, J., 1991. *Arabian Mammals. A Natural History.* London: Academic Press.

König, P., 1986. Zonation of vegetation in the mountainous region of south-western Saudi Arabia ('Asir, Tihama). *In*: H. Kürschner, ed., *Contribution to the Vegetation of Southwest Asia.* Wiesbaden: Dr. Ludwig Reichart Verlag. pp. 137–166.

Krupp, F., 1983. Fishes of Saudi Arabia. Freshwater fishes of Saudi Arabia and adjacent regions of the Arabian Peninsula. *Fauna of Saudi Arabia* 5: 569–634.

Krupp, F., Abuzinada, A.H. & Nader, I.A., 1996. *A Marine Wildlife Sanctuary for the Arabian Gulf. Environmental Research & Conservation Following the 1991 Gulf War Oil Spill.* Brussels: European Commission. (Also – Riyadh: National Commission for Wildlife Conservation and Development; Frankfurt a.M.: Forschungsinstitut Senckenberg).

Kürschner, H. & Ghazanfar, S.A., 1998. Bryophytes and Lichens. *In*: S.A. Ghazanfar. & M. Fisher., (eds.). *Vegetation of the Arabian Peninsula.* Dordrecht: Kluwer Academic Publishers. pp. 99–124.

Kürschner, H., 1998. Biogeography and Introduction to Vegetation. *In*: S.A. Ghazanfar & M. Fisher., (eds.). *Vegetation of the Arabian Peninsula.* Dordrecht: Kluwer Academic Publishers. pp. 63–98.

Lancaster, N., 1980. The formation of seif dunes from barchans – supporting evidence for Bagnold's model from the Namib Desert. *Zeitschrift für Geomorphologie* 24: 160–176.

Land Management Staff, 1981. *Schematic Soil Map.* Land Management Department, Ministry of Agriculture and Water, Saudi Arabia.

Larsen, C.E., 1983. *Life & Land Use on Bahrain: the Geoarchaeological Context to an Ancient Society.* Chicago: University of Chicago Press.

Larsen, T.B., 1984. The zoogeographical composition and distribution of Arabian butterflies (Lepidoptera; Rhopalocera). *Journal of Biogeography* 11: 119–158.

Laurent, D., 1992. *Kingdom of Saudi Arabia: Atlas of Industrial Minerals.* Jiddah: Directorate General of Mineral Resources.

Lawrence, T.E., 1926. *Seven Pillars of Wisdom.* Oxford: Subscribers' Edition (a private edition was published in 1922).

Lee, D.M., 1984. The Climate of Saudi Arabia. A Brief Introduction to the Kingdom of Saudi Arabia. Publication No. 9. Meteorology and Environmental Protection Administration (MEPA). p. 14.

Lee, S.Y., Dixon, J.B. & aba-Huysayn, M.M., 1983. Mineralogy of Saudi Arabian soils: Eastern Region. *Soil Science Society of America Journal* 47: 321–326.

Lees-smith, D.T., 1986. Composition and origins of the south-west Arabian avifauna: a preliminary analysis. *Sandgrouse,* 7: 71–92.

Lettau, K. & Lettau, H., 1969. Bulk transport of sand by barchans of the Pampa de la Joya in southern Peru. *Zeitschrift für Geomophologie* 13: 182–195.

Leviton, A.E., 1986. Description of a new species of *Coluber* (Reptilia: Serpentes: Colubridae) from the southern Tihamah of Saudi Arabia, with comments on the biogeography of southwestern Arabia. *Fauna of Saudi Arabia* 8: 436–446.

Linsley, R.K., Kohler, M.A. & Paulus, J.L.N., 1975. *Hydrology for Engineers.* New York: McGraw-Hill.

Lioubimtseva, E., 2004. Climate change in arid environments: revisiting the past to understand the future. *Progress in Physical Geography* 28: 525–538.

Lowell, J.D. & Genik, G.J., 1972. Sea-floor spreading and structural evolution of the southern Red Sea. *American Association of Petroleum Geologists Bulletin* 56: 247–259.

Magin, C.D. & Greth, A., 1994. Distribution, status and proposals for conservation of mountain gazelle *Gazella gazella cora* in south-west Saudi Arabia. *Biological Conservation* 70: 69–75.

Maitland, A., 2003. Obituary. Sir Wilfred Thesiger 1910–2003. *The Geographical Journal* 170: 92–94.

Maizels, J., 1988. Paleochannels: Plio-Pleistocene Raised Channel Systems of the Western Sharqiyah. *In*: R.W. Dutton ed. The Scientific Results of The Royal Geographical Society's Oman Wahiba Sands Project, 1985–1987. *Journal of Oman Studies, Special Report* No. 3: 95–112.

Maltzan, H. von, 1865. *Meine Wallfahrt nach Mekka. Reise in der Küstengegend und im Inneren von Hedschas* (My pilgrimage to Mecca: travels in the coastal region and the interior of Hejjas), 2 vols, Leipzig. (reprint: Hildesheim 2004).

Mandaville, J.P., 1984. Studies in the Flora of Arabia XI: Some historical and geographical aspects of a principal floristic frontier. *Notes from the Royal Botanic Garden Edinburgh* 42: 1–15.

Mandaville, J.P., 1986. Plant life in the Rub' Al-Khali (the Empty Quarter), south-central Arabia. *Proceedings of the Royal Society of Edinburgh*, 89B: 147–157.

Mandaville, J.P., 1990. *Flora of Eastern Arabia*. London: Kegan Paul International.

Martins, R.P. & Hirschfeld, E., 1998. Comments on the limits of the Western Palearctic in Iran and the Arabian Peninsula. *Sandgrouse* 20: 108–134.

Masry, A.H., 1974. Prehistory in Northeastern Arabia: the Problem of Inter-regional Interaction. Coconut Grove, Field Research Projects.

Masry, A.H., 1977. The historic legacy of Saudi Arabia. *Atlal* 1: 9–19.

McClure, H.A., 1978. Early Paleozoic glaciation in Arabia. *Paleogeography, Paleoclimatology, Paleoecology* 25: 315–326.

McClure, H.A., 1978. Ar Rub' al Khali. *In*: S.S. Al-Sayari, & J.G. Zötl, (eds.). *Quaternary Period in Saudi Arabia*. Vienna: Springer-Verlag. pp. 252–263.

McClure, H.A., 1976. Radiocarbon chronology of late Quaternary lakes in the Arabian desert. *Nature* 253: 755–756.

McLaren International Consultants, 1979. Report on Agricultural Development of the Arabian Shield: South. Ministry of Agriculture and Water, Riyadh.

Meig, P., 1953. World distribution of arid and semi-arid homoclimates. In *Reviews of Research on Arid Zone Hydrology*. UNSCO, Paris, Arid Zone Program, 1: 203–210.

Meinertzhagen, R., 1954. *Birds of Arabia*. London: Henry Sotherton.

Membery, D.A., 1983. Low level wind profiles during the Gulf Shamal. *Weather* 38: 18–24.

Menzies, M.A., Baker, J., Bosence, D.W.K., Dart, C., Davison, I., Hueford, A., al-Kadasi, M.A., McClay, K.R., Nichols, G.J., al-Subbary, A.K. & Yelland, A., 1992. The timing of crustal magmatism, uplift and crustal extension – preliminary observations from Yemen. *In*: B. Storey, ed., *Magmatism & Continental Break Up. Geological Society of London, Special Publication* 68. pp. 293–304.

MEPA, 1986. The State of the Environment. Part II, Terrestrial Environments. Jeddah, Saudi Arabia.

Metal Bulletin, 1998. Saudi zinc mining project moves ahead. *Metal Bulletin*, No. 8246, January 22, p. 7.

Middleton, N.J., 1986. Dust storms in the Middle East. *Journal of Arid Environments* 10: 83: 96.

Miller, A.G. & Cope, T.A., 1996. *Flora of the Arabian Peninsula & Socotra*. Volume 1. Edinburgh: Edinburgh University Press.

Miller, A.G. & Nyberg, J.A., 1991. Patterns of endemism in Arabia. *Flora et Vegetatio Mundi* 9: 263–279.

Miller, A.G., Hedge, I.C. & King, R.A., 1982. Studies in the flora of Arabia I: a botanical bibliography of the Arabian peninsula. *Notes Royal Botanical Garden Edinburgh* 40: 43–61.

Miller, R.P., 1937. Drainage lines in bas-relief. *Journal of Geology* 8: 432–438.

Ministry of Agriculture and Water, 1984. *Water Atlas of Saudi Arabia*. Riyadh. p. 111.

Ministry of Agriculture and Water, 1985. *General Soil Map of the Kingdom of Saudi Arabia*. Ministry of Agriculture and Water – Land Management Department, Riyadh. p. 66.

Ministry of Agriculture and Water, 1988. *Climatic Atlas of Saudi Arabia*. Riyadh. p. 117.

Ministry of Agriculture and Water, 1995. The Land Resources. Technical Annex D. A Catalogue of Vegetation Communities in Saudi Arabia. Land Management Department. Riyadh.

Ministry of Agriculture and Water, 1995a. The Land Resources of the Kingdom of Saudi Arabia. Riyadh.

Ministry of Agriculture and Water, 1995b. The Land Resources. Technical Annex C. Agro-climatic zones in Saudi Arabia. Land Management Department, Riyadh.

Ministry of Communication, 1992. Repair to the damages in the lower part of the Deleh Descent. Hydrological Study. p. 37 [also spelled Dellah].

Ministry of Defence & Aviation, 1989. *The State of the Environment in the Kingdom of Saudi Arabia*. Volume 2. Jeddah: MEPA.

Ministry of Higher Education, 1999. *Atlas of the Kingdom of Saudi Arabia*. Riyadh. (in Arabic).

Ministry of Planning, 1995. Kingdom of Saudi Arabia. Sixth Development Plan. 1415–1420 A.H. (1995–2000 A.D.).

Mohammad, F.Z. & Abo-Ghobar, H.M., 1992. Estimation of rainfall erosivity indices for the Kingdom of Saudi Arabia. *Journal of King Saudi University* 4: (Agricultural Science) – no page numbers.

Monnier, O. & Guilcher, A., 1993. Le Sharm Abhur, ria récifale du Hedjaz, mer Rouge: géomorphologie et impact de l'urbanisation. *Annales de Géographie* 569: 1–16.

Monroe, E., 1973. Philby of Arabia. London: Faber and Faber.

Muller, R.A., 1981. *A Computer Program for the Continuous Monthly Water Budget*. Department of Geography and Anthropology. Louisiana State University.

Mustafa, M.A., Akabawi, K.A. & Zoghet, M.F., 1989. Estimates of reference crop evapotranspiration for life zones of Saudi Arabia. *Journal of Arid Environments* 17: 293–300.

Nader, I.A., 1990. Checklist of the mammals of Arabia. *Fauna of Saudi Arabia* 11: 330–380.

Nasrallah, H.A. & Balling, R.C., 1993. Spatial and temporal analysis of Middle Eastern temperature changes. *Climatic Change* 25: 153–161.

Naval Intelligence Division, 1946. *Western Arabia & the Red Sea*. Geographical Handbook Series, B.R. 527.

Novikova, N.M., 1970. Drawing up a preliminary vegetation map of Arabia [in Russian]. *Geobotanicheskos Kartografirovanie*, 61–71. (English translation, The British Library, Russian Translation Service 12072.)

Ormond, R.F.G., Dawson Shepherd, A.R., Price, A.R.G. & Pitts, R.J., 1988. The distribution and character of mangroves in the Red Sea. *In*: C.D. Field & M. Vannucci, (eds.). *Symposium on New Perspectives in Research & Management of Mangrove Ecosystems*. Proceedings, November 11–14, 1986, Columbo, Sri Lanka, UNDO/UNESCO. pp. 125–130.

Overstreet, W.C., Stoeser, D.B., Overstreet, E.F. & Goudarzi, G.H., 1977. Tertiary laterite of the As Sarat Mountains, Asir Province, Kingdom of Saudi Arabia. Directorate General of Mineral Resources, *Mineral Resources Bulletin*, 21.

Palgrave, W.C., 1865. *Narrative of a Year's Journey Through Central & Eastern Arabia* (1862–63) 2 vols. London: MacMillan.

Pellaton, C., 1979. Geological map of the Yanbu la Bahr quadrangle, sheet 24C, Kingdom of Saudi Arabia. Saudi Arabian Directorate General of Mineral Resources Geological Map GM-48, scale 1:250,000.

Peters, W.D., Pint, J.J. & Kremla, N., 1990. Karst landforms in the Kingdom of Saudi Arabia. *The NSS Bulletin*, 52: 21–32.

Philby, H. St. John, 1920a. Southern Nadj. *Geographical Journal* 55:161–185.

Philby, H. St. John, 1920b. Across Arabia: from the Persian Gulf to the Red Sea. *Geographical Journal* 56:446–463.

Philby, H. St. John, 1933a. *The Empty Quarter*. London: Constable.

Philby, H. St. John, 1933b. Rub' al Khali: An Account of Exploration in the Great South Desert of Arabia under the Auspices and Patronage of His Majesty 'Abdul 'Aziz ibn Sa'ud, King of the Hejaz and Nejd and Its Dependencies. *Geographical Journal* 81:1–21

Philby, H. St. John, 1949. Two notes from central Arabia. *Geographical Journal* 113: 86–93.

Pint, J. & Peters, D., 1985. The caves of Ma'aqala. *National Speliological Society Newsletter,* September, 277–282.

Pint, J., 2000. Saudi Arabia's desert caves. *Aramco World* 41: 26–39.

Pint, J., 2003. *The Desert Caves of Saudi Arabia.* London: Stacey International Publishers.

Powers, R.W., Ramirez, L.F., Redmon, C.D. & Elberg, E.L., 1966. Geology of the Arabian Peninsula: Sedimentary Geology of Saudi Arabia. *United States Geological Survey. Professional Paper,* 560-D, p. 127.

Powers, R.W., Ramirez, L.F., Redmond, C.D. & Elberg, Jr., E.L., 1966. Geology of the Arabian Peninsula, Sedimentary Geology of Saudi Arabia, *U.S. Geological Survey., Professional Paper* 560-D.

Prescott, J.R., Robertson, G.B., Shoemaker, C., Shoemaker, E.M &Wynn, J., 2004. Luminescence dating of the Wabar meteorite craters, Saudi Arabia. *Journal of Geophysical Research,* 109: E01008.1-E01008.8.

Prive-Gill, C., Thomas, H., & Lebret, P., 1999. Fossil wood of *Sindora* (Leguminosae, Caesalpiniaceae) from the Oligo-Miocene of Saudi Arabia: paleobiogeographical considerations. *Review of Palaeobotany & Palynology* 107: 191–199.

Radford, C., 1920. Joseph Pitts of Exeter (?1663 –?1739). *Transactions, Devonshire Society* 52: 223–238.

Rauert, W., Geyh, M.A. & Henning, G.T., 1988. Results of ¹⁴C & U/Th-datings of sinter samples from the caves of the as Summan Plateau. Institute of Hydrology, G.S.F., Munich & Das Niedersächsische Landesamt für Bodenforschung, Hannover.

Rodriguez-Iturbe, I. & Valdes, J.B., 1979. The geomorphic structure of hydrological response. *Water Resources Research* 15: 1409–1420.

Roobol, M.J. & Bankher, K.H., 2000. Selected bibliography of reports and publications on the geological hazards in the Kingdom of Saudi Arabia. To accompany Open-File Report SGS-OF-2002-9 Jiddah: Saudi Geological Survey.

Roobol, M.L., Sabahi, A., Kattan, F., Sl-Ahmadi, M. & Babkair, M.S., 1999. A Tertiarly Laterite Profile Beneath Harrat As Sarrat. Deputy Ministry for Mineral Resources, Technical Report BGRM-TR-99–3.

Roobol, M.J., Shouman, S.A. & Al-Solami, A.M., 1985. Earth tremors, ground fractures and damage to buildings at Tabah. Open-File Report DGMR-OF-05-13. Jiddah: Ministry of Petroleum and Mineral Resources, Deputy Ministry for Mineral Resources.

Rowaihy, M.N., 1985. Geologic map of the Haql Quadrangle Sheet 29A, Kingdom of Saudi Arabia: Jiddah: Directorate General of Mineral Resources.

Sabtan, A., 2005. Performance of a steel structure on Ar-Rayyas Sabkha soils. *Geotechnical and Geological Engineering* 23: 157–174.

Sadleir, G.F., 1977. *Diary of a Journey Across Arabia* (1819). 1st ed. Reprinted. With a New Introduction by F.M. Edwards. Cambridge: Oleander Press. (Note the misspelling of Sadlier's surname in this reprint).

Saint Jalme, M. & van Heezik, Y., (eds.), 1996. *Propagation of the houbara bustard.* London: Kegan Paul International.

Saint Jalme, M., Combreau, O., Seddon, P.J., Paillat, P., Gaucher, P. & van Heezik, Y., 1996. Restoration of the *Chlamydotis undulata macqueenii* (houbara bustard) populations in Saudi Arabia: a progress report. *Restoration Ecology* 4: 1–8.

Salih, A.M. & Sendil, V.M., 1985. Evapotranspiration under extremely arid climates. *Journal of Irrigation, Drainage Division ASCE,* 110: 289–303.

Salpeteur, I. & Sabir, H., 1989. Orientation studies for gold in the central pediplain of the Saudi Arabian shield. *Journal of Geochemical Exploration* 34: 289–215.

Saner, S., Al-Hinai, K. & Perincek, D., 2005. Surface expressions of the Ghawar structure, Saudi Arabia. *Marine and Petroleum Geology* 22: 657–670.

Sanlaville, P., 1992. Changements climatiques dans la péninsule Arabique durant le Pléistocène Supérieur et l'Holocène. *Paléorient* 18: 5–26.

Schmidt, D.L., Hadley, D.G. & Brown, G.F., 1981. Middle Tertiary continental drift and evolution of the Red Sea in southeastern Saudi Arabia. *United States Geological Survey. Open File Report*, No. 83–641, pp. 1–56.

Schulz, E. & Whitney, J.W., 1985. Vegetation of the northern Arabian Shield and adjacent sand seas. *U.S. Geological Survey Open-File Report* 85–116. p. 52.

Schulz, E. & Whitney, J.W., 1986a. Vegetation in north-central Saudi Arabia. *Journal of Arid Environments*, 10: 175–186.

Schulz, E. & Whitney, J.W., 1986b. Upper Pleistocene and Holocene lakes in the An Nafud, Saudi Arabia. *Hydrobiologia*, 43, 175–190.

Scotese, C.R. (*no date*). Paleomap Project. http://www.scotese.com/info.htm [06/05/07].

Seddon, P.J., 1995. Master Management Plan for the Mahazat as-Sayd protected area. Unpublished report. NCWCD, Riyadh.

Seddon, P.J., 1996. Introduction. *In:* M.Sainte Jalme & Y. Van Heezik, (eds.). *Propagation of the Houbara Bustard.* London, Kegan Paul International pp. 1–2.

Seddon, P.J., 2000. Trends in Saudi Arabia: increasing community involvement and a potential role for eco-tourism. *Parks* 10: 11–24.

Seddon, P.J. & Maloney, R.F., 1996. Reintroduction of houbara bustards into central Saudi Arabia: a summary of results between 1991 & 1996. *The Phoenix* 13: 14–16.

Seddon, P.J. & Van Heezik, Y., 1996. Seasonal changes in houbara bustard *Chlamydotis undulata macqueenii* numbers in Harrat al-Harrah, Saudi Arabia: implications for managing a remnant population. *Biological Conservation* 75: 139–146.

Sen, Z., 2004. The Saudi Geological Survey (SGS) Hydrograph Method for Use in Arid Areas. Saudi Geological Survey Technical Report, SGS-TR-2004-5.

Shaltout, K.H., el-Hahwany, E.F. & el-Kady, H.F., 1996. Consequences of protection from grazing on diversity and abundance of the coastal vegetation in eastern Saudi Arabia. *Biodiversity & Conservation* 5: 27–36.

Sharaf, M.A., Farag, M.H. & Gazzaz, M., 1988. Groundwater chemistry of Wadi Uoranah – al Abdiah area, Western Province, Saudi Arabia. *Journal of King Abdulaziz University, Earth Sciences* 1:103–112.

Sharief, F.A., Khan, M.S. & Magara, K., 1991. Outcrop-subcrop sequence & diagenesis of the Upper Jurassic Arab-Hith Formations, central Saudi Arabia. *Journal of King Abdulaziz University, Earth Sciences* 4: 105–137.

Sharland, P.R., Archer, R., Casey, D.M., Davies, R.B., Hall, S.H., Heward, A.P., Horbury, A.D. & Simmons, M.D., 2001. *Arabian Plate Sequence Stratigraphy.* GeoArabia Special Publication No. 2, p. 371.

Sheppard, C., Price, A. & Roberts, C., (eds.), 1992. *Marine Ecology of the Arabian Region.* London: Academic Press.

Shihata, S.A. & Abu-Rizaiza, O.S., 1998. Effect of Ground water Rise on the Performance of Flexible Pavements in Jeddah. *Proceedings 3rd International Road Federation, Middle East Regional Meeting.* Volume 2: Maintenance Management Systems Pavement and Performance, Riyadh, Saudi Arabia, 1.539–1.622.

Shogdar, N.B.J., 1998. Palaeosols and Duricrusts in the Area North of Jeddah and Rabigh. Jeddah: Privately Published by the Author.

Simmons, J.C., 1987. *Passionate Pilgrims. English Travellers to the World of the Desert Arabs.* New York: William Morrow and Company.

Sirocko, F., 1996. The evolution of the monsoon climate over the Arabian Sea during the last 24,000 years. *Palaeoecology of Africa* 24, 53–71.

Smith, G., 1996. The concentration and extent of degradation of petroleum components from intertidal and subtidal sediments in Saudi Arabia following the Gulf War oil spill. *In*, E. Krupp, A.H. Abuzinada & I.A. Nader, 1996. *A Marine Wildlife Sanctuary for the Arabian Gulf. Environmental Research & Conservation Following the 1991 gulf War Oil Spill*. Riyadh: National Commission for Wildlife Conservation & Development.

Soon, W. & Baliunas, S., 2003. Proxy climatic and environmental changes of the past 1000 years. *Climate Research* 23: 89–110.

Sorman, A.U., 1995. Estimation of peak discharge using GIUH model in Saudi Arabia. *Journal of Water Resources Planning and Management* 121: 287–293.

Stark, F., 1936. *The Southern Gates of Arabia: a Journey in the Hadhramaut*. London: John Murray.

Stern, R.J., Avigad, D., Miller, N.R. & Beyth, M., 2006. Evidence for the snowball Earth hypothesis in he Arabia-Nubian Shield and the East African Orogen. *Journal of African Earth Sciences* 44: 1–20.

Stern, R.J., 1994. Arc assembly and continental collision in the Neoproterozoic East African Orogen – implications for the consolidation of Gondwanaland. *Annual Review of Earth and Planetary Science* 22: 319–351.

Subyani, A. & Sen, Z., 1991. Study of recharge outcrop relations of the Wasia aquifer of central Saudi Arabia. *Journal of King Abdulaziz University, Earth Science* 4: 137–147.

Sulayem, M.S., Strauss, M., Joubert, E., & Wacher, T., 1997. Action plan for the Uruq Bani Maarid protected area. Unpublished report. NCWCD, Riyadh.

Symens, P. & Suhaibani, A., 1994. The impact of the 1991 Gulf War oil spill on bird populations in the northeastern Arabian Gulf – a review. *Courier Forschungsinstitut Senckenberg* 166: 47–55.

Tardy,Y., Kobilisek, B., & Paquet, H., 1991. Mineralogical composition and geographical distribution of African and Brazilian periatlantic laterites. The influence of continental drift and tropical paleoclimates during the last 150 million years and implications of India and Australia. *Journal of African Earth Sciences* 12: 283–295.

Tchernov, E., 1981. The biostratigraphy of the Middle East. *In:* J.Cauvin & P. Sanlaville, (eds.). *Prehistoire du Levant*. Paris: Centre National de la Recherche Scientifique. pp. 67–98.

Templeton, A.R., 2002. Out of Africa again and again. *Nature* 416: 45–51.

Thesiger, W., 1946. A new journey in southern Arabia. *The Geographical Journal* 108: 129–45.

Thesiger, W., 1948a. Across the Empty Quarter. *The Geographical Journal* 111: 1–21.

Thesiger, W., 1948b. A journey through the Tihama, the Asir and the Hejaz mountains. *The Geographical Journal* 90: 188–200.

Thesiger,W., 1949. A further journey across the Empty Quarter.*The Geographical Journal* 113: 21–46.

Thesiger, W., 1950. Desert borderlands of Oman. *The Geographical Journal* 111: 137–171.

Thesiger, W., 1959. *Arabian Sands*. London: Penquin Books.

Thomas, B., 1931. Across the Rub' al Khali. *Journal of the Royal Central Asian Society*, 18: 489–504.

Thomas, B., 1932. *Arabia Felix. Across the Empty Quarter of Arabia*. London: Jonathan Cape.

Thomas, H., Geraads, D., Janjou, D., Vaslet, S., Memesh, A., Billiou, D., Bocherens, H., Dobigny, G., Eisenmann, V., Gayet, M., Lapparent de Broin, F., Petter, G. & Halawani, M., 1998. First Pleistocene fauna from the Arabian Peninsula: An Nafud desert, Saudi Arabia. *Comptes rendus Academie des Sciences, Paris, Sciences de la Terre et des Planèts* 326: 145–152.

Thomson. A., 2000. *Origins of Arabia*. London: Stacey International.

Thouless, C., Grainger, J.C., Shobrak, M. & Habibi, K., 1991. Conservation status of gazelles in Saudi Arabia. *Biological Conservation* 58: 85–98.

Thouless, C., 1991. Conservation in Saudi Arabia. *Oryx* 25: 222–228.

Tidrick, K., 1989. *Heart Beguiling Araby. The English Romance with Arabia*. London: I.B. Taurus.

Tsoar, H., 1978. *The Dynamics of Longitudinal Dunes*. Final Technical Report. European Research Office. United States Army. London. p. 171.

United States Department of Agriculture, 1975. *Soil Taxonomy: A Basic System of Land Classification for Making & Interpreting Soil Surveys.* Soil Conservation Service, United States Department of Agriculture, Handbook 436, p. 754.

Unrug, R., 1997. Rodinia to Gondwana: the geodynamic map of Gondwana supercontinent. *GSA Today* 7: 1–6.

Uvarov, B.P., 1951. Locust research and control 1929–1950. *Colonial Research Publication,* No.10. London: Colonial Office.

van Heezik, Y.M. & Seddon, P.J., 1996. Status of the houbara bustard & houbara habitat protection in Saudi Arabia. *In:* van Heezik, Y.M. & Seddon, P.J. (eds.). *Restoration of bustard populations: captive breeding, release, monitoring & habitat management.* Proceedings of the symposium "Restoration of houbara bustard populations in Saudi Arabia". NCWCD Publication No. 27. English Series. Riyadh. pp. 2–8.

Vasler, D., Janjou, D., Robelin, C., al-Muallem, M.S., Halawini, M.A., Bross, J-M., Breton, J-P., Courbouleix, S., Roobol, M.J. & Dagain, J-P., 1994. Explanatory notes to the geologic map of the Tayma quadrangle, Sheet 27C, Kingdom of Saudi Arabia. Jeddah: Ministry of Petroleum & Mineral Resources.

Vassiliev, A., 1998. *The History of Saudi Arabia.* London: Saqi Books.

Vazquez, M.A., Allen, K,W. & Kattan, Y.M., 2000. Long-term effects of the 1991 Gulf War and the hydrocarbon levels in clams at selected areas of the Saudi Arabian Gulf coastline. *Marine Pollution Bulletin* 40: 440–448.

Vesey-Fitzgerald, D.F., 1955. Vegetation of the Red Sea coast south of Jedda, Saudi Arabia. *Journal of Ecology* 43:477–489.

Vesey-Fitzgerald, D.F., 1957a. The vegetation of central and eastern Arabia. *Journal of Ecology* 45: 779–798.

Vesey-Fitzgerald, D.F., 1957b. The vegetation of the Red Sea coast north of Jedda, Saudi Arabia. *Journal of Ecology* 45: 547–562.

Viani, B.E., Al-Mashhady, A.S. & Dixon, J.B., 1983. Mineralogy of Saudi Arabian soils: central alluvial basins. *Soil Science of America Journal* 47: 149–157.

Vincent, P. & Kattan, F., 2006. Yardangs in the Cambrian Saq Sandstones, North-West Saudi Arabia. *Zeitschrift für Geomorphologie* 52: 305–320.

Vincent, P. & Sadah, A., 1996. Downslope changes in the shape of pedisdiment debris. *Sedimentary Geology* 95: 207–219.

Vincent, P. & Sadah, A., 1996. The profile form of rock-cut pediments in western Saudi Arabia. *Journal of Arid Environments* 32: 121–139.

Vincent, P., 2004. Jeddah's environmental problems, *Geographical Review* 93: 394–413.

Voggenreiter, W., Hötzl, H. & Mechie, J., 1988. Low-angle detachment origin for the Red Sea rift system. *Technophysics* 150: 51–75.

Walen, N.M. & Schatte, K.E., 1997. Pleistocene sites in southern Yemen. *Arabian Archaeology and Epigraphy* 8: 1–10.

Wallace, A.R., 1876. *The Geographical Distribution of Animals.* London: Macmillan.

Wallach, J.,1996. *Desert Queen. The Extraordinary Life of Gertrude Bell.* London: Weidenfeld & Nicolson.

Walters, M.O., 1989. Transmission losses in arid regions. *Journal of Hydraulic Division. American Society of Civil Engineers* 116: 129–138.

Warner, T.T., 2004. Desert Meteorology. Cambridge: Cambridge University Press.

Watson, A., 1990. The control of blowing sand and mobile desert dunes. *In:* A. Goudie, ed., *Techniques for Desert Reclamation.* Chichester: John Wiley. pp. 35–85.

Watson, A., 1985. The control of wind blown sand and moving dunes: a review of the methods of sand control in deserts, with observations from Saudi Arabia. *Quarterly Journal of Engineering Geology London* 18: 237–252.

Watts, D. & Al-Nafie, A.H., 2003. *Vegetation & Biogeography of the Sand Seas of Saudi Arabia.* London: Kegan Paul.

Weijermars. R & Asif Khan, M., 2000. Mid-crustal dynamics and island-arc accretion in the Arabian Shield: insight from the Earth's natural laboratory. *Earth-Science Reviews* 49: 77–120.

White, F. & Leonard, J., 1991. Phytogeographical links between Africa and Southwest Asia. *Flora et Vegetatio Mundi* 9: 229–246.

White, F., 1983. *The Vegetation of Africa. A Descriptive Memoir to Accompany the UNESCO, AETFAT, UNSO, Vegetation Map of Africa.* Paris: UNESCO.

Whitney, J.W., 1983 (1404 AH). Erosional history and surficial geology of western Saudi Arabia. Technical Record, USGS-TR-04-1. Jeddah: Ministry of Petroleum & Minerals, Deputy Ministry for Mineral Resources.

Whitney, J.W., Faulkender, D.J. & Rubin, M., 1983. The environmental history and present conditions of Saudi Arabia's northern sand seas. United States Department of the Interior. *Geological Survey. Open-File Report* 83–749, p. 39.

Whybrow, P.J. & McClure, H.A., 1981. Fossil mangrove roots and paleoenvironments of the Miocene of the eastern Arabian Peninsula. *Palaeogeography, Palaeoclimatology, Palaeoecology* 32: 213–225.

Wilkinson, J., 1978. Islamic water law with special reference to oasis settlement. *Journal of Arid Environment* 1: 87–96.

Williams, V.S., AL-Rehaili, M., Khiyami, H., Showail, A.H., Basahel, H., Turkistani, A. & Al-Zarani, M., 2000. Preliminary report on neotectonic studies of the Red Sea coast near Umm Lajj. Saudi Geological Survey. Unpublished.

Winstone, H.V.F., 1976. *Captain Shakespear: a Portrait.* London: Jonathan Cape.

Winstone, H.V.F., 2003. *Lady Anne Blunt. A Biography.* London: Barzan Publishing.

Zarins, J., Murad, A. & Al-Yish, K.S., 1981. Second Preliminary Report of the South-western Province. *Altai* 5: 9–59.

Zarins, J., Whalen, N., Ibrahim, M., Mursi, A. & Khan, M., 1980. Preliminary Report on the Central and South-western Provinces Survey. *Altai* 4: 9–36.

Zohary, M., 1957. A contribution to the flora of Saudi Arabia. *The Journal of the Linnaean Society of London, Botany* 55: 623–643.

Zohary, M., 1973. *Geobotanical Foundations of the Middle East.* 2 Vols. Stuttgart: Fisher Verlag.

Zuener, F.E., 1954. 'Neolithic' sites from the Rub al-Khali, southern Arabia. *Man* 52: 133–136.

Index